Renin-Angiotensin System in Cardiovascular Biology

Guest Editors

Mária Szekeres
György L. Nádasy
András Balla

Basel • Beijing • Wuhan • Barcelona • Belgrade • Novi Sad • Cluj • Manchester

Guest Editors

Mária Szekeres
Semmelweis University
Budapest
Hungary

György L. Nádasy
Semmelweis University
Budapest
Hungary

András Balla
Semmelweis University
Budapest
Hungary

Editorial Office
MDPI AG
Grosspeteranlage 5
4052 Basel, Switzerland

This is a reprint of the Special Issue, published open access by the journal *Biomedicines* (ISSN 2227-9059), freely accessible at: https://www.mdpi.com/journal/biomedicines/special_issues/2AI3LK6P1T.

For citation purposes, cite each article independently as indicated on the article page online and as indicated below:

Lastname, A.A.; Lastname, B.B. Article Title. *Journal Name* **Year**, *Volume Number*, Page Range.

ISBN 978-3-7258-3109-8 (Hbk)
ISBN 978-3-7258-3110-4 (PDF)
https://doi.org/10.3390/books978-3-7258-3110-4

© 2025 by the authors. Articles in this book are Open Access and distributed under the Creative Commons Attribution (CC BY) license. The book as a whole is distributed by MDPI under the terms and conditions of the Creative Commons Attribution-NonCommercial-NoDerivs (CC BY-NC-ND) license (https://creativecommons.org/licenses/by-nc-nd/4.0/).

Contents

Preface vii

György L. Nádasy, András Balla and Mária Szekeres
From Living in Saltwater to a Scarcity of Salt and Water, and Then an Overabundance of Salt—The Biological Roller Coaster to Which the Renin–Angiotensin System Has Had to Adapt: An Editorial
Reprinted from: *Biomedicines* 2023, 11, 3004, https://doi.org/10.3390/biomedicines11113004 . . 1

Dana Pop, Alexandra Dădârlat-Pop, Raluca Tomoaia, Dumitru Zdrenghea and Bogdan Caloian
Updates on the Renin–Angiotensin–Aldosterone System and the Cardiovascular Continuum
Reprinted from: *Biomedicines* 2024, 12, 1582, https://doi.org/10.3390/biomedicines12071582 . . 10

Elena Cojocaru, Cristian Cojocaru, Cristiana-Elena Vlad and Lucian Eva
Role of the Renin-Angiotensin System in Long COVID's Cardiovascular Injuries
Reprinted from: *Biomedicines* 2023, 11, 2004, https://doi.org/10.3390/biomedicines11072004 . . 22

Nikolina Kolobarić, Nataša Kozina, Zrinka Mihaljević and Ines Drenjančević
Angiotensin II Exposure In Vitro Reduces High Salt-Induced Reactive Oxygen Species Production and Modulates Cell Adhesion Molecules' Expression in Human Aortic Endothelial Cell Line
Reprinted from: *Biomedicines* 2024, 12, 2741, https://doi.org/10.3390/biomedicines12122741 . . 37

Déborah Victória Gomes Nascimento, Darlyson Ferreira Alencar, Matheus Vinicius Barbosa da Silva, Danilo Galvão Rocha, Camila Ferreira Roncari, Roberta Jeane Bezerra Jorge, et al.
Cardiovascular and Renal Effects Induced by Alpha-Lipoic Acid Treatment in Two-Kidney-One-Clip Hypertensive Rats
Reprinted from: *Biomedicines* 2024, 12, 1751, https://doi.org/10.3390/biomedicines12081751 . . 54

Julius Abiola, Anna Maria Berg, Olapeju Aiyelaagbe, Akindele Adeyi and Simone König
Dabsylated Bradykinin Is Cleaved by Snake Venom Proteases from *Echis ocellatus*
Reprinted from: *Biomedicines* 2024, 12, 1027, https://doi.org/10.3390/biomedicines12051027 . . 69

Adrian Martyniak, Dorota Drożdż and Przemysław J. Tomasik
Classical and Alternative Pathways of the Renin–Angiotensin–Aldosterone System in Regulating Blood Pressure in Hypertension and Obese Adolescents
Reprinted from: *Biomedicines* 2024, 12, 620, https://doi.org/10.3390/biomedicines12030620 . . . 80

Attila Nagy, Réka Májer, Judit Boczán, Sándor Sipka, Jr., Attila Szabó, Enikő Edit Enyedi, et al.
Enalapril Is Superior to Lisinopril in Improving Endothelial Function without a Difference in Blood–Pressure–Lowering Effects in Newly Diagnosed Hypertensives
Reprinted from: *Biomedicines* 2023, 11, 3323, https://doi.org/10.3390/biomedicines11123323 . . 90

Robin Shoemaker, Marko Poglitsch, Dolph Davis, Hong Huang, Aric Schadler, Neil Patel, et al.
Association of Elevated Serum Aldosterone Concentrations in Pregnancy with Hypertension
Reprinted from: *Biomedicines* 2023, 11, 2954, https://doi.org/10.3390/biomedicines11112954 . . 103

**Danyelle Siqueira Miotto, Francine Duchatsch, Aline Dionizio,
Marília Afonso Rabelo Buzalaf and Sandra Lia Amaral**
Physical Training vs. Perindopril Treatment on Arterial Stiffening of Spontaneously Hypertensive Rats: A Proteomic Analysis and Possible Mechanisms
Reprinted from: *Biomedicines* **2023**, *11*, 1381, https://doi.org/10.3390/biomedicines11051381 . . **120**

**Sergei M. Danilov, Mark S. Jain, Pavel A. Petukhov, Olga V. Kurilova, Valery V. Ilinsky,
Pavel E. Trakhtman, et al.**
Blood ACE Phenotyping for Personalized Medicine: Revelation of Patients with Conformationally Altered ACE
Reprinted from: *Biomedicines* **2023**, *11*, 534, https://doi.org/10.3390/biomedicines11020534 . . . **137**

Preface

Cardiovascular diseases are the most common chronic diseases with the most frequent causes of mortality worldwide. The aim of this Special Issue is to present some latest updates on the role of the renin–angiotensin system (RAS) in cardiovascular biology, from cellular mechanisms to organ functions and novel therapeutic mechanisms. The RAS has an important role in the regulation of blood pressure and salt-water homeostasis. Angiotensin II is the main regulator, which operates mostly by stimulating AT1 receptors, a member of the G protein-coupled receptor family, and activating mainly calcium signaling mechanisms inducing, e.g., smooth muscle contractions, vasoconstriction, and blood pressure elevation. Long-term overactivation of the RAS may result in inflammatory mechanisms, vascular hypertrophy and remodeling, water retention, hypertension, and atherosclerosis.

This Special Issue provides original work and review articles by outstanding research groups based on the latest scientific knowledge on the area of RAS in cardiovascular biology.

Our Readers are invited to obtain a deeper insight into this interesting research area to update their knowledge on cardiovascular diseases and treatment proposals.

Mária Szekeres, György L. Nádasy, and András Balla
Guest Editors

Editorial

From Living in Saltwater to a Scarcity of Salt and Water, and Then an Overabundance of Salt—The Biological Roller Coaster to Which the Renin–Angiotensin System Has Had to Adapt: An Editorial

György L. Nádasy [1], András Balla [1,2] and Mária Szekeres [1,3,*]

[1] Department of Physiology, Faculty of Medicine, Semmelweis University, 37-47 Tűzoltó Street, 1094 Budapest, Hungary; nadasy.gyorgy@med.semmelweis-univ.hu (G.L.N.); balla.andras@med.semmelweis-univ.hu (A.B.)
[2] Laboratory of Molecular Physiology, Eötvös Loránd Research Network, Research Centre for Natural Sciences, 2 Magyar Tudósok Körútja, 1117 Budapest, Hungary
[3] Department of Morphology and Physiology, Faculty of Health Sciences, Semmelweis University, 17 Vas Street, 1088 Budapest, Hungary
* Correspondence: szekeres.maria@semmelweis.hu

Angiotensin II (Ang II) is a hormone with much more complex actions than is typical for other agonists with heterotrimeric G protein-coupled receptors (GPCRs). Its principal receptor, angiotensin type 1 receptor (AT$_1$R), is distributed in many cells of the body in diverse organs, controlling the water and salt balance, vascular contractility and blood pressure. The main functions are discussed below.

1. Pleiotropic Effects of Ang II

AT$_1$Rs can attach promiscuously to G$_{q/11}$, Gi/o and G$_{12/13}$ intracellular signal proteins in various cells of the body. The activation of the mitogen-activated protein kinase (MAPK) cascade and the release of epithelial growth factor (EGF) and vascular endothelial growth factor (VEGF) lend a trophic character to the hormone [1–3]. In the blood vessel wall, it induces inflammatory cell transformation, chronic inflammation with cytokine release, white cell migration, fibrosis and apoptosis of cells. It can be considered an inflammatory transmitter with a strong resemblance to chemokines [4]. It is also a neural transmitter. In the brain stem, diencephalon and limbic system neural circuits controlling salt and water balance even behavior elements containing Ang II-releasing neurons [5,6]. These complex actions are very far from the original epithelial salt-conserving function of this archaic agonist. In the mammalian body, several functions intermingle inseparably in the diverse actions of Ang II. Corrections needed to restore the salt balance, volume balance, blood pressure, vascular mechanics, cellular longevity and drinking behavior in optimal cases may be parallel, but it could be that the adjustment of one important parameter induces disadvantageous changes in another. In the case of fast environmental changes, with no time for genetically fine-tuning the whole system, *Ang II will emerge as a pathological factor*.

Three papers which have been published in this Special Issue deal with such pleiotropic functions of the renin–angiotensin system (RAS). In an excellent study by Miotto et al. [7] from the State University of Sao Paulo, Brazil, the aortic wall of perindopril-treated spontaneously hypertensive rats (SHR) was analyzed. Using the best available high throughput proteomic methods, they found that chronic treatment with the angiotensin converting enzyme (ACE) inhibitor perindopril in the artic wall altered 38 subcategories of cellular protein components, among them "supramolecular polymer", "heterotrimeric G protein complex", "actin cytoskeleton", "supramolecular fiber", "intermediate filament", "membrane raft" and "oxidoreductase" complexes. The reduced aortic stiffness they observed seems to be connected to the elevated expression of the Ehd2 (EH domain-containing) protein, a component involved in the endothelial nitric oxide synthase (eNOS)–nitric oxide

---

Citation: Nádasy, G.L.; Balla, A.; Szekeres, M. From Living in Saltwater to a Scarcity of Salt and Water, and Then an Overabundance of Salt—The Biological Roller Coaster to Which the Renin–Angiotensin System Has Had to Adapt: An Editorial. *Biomedicines* **2023**, *11*, 3004. https://doi.org/10.3390/biomedicines11113004

Received: 12 October 2023
Revised: 3 November 2023
Accepted: 7 November 2023
Published: 9 November 2023

Copyright: © 2023 by the authors. Licensee MDPI, Basel, Switzerland. This article is an open access article distributed under the terms and conditions of the Creative Commons Attribution (CC BY) license (https://creativecommons.org/licenses/by/4.0/).

(NO) endothelial vasodilatory pathway. They succeeded in proving that the rigidity-reducing effect of chronic exercise has a different mechanism [7]. The extremely complex nature of the renin–angiotensin system (RAS) is shown by another paper in this Special Issue. Danilov et al. [8] found that the steric structure of the ACE protein is dependent not only on its amino acid sequence, but also on the presence of certain blood constituents [8]. Pathological connections of the RAS are far-reaching. Interspecies conservatism of the structure of ACE2, its receptor protein, made it possible for the SARS-CoV-2 virus to adapt from the original bat host to humans with surprising speed. A virus binding to the attacked cells will be followed by its internalization, together with the receptor. The elements of RAS will be involved in several subsequent pathological events. Such long post-COVID cardiovascular injuries have been reviewed by Cojocaru et al. in a recent Special Issue [9].

2. The Natural History of Water and Salt Balance

If we want to understand the role of Ang II as an important pathological factor, we have to explore how its problematic, divergent characteristics developed. The types of life occupying the disturbingly thin (12 miles) biosphere on the surface of the Earth are inherently attached to highly organized macromolecular interactions in watery solutions. Water, one of the most abundant "liquid stones" of the Earth's crust, diluted salty substances from the solid rocks. Early forms of life adapted to that salty water, the salinity of which increased with time. Our planet can be considered extremely fortunate, having been able to keep water and maintain it in its fluid form on its surface for such a long time. This has ensured the continuous existence and development of life. With the rising of the continents, dry land appeared, forming new habitats for living creatures that they were eager to colonize. On continents, water needed for life processes could be ensured through the atmospheric water circulation: rain, molten snow and ice watered the soil and fed rivers and lakes. However, this water was only of very limited salt content. Sophisticated mechanisms developed to keep the water and salt balance of the organisms even under such conditions [10–12].

Marine invertebrates succumbed to the slowly elevating salinity of surrounding sea-water, their extracellular osmolarity elevated in equilibrium with it. Their angiotensin II hormone controlled the extracellular fluid pressure and volume. In marine fish, however, like in all vertebrates, extracellular fluid osmolarity conserved the lower salinity of the ancient oceans. This was a mere third of present day's ocean salinity values. Keeping extracellular sodium values constant simplified the genetic maintenance of the elaborate membrane ionic processes on which their sophisticated and, in the case of vertebrates, very successful neural and muscle functions depended. For marine vertebrates, the removal of salt from the body turned out to be the most important task. In fishes, it is performed due to the esophageal desalination of swallowed sea water and sodium excretion through the epithelial ionocytes of the gills [13]. The RAS controls extracellular fluid volume (thirst, determining salty sea-water swallowing), as well as the contraction of the newly acquired blood vessels that form a closed "intravascular" compartment lined with a smooth, continuous, low-friction endothelial sheath and, in case of larger vessels, additionally surrounded by a contractile smooth muscle media. Adjustments in blood volume and vascular and heart contractilities ensure proper intravascular blood flow and pressure. All these parameters are controlled (among others) by the renin–angiotensin system, through the action of the renin enzyme secreted into the blood by the mesonephros and of the peripheral ACE producing the active angiotensin II octapeptide. The kidneys of fishes are mesonephros, where nephrons do not have a Henle loop [14]. The mesonephros is present in the human embryo. It is fully developed at the 8th week and disappears at embryonal week 16, giving way to the metanephros [15]. Fish species living in freshwater rivers and lakes are confronted with just the opposite task: while water was still in abundancy (if a draught did not happen), there was a scarcity of salt. The lack of salt should be prevented by proper sodium reuptake processes in the kidneys and uptake from the low-sodium freshwater habitats. The very long shores of rivers, lakes, rivulets and swamps made it a successful strategy to live partially (amphibians) or permanently (reptiles, birds, mammals) on dry land where, however, both water and salt were scarce. From that point onward, ensuring both water and salt requirements of the body, and economizing them, became one of the most important tasks to ensure survival. Quadrupeds living on dry land for a substantial part of their life could only manage it through their sophisticated kidneys, the metanephros, with very effective sodium reuptake mechanisms in the Henle loop, distal and connecting tubules, as well as in the collecting duct, controlled both directly and indirectly (mineralocorticoid) through metanephric JGA renin and peripheral angiotensin II. The metanephros in the human fetus starts to develop at week 5 and remains thereafter in the final form of the kidney [15]. While renin-containing granules are distributed in the cells along the length of the afferent arteriole in fish mesonephros and early fetal metanephros, renin production is concentrated in the juxtaglomerular apparatus in adult mammals: granular cells of the afferent arteriole in the vicinity of the macula densa of the distal tubules store and release the renin controlled by (among other factors) the amount of salt in the distal tubules [16]. An effective osmoregulation ensures corresponding water conservation.

3. The Natural History of the Elements of RAS

As we can see, the renin–angiotensin system plays an important role in maintaining the salt and water balance of very diverse animal species under very diverse conditions. It may be surprising that the components of the renin–angiotensin system are highly conserved proteins and peptides, and that their basic molecular structure has been maintained during their evolution. Vertebrate angiotensin peptides differ from each other in no more than 2–3 amino acid residues [17]. Functionable ACE is present in Gram-negative bacteria, annelid worms and molluscs [18,19]. Vertebrate development has been associated with the duplication of the ACE gene, where a more active form of the enzyme developed and the ancestral unduplicated form of the protein retained its role in testicular development (tACE) [20]. The angiotensin II stimulation of adrenocortical cells to produce mineralocorticoid appeared in early fish (shark) [21]. Renin production in the kidney, peripheral ACE, angiotensin I and II production and angiotensin receptors are present in all vertebrate classes (with the potential exemption of some lower fishes) [17]. Angiotensin II contracts fish blood vessels either directly or through catecholamine release and elevates blood pressure, which has a central dipsogenic effect [16]. The angiotensin receptors (AT_1R and AT_2R) have substantial homology with chemokine receptors, apelin receptors and even opioid receptors, which can explain their involvement in inflammatory, cell differentiation and neurotransmitter processes [22]. The ancestry of RAS in sea life can be judged from the fact that extracts of several marine microbial fungal strains have been found to effectively inhibit the mammalian AT_1 receptor (also ET_A and ET_B receptors). At the same time, this observation demonstrates what pharmacological treasures can be present in only superficially examined ecosystems of our planet [23]. It is interesting to note here that the phylogenetic conservatism of RAS proteins had a very serious consequence: the large similarity of the bat and human ACE2 proteins made it possible for the COVID-19 virus, which used this particular protein as a cell surface anchor, to be transmitted from bats to humans and to adapt to the human molecule within only a few mutations ("variants of concern"), inducing a pandemic which recently killed millions of human beings [24].

4. Minor Variabilities of the Human RAS Genes

Minor variabilities of the human RAS genes can be important factors in hypertension and cardiovascular disease development. The time is approaching when—similar to present-day cancer molecular diagnostics—patient DNA analysis will contribute to the examination of pathomechanisms and the determination of individual optimal therapy. From the point of view of human biology, it would be very interesting to reveal which molecular alterations have accumulated under which geographical and ethnographic conditions. Two polymorphisms of the human ACE gene (*ACE I/D* and *ACE2 rs2106809*) have been reported which increase Ang II levels and elevate blood pressure, but at the same time have protective effects against severe malaria. Their frequency is elevated in malaria-endemic geographical areas [25]. The *ACE I/D* also increases the risk of hypertrophic cardiomyopathy [26], together with other variabilities of ACE [27]. Polymorphisms of the ACE and AT_1R and AT_2R genes seem to be associated with preeclampsia and pregnancy-induced hypertension [28]. Angiotensinogen is also a fairly conservative protein: several sequence features are maintained in 57 vertebrate species. In the human genome, however, minor variabilities are very frequent: 690 angiotensinogen variants have been identified in 1092 human genomes, the most frequent variations being somatic single nucleotide polymorphisms [29]. The potential pathological significance has yet to be investigated.

5. The Diverse Functions of AT1R

The AT_1R is distributed in diverse tissues, inducing diverse functions at the level of the organism. However, reflecting its long and complex development, this receptor has complex functions even at the molecular level. The pleiotropic effects of AT_1R stimulation, almost unique among heterotrimeric G protein-coupled receptors, have been reviewed [1,2]. This receptor can associate, in addition to $G_{q/11}$, $G_{i/o}$ and $G_{12/13}$ signal proteins. The termination of activation can be accomplished through the internalization of the receptor molecule via β-arrestin-mediated mechanisms. The substantial activation of the MAPK cascade reveals trophic effects and, interestingly, several dual-specificity MAPK phosphatases negatively regulating their activities are also upregulated in a negative feedback manner [3]. Earlier in the H295R human adrenocarcinoma cell line and in primary rat adrenocortical glomerular cells, we mapped Ang II-induced expression kinetics of *CYP11B2* and *BDNF* genes [30]. Later, in chronic in vivo studies, we found that the infusion of angiotensin II into rats resulted in reduced morphological lumen and the elevated thickness of the wall of resistance arteries, which only slowly and partially recovered after the cessation of the infusion [31]. In addition, aberrations in the geometry of the intramural coronary resistance artery network were observed [32]. A further complication is that the activation of AT_1R by Ang II is accompanied by the release of endocannabinoids, as was found in a cell expression system [33] and in vascular tissues such as the skeletal muscle (gracilis) arteriole, coronary vessels or aorta [34,35]. The Ang II-stimulated release of 2-arachidonoylglycerol was directly detected in aortic vascular smooth muscle cells [36]. In cardiac tissue, we also found that an Ang II-induced release of endocannabinoids may counteract the positive inotropic effect of Ang II to decrease metabolic demand [37]. Ang II also activates chronic inflammatory cellular pathways, inducing vascular damage and accelerating vascular aging (for a review, see [4]). In addition, according to some disputed views, ACE and Ang II even might restrict longevity throughout the animal world [38].

6. The Role of Ang II in Maintaining Blood Pressure under Basal Conditions

In terrestrial animals, salt and water conservation, blood volume conservation and blood pressure maintenance are inherently connected to each other. In the diverse functions of the renin–angiotensin–aldosterone system, this fact is reflected. The four most important actions of the renin–angiotensin system are all directed to elevate and maintain blood pressure: 1. Cellular (epithelial) sodium reuptake. 2. Stimulation of mineralocorticoid production (with similar effects). 3. Vascular smooth muscle contraction. 4. Inducing thirst. Among the physiological circuits controlling and maintaining normal blood pressure, those with the involvement of this octapeptide are among the most important. The renin gene and AT_{1a} receptor gene knock-out mice (homozygous) had blood pressure reductions of around 20–30 mmHg [39]. The RAS is active under basal conditions at "normal" levels of water and salt intake and fluid loss. Ang II is continuously produced in many tissues; the overdose of ACE inhibitor was found to decrease blood pressure to as low as 88/42 (57) mmHg (systolic pressure/diastolic pressure (mean pressure)) in a set of toxicology patients [40].

7. The Role of Ang II in Maintaining Blood Pressure under Extreme Conditions

The significance of having an effective renin–angiotensin system for human survival is even more prominent under extreme conditions. The mammalian brain (and to some degree heart) are extremely sensitive to any reduction in the continuous supply of oxygen through continuous blood flow. A drop in the mean arterial blood pressure below about 70 mmHg interferes with the cerebral blood flow and disturbs sensation, motor abilities, decision making and communication, and substantially reduces muscle force and physical and mental performance, in a real situation massively reducing the probability of survival, maybe even in a few minutes. Historically, our ancestors were subjected to an impressive array of dangerous blood pressure lowering effects, such as bleeding, crushing wounds, toxicoses, exsiccosis, water deprivation, excessive heat, infections, diarrhea, vomiting, profuse sweating, etc. Several blood pressure elevating mechanisms help to prevent this from occurring, and the renin–angiotensin system is one of the most powerful among them. In irreversible circulatory shock, plasma renin values enormously elevate, but even such extreme values cannot produce the blood pressure needed for adequate microcirculation. High plasma renin values can thus be considered a bad prognostic marker for critically ill patients [41].

8. Salt Appetite, Salt Preference and the Overconsumption of Salt

Behavioral elements are important in processes that control body water and salt. Ang II is a transmitter in hypothalamic, brain stem and limbic neuronal circuits controlling thirst, vasopressin release and salt preference. Salt preference increases when sodium deprivation is present. Several mammals have a salt appetite, and humans are among them. A salt appetite might have had an additional advantage in early humans turning hunters: even unsalted meat has a fairly salty taste [5,6].

Salt appetite and salt preference could have been important behavioral elements in keeping a salt (and fluid) balance in our early ancestors. We have good reason to think that salt preference, controlling salt intake, has been genetically adjusted to conditions when salt was scarce and the dangers of salt deficiency were larger than the dangers of consuming more salt than was optimal. However, with historical development, humans learned how to produce that commodity by evaporating sea water and by mining salt deposits from layers of solid rock. Hallstatt salt mining in the Austrian Alps dates back 7000 years, and in the bronze age in Europe there was substantial commercial salt transportation [42]. In Medieval times, the consumption of commercial salt turned habitual even among the poor, and salt taxation was one of the most stable entries in governmental income lists. In West and North European cities, foodstuffs preserved using salt became components of the everyday food of city dwellers, ensuring the development of high concentrations of non-agricultural urban residents with specific industrial, commercial and intellectual skills. Even lengthy deep-sea ship travels and military operations depended on salt-preserved food. With improved financial situations, limits on salt consumption practically disappeared and the genetically inherited salt taste dictating salt consumption started to elevate. This is an important element in the hypertension pandemic we are observing in the present day [43].

In developed countries, elevated salt consumption is—together with other factors—an important component of the increasing prevalence of the essential hypertension disease. Daily salt intake in practically all cultures now exceeds the estimated physiological requirements of 10–20 mmol/day [44]. The global sodium intake of adults runs at 4310 mg/day (10,780 mg/day salt), twice the recommended limit of the World Health Organization of 2000 mg/day (<5000 mg salt/day) [45] or the sodium limit advised by the US Institute of Medicine (2300 mg/day, 100 mmol/24 h) [46].

There is a classical negative feedback system between salt intake and the activity of the renin–angiotensin–aldosterone system, a system that works in the direction of salt accumulation [47]. In salt-sensitive individuals, anti-natriuretic effects are not sufficiently suppressed by salt consumption, which results in marked blood pressure elevation [48,49]. Reducing salt intake has a blood-pressure-lowering effect. According to the meta-analysis by He and McGregor [50], in the range of 3–12 g/day, the lower the salt intake, the lower the blood pressure. Salt restriction or the consumption of salt substitutes can be an alternative to drug treatment in certain hypertension cases [51].

9. The RAS as a Pharmacological Target in Large Populations

The enormous significance of these eight biologically inherited molecules of the RAS for human health can be judged from some statistics at hand. In the US, anti-hypertensive drug use increased from 20 to 27% of the population between 1999 and 2000 to 2011 and 2012, with 31% and 66% of the non-institutionalized population being involved in age groups 40–64 and >65 years, respectively. Of these, 12% took ACE inhibitors and 5.8% AT_1R inhibitors [52]. In England, in 2018, 22% of primary care patients received antihypertensive prescriptions. ACE inhibitors being ordered increased from less than 0.5% in 1988 to over 9% in 2018, while angiotensin receptor blockers, which started to be produced in 1995, reached 4% in 2018 [53]. Among hypertensive treatments, in the years after 2011, ACE inhibitors represented 38% and angiotensin receptor blockers 2.1% [54]. Similarly high numbers, but with a different ratio between drug groups, have been reported from Denmark. Among 352 antihypertensive drug users per 1000 inhabitants, 70 were taking ACE inhibitors and 65 angiotensin receptor blockers [55]. Looking at such statistics, we can conclude that in a substantial part of the population of developed countries, especially in the elderly, we cannot talk any longer about the "physiological control of blood pressure and salt balance": there is a combined pharmaco-physiological control, instead.

10. Conclusions

We conclude that our RAS and salt preference behavior has not genetically adapted to the hedonistic salt use which is broadly accessible today. Public information should be more widely distributed about the risks of high salt consumption. Diagnostic tests and even genetic tests should be worked out to identify, within the supposedly heterogenous group of "essential hypertensives", those for whom moderate or aggressive salt restriction can be an important element of therapy. Further, with such tests, a more targeted, individualized pharmacotherapy can be hopefully planned in the not-too-far future.

Author Contributions: Conceptualization, G.L.N.; data curation, G.L.N., A.B. and M.S.; writing—original draft preparation, G.L.N.; writing—review and editing, G.L.N., A.B. and M.S.; All authors have read and agreed to the published version of the manuscript.

Funding: This research received no external funding.

Institutional Review Board Statement: Not applicable.

Informed Consent Statement: Not applicable.

Data Availability Statement: Not applicable.

Acknowledgments: Studies related to RAS in the Department have been supported by the Hungarian National Science Foundation TO-32019, NK-72661, NK-100883, K-116954, K139231; The Hungarian Society of Hypertension, Research Grant 2023 (M.S.); The Hungarian Kidney Foundation; the National Development Agency, Hungary TÁMOP 4.2.1.B-09/1/KMR-2010-0001; Hungarian National Research, Development and Innovation NVKP_16-1-2016-0039 and VEKOP-2.3.2-16-2016-00002. The authors are grateful to László Hunyady for helpful advice and support.

Conflicts of Interest: The authors declare no conflict of interest.

References

1. Hunyady, L.; Catt, K.J. Pleiotropic AT1 receptor signaling pathways mediating physiological and pathogenic actions of angiotensin II. (Review). *Mol. Endocrinol.* **2006**, *20*, 953–970. [CrossRef]
2. Toth, A.; Turu, G.; Hunyady, L.; Balla, A. Novel mechanisms of G-protein.coupled receptors functions: QT1 angiotensin receptor acts as a signaling hub and focal point of receptor cross-talk. *Best Pract. Red Clin. Endocrynol. Metabol.* **2018**, *32*, 69–82. [CrossRef]

3. Gem, J.B.; Kovacs, K.B.; Szalai, L.; Szakadati, G.; Porkolab, E.; Szalai, B.; Turu, G.; Toth, A.D.; Szekeres, M.; Hunyady, L.; et al. Characterization of Type 1 Angiotensin receptor activation induced dual-specificity MAPK phosphatase gene expression changes in rat vascular smooth muscle cells. *Cells* **2021**, *10*, 3538. [CrossRef]
4. Ungvari, Z.; Tarantini, S.; Donato, A.J.; Galvan, V.; Csiszar, A. Mechanism of vascular aging. *Circ. Res.* **2018**, *123*, 849–867. [CrossRef]
5. Iovino, M.; Guastamacchia, E.; Giagulli, V.A.; Licchelli, B.; Triggiani, V. Vasopressin secretion control: Central neural path-ways, neurotransmitters and effects of drugs. *Curr. Pharm. Design* **2012**, *18*, 4714–4724. [CrossRef]
6. Iovino, M.; Messana, T.; Lisco, G.; Vanacore, A.; Giagulli, V.A.; Guastamacchia, E.; DePergola, G.; Triggiani, V. Signal trans-duction of mineralocorticoid and angiotensin II receptors in the central control of sodium appetite: A narrative review. *Internatl J. Mol. Sci.* **2021**, *22*, 11735. [CrossRef]
7. Miotto, D.S.; Duchatsch, F.; Dionizio, A.; Buzalaf, M.A.R.; Amaral, S.L. Physical training vs. perindopril treatment on arterial stiffening of spontaneously hypertensive rats: A proteomic analysis and possible mechanisms. *Biomedicines* **2023**, *11*, 1381. [CrossRef] [PubMed]
8. Danilov, S.M.; Jain, M.S.; Petukhov, P.A.; Kurilova, O.V.; Ilinsky, V.V.; Trakhtman, P.E.; Dadali, E.L.; Samokhodskaya, L.M.; Kamalov, A.A.; Kost, O.A. Blood ACE phenotyping for personalized medicine: Revelation of patients with conforma-tionally altered ACE. *Biomedicines* **2023**, *11*, 534. [CrossRef] [PubMed]
9. Cojocaru, E.; Cojocaru, C.; Vlad, C.-E.; Eva, L. Role of the Renin-Angiotensin System in long COVID's cardiovascular injuries. *Biomedicines* **2023**, *11*, 2004. [CrossRef] [PubMed]
10. Colbert, E.H. *Evolution of the Vertebrates*, 2nd ed.; Wiley: New York, NY, USA, 1969.
11. Takei, Y. Comparative physiology of body fluid regulation in vertebrates with special reference to thirst regulation. *Jpn. J. Physiol.* **2000**, *50*, 171–186. [CrossRef]
12. Smith, H.W. *From Fish to Philosopher; The Story of Our Internal Environment*; Little, Brown: Boston, MA, USA, 1953.
13. Takei, Y.; Wong, M.K.-S.; Pipil, S.; Ozaki, H.; Suzuki, Y.; Iwasaki, W.; Kusakabe, M. Molecular mechanisms underlying active desalination and low water permeability in the esophagus of eels acclimated to seawater. *Am. J. Physiol. Regul. Physiol.* **2017**, *312*, R231–R244. [CrossRef] [PubMed]
14. Dantzler, W.H. Comparative aspects of renal function. In *The Kidney: Physiology and Pathophysiology*, 2nd ed.; Seldin, D.W., Giemiesch, G., Eds.; Raven Press: New York, NY, USA, 1992; pp. 885–942.
15. De Martino, C.; Zamboni, L.L. A morphologic study of the mesonephros of the human embryo. *J. Ultrastruct. Res.* **1966**, *16*, 399–427. [CrossRef] [PubMed]
16. Nishimura, H. Renin-angiotensin system in vertebrates: Phylogenetic view of structure and function. (Review). *Anat. Sci. Int.* **2017**, *92*, 215–247. [CrossRef] [PubMed]
17. Nishimura, H. Angiotensin receptors—evolutionary overview and perspectives. Review. *Comp. Biochem. Physiol. Part. A* **2001**, *128*, 11–30. [CrossRef]
18. Riviere, G.; Michaud, A.; Corradi, H.R.; Sturrock, E.D.; Acharya, K.R.; Cogez, V.; Bohin, J.P.; Vieau, D.; Corvol, P. Character-ization of the first angiotensin-converting-like enzyme in bacteria: Ancestor ACE is already active. *Gene* **2007**, *399*, 81–90. [CrossRef]
19. Riviere, G. (Angiotensin converting enzyme: A protein conserved during evolution. Review) (French). *J. Societ Biol.* **2009**, *203*, 281–293.
20. Moskowitz, D.W.; Johnson, F.E. The central role of angiotensin I-converting enzyme in vertebrate pathophysiology. *Curr. Top. Med. Chem.* **2004**, *4*, 1443–1454. [CrossRef]
21. Armour, K.J.; O'Toole, L.B.; Hazon, N. Mechanisms of ACTH- and angiotensin II stimulated 1 alpha-hydroxycorticosterone secretion in the dogfish, *Scyliorhynus canicula*. *J. Mol. Endocrinol.* **1993**, *10*, 235–244. [CrossRef]
22. Lio, P.; Vannucci, M. Investigating the evolution and structure of chemokine receptors. *Gene* **2003**, *317*, 29–37. [CrossRef]
23. Bolanos, J.; DeLeon, L.F.; Ochoa, E.; Darias, J.; Raja, H.A.; Shearer, C.A.; Miller, A.N.; vanderheyden, P.; Porras-Altaro, A.; Caballero-George, C. Phylogenetic diversity of sponge-associated fungi from the Cariibean and the Pacific of Panama and their in vitro effect on angiotensin and endothelin receptors. *Marine Biotechnol.* **2015**, *17*, 533–564. [CrossRef]
24. Telenti, A.; Hodcroft, E.B.; Robertson, D.L. The evolution biology of SARS-CoV-2 variants. (Review). *Cold Spring Harbor Perspect Med.* **2022**, *12*, 27. [CrossRef] [PubMed]
25. De, A.; Tiwari, A.; Pande, V.; Sinha, A. Evolutionary trilogy of malaria, angiotensin II and hypertension: Deeper insights and the way forward. *J. Hum. Hypertens.* **2021**, *36*, 344–351. [CrossRef]
26. Yuan, Y.; Meng, L.; Zhou, Y.; Lu, N. Genetic polymorphism of angiotensin-converting enzyme and hypertrophic cardio-myopathy risk. *Medicine* **2017**, *96*, 48. [CrossRef] [PubMed]
27. Bleumink, G.S.; Schut, A.F.; Sturkenboom, M.C.; Deckers, J.W.; van Duijn, C.M.; Stricker, B.H. Genetic polymorphisms and heart failure. *Genet Med.* **2004**, *6*, 465–474. [CrossRef] [PubMed]
28. Wei, W.; Wang, X.; Zhou, Y.; Shang, X.; Yu, H. The genetic risk factors for pregnancy-induced hypertension: Evidence from genetic polymorphisms. Review. *FASEB J.* **2022**, *36*, e22413. [CrossRef]
29. Kumar, A.; Sarde, S.J.; Bhandari, A. Revising angiotensinogen from phylogenetic and genetic variants perspectives. *Biochem. Biophys. Res. Commun.* **2014**, *446*, 504–518. [CrossRef]

30. Szekeres, M.; Nádasy, G.L.; Turu, G.; Süpeki, K.; Szidonya, L.; Buday, L.; Chaplin, T.; Clark, A.J.L.; Hunyady, L. Angiotensin II-induced expression of brain-derived neurotrophic factor in human and rat adrenocortical cells. *Endocrinology* **2010**, *151*, 1695–1703. [CrossRef]
31. Nadasy, G.L.; Varbiro, S.; Szekeres, M.; Kocsis, A.; Szekacs, B.; Monos, E.; Kollai, M. Biomechanics of resistance artery wall remodeling in angiotensin-II hypertension and subsequent recovery. *Kidney Blood Press. Res.* **2010**, *33*, 37–47. [CrossRef]
32. Monori-Kiss, A.; Antal, P.; Szekeres, M.; Varbiro, S.; Fees, A.; Szekacs, B.; Nadasy, G.L. Morphological remodeling of the intramural coronary resistance artery network geometry in chronically Angiotensin II infused hypertensive female rats. *Heliyon* **2020**, *6*, e03807. [CrossRef]
33. Turu, G.; Simon, A.; Gyombolai, P.; Szidonya, L.; Bagdy, G.; Lenkei, Z.; Hunyady, L. The role of diacylglycerol lipase in con-stitutive and angiotensin AT1 receptor-stimulated cannabinoid CB1 receptor activity. *J. Biol. Chem.* **2007**, *282*, 7753–7757. [CrossRef]
34. Szekeres, M.; Nadasy, G.L.; Turu, G.; Soltesz-Katona, E.; Toth, Z.E.; Balla, A.; Catt, K.J.; Hunyady, L. Angiotensin II induces vascular endocannabinoid release which attenuates its vasoconstrictor effect via CB1 cannabinoid receptors. *J. Biol. Chem.* **2012**, *287*, 31540–31550. [CrossRef] [PubMed]
35. Szekeres, M.; Nadasy, G.L.; Soltesz-Katona, E.; Hunyady, L. Control of myogenic tone and agonist induced contraction of intramural coronary resistance arterioles by cannabinoid type 1 receptors and endocannabinoids. *Prostaglandins Other Lipid Mediat.* **2018**, *134*, 77–83. [CrossRef] [PubMed]
36. Szekeres, M.; Nadasy, G.L.; Turu, G.; Soltesz-Katona, E.; Benyo, Z.; Offermanns, S.; Ruisanchez, E.; Szabo, E.; Takats, Z.; Batkai, S.; et al. Endocannabinoid-mediated modulation of Gq/11 protein-coupled receptor signal-ing-induced vasoconstriction and hypertension. *Mol. Cell Endocrinol.* **2015**, *403*, 46–56. [CrossRef] [PubMed]
37. Miklos, Z.; Wafa, D.; Nadasy, G.L.; Toth, Z.E.; Besztercei, B.; Dornyei, G.; Laska, Z.; Benyo, Z.; Ivanics, T.; Hunyady, L.; et al. Angiotensin II-induced cardiac effects are modulated by endocannabinoid-mediated CB$_1$ receptor activation. *Cells* **2021**, *10*, 724. [CrossRef] [PubMed]
38. Kumar, S.; Dietrich, N.; Kornfeld, K. Angiotensin Converting Enzyme (ACE) inhibitor extends *Caenorhabditis elegans* life span. *PLoS Genet.* **2016**, *12*, e1005866. [CrossRef] [PubMed]
39. Takimoto-Ohnishi, E.; Murakami, K. Renin–angiotensin system research: From molecules to the whole body. *J. Physiol. Sci.* **2019**, *69*, 581–587. [CrossRef]
40. Christie, G.A.; Lucas, C.; Bateman, D.N.; Waring, W.S. Redefining the ACE-inhibitor dose-response relationship: Substantial blood pressure lowering after massive doses. *Eur. J. Clin. Pharmacol.* **2006**, *62*, 989–993. [CrossRef]
41. Gleeson, P.J.; Crippa, I.A.; Mongkolpun, W.; Cavicchi, F.Z.; Meerhaghe, T.V.; Brimioulle, S.; Taccone, F.S.; Vincent, J.-L.; Creteur, J. Renin as a marker of tissue-perfusion and prognosis in critically ill patients. *Crit. Care Med.* **2019**, *47*, 152–158. [CrossRef]
42. Reschreiter, H.; Kowarik, K. Bronze age mining in Hallstatt. A new picture of everyday life in the salt mines and beyond. *Archeol. Austriat* **2019**, *103*, 99–136.
43. Roberts, W.C. High salt intake, its origins, its economic impact, and its effect on blood pressure. *Am. J. Cardiol.* **2001**, *88*, 1338–1346. [CrossRef]
44. Brown, I.; Tzoulaki, I.; Candeias, V.; Elliott, P. Salt intakes around the world: Implications for public health. *Int. J. Epidemiol.* **2009**, *38*, 791–813. [CrossRef] [PubMed]
45. World Health Organization. Salt Reduction. Available online: https://www.who.int/news-room/fact-sheets/detail/salt-reduction (accessed on 1 October 2023).
46. Storn, B.I.; Anderson, C.A.; Ix, J.H. Sodium reduction in populations insights from the Institute of Medicine Committee. *JAMA* **2013**, *310*, 31–32.
47. Schweda, F. Salt feed-back on the renin-angiotensin-aldosterone system (Review). *Pflugers Arch. Eur. J. Physiol.* **2015**, *467*, 565–576. [CrossRef] [PubMed]
48. Balafa, O.; Kalaitzidis, R.G. Salt sensitivity and hypertension. *J. Hum. Hypertens.* **2021**, *35*, 184–192. [CrossRef]
49. Kanbay, M.; Chen, Y.; Solak, Y.; Sanders, P.W. Mechanisms and consequences of salt sensitivity and dietary salt intake. *Curr. Opin. Nephrol. Hypertens.* **2011**, *20*, 37–43. [CrossRef]
50. He, F.J.; McGregor, G.A. Effect of longer-term salt reduction on blood pressure. (Review). *Cochrane Database Syst. Rev.* **2004**, *3*, CD004937.
51. Zidek, W. Can salt substitution or reduction replace pharmaceuticals for arterial hypertension? Review. *Innere Medizin* **2022**, *63*, 1097–1104. [CrossRef]
52. Kantor, E.D.; Rehm, C.D.; Haas, J.S.; Chan, A.T.; Giovannucci, E.L. Trends in drug use among adults in the United States from 1999–2012. *JAMA* **2015**, *314*, 1818–1831. [CrossRef]
53. Rouette, J.; McDonald, E.G.; Schuster, T.; Brophy, J.M.; Azoulay, L. Treatment and prescribing trends of antihypertensive drugs in 2.7 million UK primary care patients over 1 years: A population-based cohort study. *BMJ Open* **2022**, *12*, e057510. [CrossRef]

54. Jiao, T.; Platt, R.; Douros, A.; Filion, K.B. Prescription patterns for the use of antihypertensive drugs for primary prevention among patients with hypertension in the United Kingdom. *Am. J. Hypertens.* **2021**, *35*, 42–53. [CrossRef]
55. Sundball, J.; Adelborg, K.; Mansfield, K.E.; Tomlinson, L.A.; Schmidt, M. Seventeen-year nationwide trends in antihypertensive drug use in Denmark. *Am. J. Cardiol.* **2017**, *120*, 2193–2200. [CrossRef] [PubMed]

Disclaimer/Publisher's Note: The statements, opinions and data contained in all publications are solely those of the individual author(s) and contributor(s) and not of MDPI and/or the editor(s). MDPI and/or the editor(s) disclaim responsibility for any injury to people or property resulting from any ideas, methods, instructions or products referred to in the content.

 *biomedicines*

Review

Updates on the Renin–Angiotensin–Aldosterone System and the Cardiovascular Continuum

Dana Pop [1,2], Alexandra Dădârlat-Pop [1,3,*], Raluca Tomoaia [1,2], Dumitru Zdrenghea [1] and Bogdan Caloian [1,2]

[1] 4th Department of Internal Medicine, Faculty of Medicine, "Iuliu Hațieganu" University of Medicine and Pharmacy, 400347 Cluj-Napoca, Romania; pop67dana@gmail.com (D.P.); raluca.tomoaia@gmail.com (R.T.); dzdrenghea@yahoo.com (D.Z.); bogdan912@yahoo.com (B.C.)
[2] Cardiology Department, Rehabilitation Hospital, 400012 Cluj-Napoca, Romania
[3] Cardiology Department, Heart Institute "N. Stăncioiu", 400001 Cluj-Napoca, Romania
* Correspondence: dadarlat.alexandra@yahoo.ro

Abstract: The cardiovascular continuum describes how several cardiovascular risk factors contribute to the development of atherothrombosis, ischemic heart disease, and peripheral arteriopathy, leading to cardiac and renal failure and ultimately death. Due to its multiple valences, the renin–angiotensin–aldosterone system plays an important role in all stages of the cardiovascular continuum, starting from a cluster of cardiovascular risk factors, and continuing with the development of atherosclerosis thorough various mechanisms, and culminating with heart failure. Therefore, this article aims to analyze how certain components of the renin–angiotensin–aldosterone system (converting enzymes, angiotensin, angiotensin receptors, and aldosterone) are involved in the underlying pathophysiology of the cardiovascular continuum and the possible arrest of its progression.

Keywords: renin–angiotensin–aldosterone system; cardiovascular risk factors; stroke; heart failure; atrial fibrillation

Citation: Pop, D.; Dădârlat-Pop, A.; Tomoaia, R.; Zdrenghea, D.; Caloian, B. Updates on the Renin–Angiotensin–Aldosterone System and the Cardiovascular Continuum. *Biomedicines* **2024**, *12*, 1582. https://doi.org/10.3390/biomedicines12071582

Academic Editor: Mária Szekeres

Received: 30 May 2024
Revised: 3 July 2024
Accepted: 15 July 2024
Published: 17 July 2024

Copyright: © 2024 by the authors. Licensee MDPI, Basel, Switzerland. This article is an open access article distributed under the terms and conditions of the Creative Commons Attribution (CC BY) license (https://creativecommons.org/licenses/by/4.0/).

1. Introduction

The cardiovascular continuum was first described by Victor J. Dzau et al. in 2006 [1] as a chain of events precipitated by several cardiovascular risk factors, which may lead to the occurrence of atherothrombosis, ischemic heart disease, and peripheral arteriopathy if left untreated. Also, major complications such as cardiac and renal failure and finally death may occur [1].

The renin–angiotensin–aldosterone system (RAAS) and its significant classical role in the occurrence of arterial hypertension was first discussed in the 1960s [2,3]. In 1976, the first converting enzyme inhibitor, captopril, was patented, and in the early 1980s it was recognized as a medication for lowering blood pressure values [4]. Afterwards, other converting enzyme inhibitors (ACEIs) were developed, and then the blockers of AT1 receptors of angiotensin II (ARBs) or sartans were discovered. Over time, the understanding of RAAS has expanded tremendously, with ACEIs and ARBs having an important role in the treatment of heart failure by blocking RAAS at different steps. Also, these drugs, especially ACEIs, have anti-atherosclerotic effects, too. RAAS is ubiquitous in multiple organ systems, especially the kidneys, systemic vasculature, and adrenal cortex, and is strongly involved in the pathogenesis of hypertension, heart failure, other cardiovascular diseases, and renal diseases. Due to its multiple valences, RAAS plays a crucial role in all stages of the cardiovascular continuum, starting with several cardiovascular risk factors, and continuing with the development and progression of atherosclerosis, ultimately leading to the onset of heart failure.

Therefore, the discussion in this article is focused on certain components of the renin–angiotensin–aldosterone system (converting enzymes, angiotensin, angiotensin receptors, and aldosterone) and their involvement in the pathogenesis of various conditions within the cardiovascular continuum, as you can see in Figure 1.

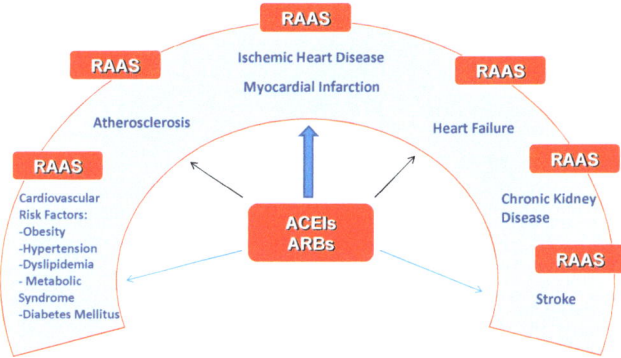

Figure 1. RAAS and the pathogenesis of various conditions within the cardiovascular continuum.

2. RAAS and Several Cardiovascular Risk Factors

2.1. Hypertension

The role of RAAS in the pathogenesis of arterial hypertension is already well known. The juxtaglomerular cells favor the cleavage of prorenin to renin. The activation of prorenin in the kidney is stimulated by several enzymes such as proconvertase 1 and cathepsin B [5]. Renin degrades angiotensinogen (AGT) synthesized in the liver into angiotensin I. Cathepsin D and tonins also have the same AGT action [5]. Angiotensin I (AG I) under the action of the "classic" converting enzyme (ACE) is transformed into angiotensin II (AG II)—Figure 2.

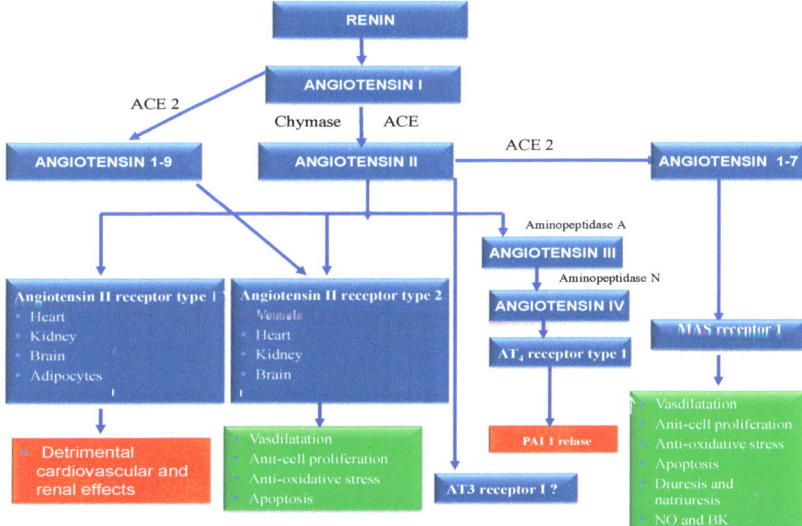

Figure 2. RAAS components—the angiotensins, the converting enzymes and their specific receptors.

However, there are other enzymes called non-EC enzymes (chymase and cathepsin G) that have the same effects [5].

Under pathological conditions, the excess of AG II causes stimulation of AG II AT1 receptors, leading to arterial vasoconstriction, sodium and water retention due to increased release of aldosterone, and finally to atrial remodeling and stimulation of the sympathetic nervous system. All these pathophysiological mechanisms are responsible for arterial hypertension development.

On the other hand, AT2 receptor stimulation results in vasodilatation and nitric oxide release, with proven antiproliferative effects [5]. At the same time, the conversion enzyme determines the degradation of bradykinin (into inactive peptides), which is known as a substance with important vasodilatory effects. The renin–angiotensin-converting enzyme-2 (ACE2) system has a central role in triggering some counter-regulatory mechanisms compared to the classic converting enzyme, with a structure somewhat similar to it (42% of the terminal amino acid structure is identical in the two enzymes).

ACE-2 acts on AG I, converting it to AG (1–9) and on AG II by hydrolyzing it to AG (1–7). Thus, ACE-2 has the role of decreasing the level of AG II where there are increases and activations of this hormone, as you can see in Figure 2.

Thus, EC-2 represents an important counter-regulatory substance of AG II, not only due to the hydrolysis of AG II, but also due to the production of AG (1–7) which, through actions on specific Mas receptors, determines the release of nitric oxide synthase (NOS) and increased sodium excretion—Figure 2. Also, AG (1–7) acts on AT2 receptors and type 2 bradykinin (BK) receptors, which implies important vasodilatory actions. In recent years, several components of the system have been identified, such as AG (1–12), angiotensin A, and alamandin—Figure 2.

Alamandin is produced by the degradation of angiotensin A by ACE-2 or directly from AG (1–7). It acts on specific receptors, such as -MrgD, which also has vasodilatory, antifibrotic, antiproliferative, and natriuretic effects [6,7]. So, we are talking about receptors whose stimulation causes favorable, cardioprotective effects (AT2, Mas, and MrgD) and about AT1 receptors with detrimental cardiovascular effects [8].

In healthy people, there is a balance between the production of ACE and ACE-2. In conditions of arterial hypertension, this balance is lost in favor of the production of ACE [8].

Considering the important role that the renin–angiotensin–aldosterone system plays in the pathogenesis of hypertension, current guidelines recommend angiotensin-converting enzyme inhibitors (ACEIs) or sartans (ARBs) as the first line of treatment in this disease in most patients if there are no contraindications [9].

2.2. Obesity

Obesity is another important risk factor often associated with high blood pressure. In people with obesity, a series of substances called adipokines are secreted from within richly represented adipose tissue. These are mainly represented by adiponectin, leptin, omentin, resistin, visfatin, TNF-α, IL-4, CRP, and PAI-1, but also by AGT. Apart from adiponectin and omentin, all other adipokines, and therefore also AGT, are involved in the emergence of endothelial dysfunction [10].

At the same time, the existence of local AG II in adipose tissue and AT1 and AT2 receptors at the adipocyte level have been described [11]. Another interesting thing demonstrated is that stimulation of AT2 receptors by AG II promotes preadipocyte differentiation, and, in humans, AG II generated by mature adipocytes suppresses the differentiation of adipocyte precursors, decreasing the proportion of small insulin-sensitive adipocytes [12]. AGII also has other actions at the adipocyte level: it determines leptin release, increases PGI2 secretion, and favors the activity and transcription rate of glycerol-3-phosphate dehydrogenase and fatty acid synthase [13]. On the other hand, in people with obesity, the presence of increased insulin resistance is accompanied by the amplification of renin activity [13].

The renin–angiotensin system, locally present, plays an important role in the development of adipocytes [13]. At the same time, the presence of important components of the renin–angiotensin system at the adipocyte level could have detrimental cardiovascular effects in people with obesity, which could lead to the creation of a vicious circle of obesity–arterial hypertension–diabetes–atherosclerosis.

2.3. Diabetes Mellitus

Diabetes mellitus is frequently associated with hypertension and obesity.

Animal studies have shown that in the presence of prediabetes, there is important activity of the renin–angiotensin system. Thus, increased values of renin, angiotensinogen, ACE, AT1 receptor, aldosterone, and AG II were detected in these studies [14]. The presence of diabetes in the context of hyperglycemia increases insulin resistance and causes endothelial dysfunction with the activation of the renin–angiotensin system and levels of AG II.

The important role of the renin–angiotensin system in diabetes was demonstrated by the evidence provided by ACEIs and ARBs in the treatment of diabetic nephropathy, in which they determined the decrease in proteinuria, renoprotective effects, and the regression of this complication [15]. Moreover, at the renal level, there are all the components of RAAS. The renal effects of the excess release of AG II cause glomerular sclerosis, tubulointerstitial fibrosis, and a reduction in nephron mass with the appearance of chronic nephropathy through multiple mechanisms.

At the retinal level, there are components of the renin–angiotensin system: renin, AGT, AG II, EC, EC-2, AG 1–7, AG II, AT1, AT2, and Mas receptors [16]. AG II has an important role in maintaining local homeostasis, with the control of vasoconstriction, the regulation of glial cell function, and the modulation of neuronal functions [17].

In a study conducted in Canada in which type 1 diabetes patients were included, it was demonstrated that the renin–angiotensin–aldosterone system is involved in the occurrence of diabetic retinopathy [18]. It contributes to the increase in vascular permeability and cell proliferation with important pro-inflammatory effects and increasing oxidative stress, favoring the development of diabetic retinopathy [19].

Under these conditions, ACEIs and ARBs have proven their effectiveness in patients with diabetic retinopathy. Thus, lisinopril administered to patients with type 1 diabetes with normal blood pressure caused a decrease in the progression of diabetic retinopathy of up to 50% [20]. At the same time, the administration of candesartan in patients with type 2 diabetes was associated with a decrease in the progression of retinopathy [21].

2.4. Dyslipidemia

The accumulation of Ox-LDL in the arteries causes the activation of the local RAAS, which further contributes to the production of LDL and its oxidation into ox-LDL. All this leads to increased oxidative stress and inflammation [22,23]. At the same time, there is recent evidence showing that megalin, an important component of the low-density lipoprotein receptor superfamily, can contribute to RASS activation, promoting atherosclerosis [24].

2.5. Obstructive Sleep Apnea

Recent data highlight a possible role of RASS in obstructive sleep apnea. Thus, in a meta-analysis published in 2023 that includes data collected from 20 studies that included 2828 participants, it is shown that patients with obstructive sleep apnea not only have higher levels of the components of the renin–angiotensin–aldosterone system, but also have higher values of blood pressure and heart rate compared to those without this pathology, even among patients without treatment-resistant hypertension [25].

3. Renin–Angiotensin–Aldosterone System and Renal Function

Excessive production of AG II and aldosterone in pathological situations is associated with the development and progression of kidney diseases, such as diabetic or non-diabetic nephropathy. The mechanisms involved in this sense would be increased glomerular capillary pressure, profibrotic effects, and proteinuria [26].

The damage to the nephrons in the context of some pathological renal injuries causes the increase in filtration and the increase in glomerular capillary pressure, with consequent damage to the glomeruli. RASS is involved in the synthesis of nephrin, a protein that is formed at the transmembrane level, in glomerular podocytes [27] which has a role in limiting the release of proteins from the glomerular level.

Activation of RASS causes a decrease in the release of nephrin with the appearance of proteinuria. AG (1–7) determines, at the renal level, the stimulation of phospholipase activity, the stimulation of sodium transport in the proximal tubule, natriuresis, diuretics, and the increase in glomerular filtration [28].

At the renal and local levels, there are components of RASS. AG II is involved in cell proliferation, increased collagen deposits, and apoptosis, and the appearance of glomerular sclerosis and tubulointerstitial fibrosis, all of which lead to glomerular sclerosis and tubulointerstitial fibrosis. Further evolution towards reducing the number of nephrons and chronic nephropathy is needed.

Stimulation of AT1 receptors at the renal level causes local vasoconstriction, increased renal sodium reabsorption, activation of pro-inflammatory cytokines, activation of oxidative stress and endothelial dysfunction, increased PAI-1 activity, and promotion of thrombosis [29]. On the other hand, stimulation of renal AT2 receptors leads to renal and systemic vasodilatation, a decrease in renal sodium reabsorption, inflammation, mitogenesis, and fibrosis [29].

RASS blockers have renoprotective effects, independent of blood pressure values, and can prevent proteinuria and its regression when it exists. These medications cause vasodilatation of the related arterioles with a reduction in intraglomerular pressure, regression of remodeling at the arterial level, and improvement of endothelial function at the level of resistance arterioles, especially in patients with nephrosclerosis [29].

4. Implications of the Renin–Angiotensin–Aldosterone System in the Development of Atherosclerosis

The presence of the risk factors mentioned above can contribute over time to the appearance of the atherosclerosis process. The RAAS is also involved in the generation and progression of the atheroma plaque. RAAS components act on all cells involved in plaque formation at the level of smooth muscle cells, endothelial cells, and macrophages [30].

At the same time, AT1, AT2, and Mas receptors exist at these levels [31]. Through the actions of AG II and AG IV, endothelial dysfunction is promoted. The RAAS plays a key role in regulating the inflammatory response by attracting inflammatory cells to the site of arterial wall insult. At the same time, local synthesis of RAAS components takes place at the level of inflammatory cells [32]. Other mechanisms by which AG II can contribute to the formation of atheroma plaque are represented by vasoconstriction, endothelial dysfunction, bradykinin degradation, effects on fibrinolysis, increased release of nitric oxide, and plasminogen activator inhibitor I.

On the contrary, bradykinin and angiotensin (1–7) contribute to maintaining a healthy endothelium [32]. Thus, it is clear that ACEs can have anti-atherosclerotic effects. Several ACEIs have proven cardioprotective effects due to their effects on the bradykinin cascade, fibrinolysis, and the decrease in the levels of AG II [33].

In the HOPE trial, patients with high cardiovascular risk were included, and were administered ramipril. Authors reported a decrease in the rate of occurrence of cardiovascular mortality, acute myocardial infarction, or stroke. The beneficial actions of ramipril were mostly independent of the reduction in blood pressure values, and also of the presence of left ventricular hypertrophy, which argues for the direct anti-ischemic effects of ramipril [34].

Treatment with perindopril in the EUROPA study, in patients with stable angina pectoris and low cardiovascular risk, led to a significant decrease in the main objectives represented by cardiovascular mortality, myocardial infarction, and cardiovascular resuscitations [35]. In the PERSPECTIVE sub-study within the EUROPA trial, the same ACE achieved more than that, namely, stabilization of the coronary plaque evaluated by intravascular ultrasonography [36].

In another study with perindopril, PREAMI, patients over 65 years of age with myocardial infarction and normal ejection fraction were evaluated, and administration of this ACE led to a significant decrease in the risk of heart failure and death [37]. In the PERTINENT study, the effects of perindopril on some markers of atherosclerosis were followed. Its administration caused significant decreases in the levels of angiotensin II, tumor necrosis factor-alpha, von Willebrand factor, and increased levels of nitric oxide, bradykinin, and nitrates/nitrites [38], but did not influence the levels of C-reactive protein and fibrinogen [39].

The addition of perindopril to amlodipine, in the ASCOT-BPLA study, in hypertensive patients determined a significant reduction in the rate of cardiovascular events, the risk of nonfatal myocardial infarction, and fatal ischemic heart disease compared to "standard" β-blocker (atenolol) antihypertensive therapy ± diuretic (thiazide) [40]. This reduction has been observed since the introduction of perindopril therapy [40].

Anti-atherosclerotic effects were also proven by quinapril therapy, which, when administered to post-PTCA patients, caused a significant reduction in the rate of restenosis compared to placebo in the TREND study [41].

The benefits of ACE inhibitor therapy in myocardial infarction were primarily demonstrated by the SMILE trial, in which administration of zofenopril within 24 h of the onset of the myocardial infarction resulted in a significant reduction in mortality after 6 weeks of treatment, which was maintained or increased after one year of treatment [42].

A meta-analysis published in 2000 that included three large trials (SAVE, TRACE, AIRE) concluded that ACE inhibitors administered 48 h from the onset of AMI caused a decrease in mortality, a reduction in the risk of heart failure and the number of hospitalizations for this condition. At the same time, there was also a reduction in the risk of reinfarction [43]. In conclusion, the trials that included patients with various forms of ischemic heart disease and who benefited from ACEI therapy (SAVE, AIRE, TRACE, SMILE, EUROPA, and QUIET) demonstrated the benefits of this class of drugs in terms of reducing cardiovascular events.

The anti-atherosclerotic effects of ARBs are controversial. Undoubtedly, ARBs also have anti-inflammatory effects, demonstrated in animal studies, that are largely mediated by blocking AT1 receptors, inhibiting the release of pro-inflammatory cytokines, such as tumor necrosis factor (TNF)-α, interleukin (IL)-6, and decreased aldosterone release [44,45]. ARBs do not cause an increase in BK synthesis and contribute to an increase in AG II levels.

Under these conditions, AG II actions are not exerted on AT1 receptors, but can be exerted on alternative pathways with the production of AG III and IV with overactivation of AT2 and AT4 receptors [33]. Stimulation of AT2 receptors by AG II leads to the release of prostaglandin E2, which mediates the release of matrix metalloproteinases from macrophages, an enzyme with an important role in the rupture of the atheroma plaque [46].

In turn, AGIV, by stimulating AT4 receptors, causes an increase in the production of nuclear factor-κB and the release of other pro-inflammatory factors that act on vascular smooth muscle cells [44,47]. Blocking AT4 receptors, however, also causes important myocardial antiapoptotic, antifibrotic, and antihypertrophic effects [44,48]. The TRANSCEND study (Telmisartan Randomized Assessment Study in ACE-I Intolerant Subjects with Cardiovascular Disease) evaluated the role of telmisartan in patients diagnosed with ischemic heart disease or diabetes with organ damage, who did not tolerate ACE inhibitors. Authors reported a significant reduction in the risk of major cardiovascular events [49]. At the same time, in the ONTARGET study (Ongoing Telmisartan Alone and in Combination with Ramipril Global End-point Trial) the effectiveness of telmisartan was similar to that of ramipril in reducing the risk of cardiovascular mortality, myocardial infarction, stroke, or hospitalizations for heart failure. It was also evaluated when it was administered to patients with vascular diseases or diabetics with increased cardiovascular risk, without heart failure [50]. Another ARB, losartan, had similar effects to captopril in post-myocardial infarction patients [51].

In the OLIVUS study (Impact of OLmesartan on progression of coronary atherosclerosis: evaluation by IntraVascular UltraSound), the administration of olmesartan to some patients with stable angina pectoris caused a decrease in the volume of the atheroma plaque [52].

Among the ARBs, only valsartan proved its effectiveness in reducing cardiovascular events after myocardial infarction [53], with the VALIANT study demonstrating that valsartan administered post-myocardial infarction causes the same decrease in mortality as captopril [53].

ACEIs (or ARBs) are indicated in the European ESC for the secondary prevention of ischemic heart disease in patients who also have heart failure, diabetes, or arterial hypertension [54].

Aldosterone-related medication has important antifibrotic actions, including the regression of myocardial remodeling and myocardial infarction. In the EPHESUS study, which included post-myocardial infarction patients with an ejection fraction of less than or equal to 40%, patients were given 50 mg of eplerenone daily, and there was a significant decrease in all-cause mortality, cardiovascular mortality, and hospitalizations for heart failure [55]. The same results were obtained in the RALES study performed with spironolactone [56]. The ESC guidelines recommend anti-aldosterone medication in patients with a history of myocardial infarction, with EF $\leq$40%, with heart failure or diabetes, and without renal failure or hyperkalemia, along with treatment with ACE inhibitors and beta-blockers [54].

5. The Renin–Angiotensin–Aldosterone System and Stroke

The excess secretion of AG II also has negative effects on cerebral vessels. It increases inflammation and promotes atherosclerosis at this level, activates the sympathetic nervous system and also the local RAAS. This is because, at the cerebral level, there is also a component of the RASS, including AT1 and AT2 receptors [57].

Both ACEIs and ARBs have proven their effectiveness in the secondary prevention of cerebral vascular accidents [58]. Thus, in the PROGRESS (the perindopril protection against recurrent stroke) study, perindopril was determined to reduce the risk of recurrent stroke, with the benefits being similar in both normotensive and hypertensive patients [59]. In a sub-study of the Heart Outcomes Prevention Evaluation (HOPE) trial, patients receiving ramipril had a 32% reduction in the risk of any type of stroke and a 61% reduction in fatal stroke compared to the placebo group [34]. The benefits were observed in all patients, regardless of BP values. Beneficial effects were also obtained following treatment with ARBs.

In the LIFE (Losartan Intervention for Endpoint Reduction in Hypertension) trial, losartan administered to hypertensive patients with left ventricular hypertrophy resulted in a 24% reduction in major vascular events and a 21% reduction in stroke compared to placebo [60]. The analysis of a subgroup of patients with previous stroke within the SCOPE trial (Study on Cognition and Prognosis in the Elderly) showed that the administration of candesartan significantly reduced the rate of recurrent stroke and cardiovascular events [61]. Also, in the ACCES (Evolution of Acute Candesartan Cilexetil Therapy in Stroke Survivors) trial, candesartan administered to hypertensive patients on the first day of stroke, determined after one year, resulted in a significant reduction in mortality and cardiovascular events [62].

6. The Renin–Angiotensin–Aldosterone System and Heart Failure

Secretion of AG II, stimulation of AT1 receptors, and release of aldosterone represent the main mechanisms involved in the pathogenesis of heart failure (HF). The main actions of aldosterone consist of cardiac effects (cardiac fibrosis), renal effects (sodium/water retention and potassium excretion), but also other effects (endothelial dysfunction).

There is a series of well-known studies in which there were patients with HF, and in which the administration of ACEIs and ARBs determined the decrease in cardiovascular mortality and the number of days of hospitalization [63–66]. There are also meta-analyses in which ACEI or ARB studies were included, which proved the effectiveness of the two drug classes in HF, especially in terms of reducing mortality [67,68].

Thus, in the CONSENSUS (Cooperative North Scandinavian Enalapril Study) [63] and SOLVD (Studies of Left Ventricular Dysfunction) [64] trials, it was demonstrated that the administration of ACEs, specifically enalapril, in patients with heart failure, increased survival and improved the NYHA class in patients with this condition. In the Val-HeFT trial (The Valsartan Heart Failure Trial), which included valsartan administered to patients with CHF, a statistically significant reduction of 13.3% in the risk of mortality and cardiovascular morbidity was determined [66]. The administration of candesartan in the CHARM trial (Candesartan in Heart Failure–assessment of reduction in mortality and morbidity) in patients with CHF who did not tolerate ACE inhibitors led to a 23% reduction in cardiovascular mortality and hospital days [65].

The latest ESC guidelines recommend ARBs in patients with reduced ejection fractions if they do not tolerate ACEIs or angiotensin receptor–neprilysin inhibitors. These drug classes reduce cardiovascular mortality and hospitalizations for HF [69].

7. The Renin–Angiotensin–Aldosterone System and Atrial Fibrillation

The RAAS is implicated in the occurrence of atrial fibrillation (AF) through several mechanisms, the most important of which would be the direct ones on the atrial structural and electrical properties and the indirect ones in the context of HF and arterial hypertension (important comorbidities of AF).

In arterial hypertension, the elevation of left atrial pressure and LV end-diastolic pressure occurs, which favors the development of AF [70]. In animal models with HF, ACEIs, and ARBs determined the regression of atrial remodeling and fibrosis and shortened the atrial effective refractory period [71]. At the same time, an analysis of human atrial myocytes taken from patients undergoing cardiac surgery reported increased tissue levels of ACE and AG II receptors in patients with AF compared to those in sinus rhythm [72].

The increase in the level of AG II activates the mechanisms of inflammation and fibrosis and the release of metalloproteinases with the excess production of collagen at the atrial level [73]. Genetic polymorphisms of ACE (ACE I/D polymorphism) and aldosterone synthase involved in the occurrence of AF have also been described [73]. At the same time, it is important to underline the fact that RAAS components are also synthesized at the local and atrial level.

The effects of ARBs respective to ACEIs in AF have been studied substantially. Thus, in the VALUE study that included hypertensive patients, the administration of valsartan was accompanied by a decrease in the risk of AF by 24% in the first 3 years, and 16% after 6 years [74].

In a systematic review and meta-analysis of 26 randomized clinical trials, it was shown that RAAS blockade had an important role in the prophylaxis of AF in patients with HF, causing a 24% reduction in the risk of its occurrence [75]. Other data show that RAAS blockade resulted in a 34% reduction in progression to permanent AF [73].

In a meta-analysis published in 2015, it was demonstrated that the use of ACEIs and ARBs in comparison with beta-blockers, calcium blockers, or diuretics, prevented the occurrence of AF in patients with systolic HF and those with arterial hypertension [73].

The role of anti-aldosteronics in the development of AF is also discussed. Thus, eplerenone was beneficial in maintaining sinus rhythm after catheter ablation in patients with permanent AF [76]. A very recent meta-analysis shows that the administration of anti-aldosteronics in patients with cardiovascular disease or at risk of developing cardiovascular disease is equally effective in preventing cardiovascular events in patients with and without HF and most likely retains its effectiveness regardless of the presence of AF [77]. They have demonstrated moderate effectiveness in preventing the onset or recurrence of AF episodes [77].

The 2023 ESH Guidelines for the management of arterial hypertension emphasize the importance of RAAS blockers, along with beta-blockers, in the prevention of AF recurrences in hypertensive patients [7].

8. Conclusions

RAAS plays a primary role in the pathogenesis of atherosclerosis, risk factors, and cardiovascular diseases, exerting significant effects throughout the entire cardiovascular continuum.

Author Contributions: Conceptualization, D.P., R.T. and A.D.-P.; investigation, D.P., R.T. and A.D.-P.; resources, D.P., R.T. and A.D.-P.; data curation, D.Z. and B.C.; writing—original draft preparation, D.P., R.T. and A.D.-P.; writing—review and editing, D.Z. and B.C.; visualization, D.Z. and B.C.; supervision, D.Z. and B.C. All authors have read and agreed to the published version of the manuscript.

Funding: This research received no external funding.

Institutional Review Board Statement: Not applicable.

Informed Consent Statement: Not applicable.

Data Availability Statement: No new data were created or analyzed in this study. Data sharing is not applicable to this article.

Conflicts of Interest: The authors declare no conflicts of interest.

References

1. Dzau, V.J.; Antman, E.M.; Black, H.R.; Hayes, D.L.; Manson, J.E.; Plutzky, J.; Popma, J.J.; Stevenson, W. The cardiovascular disease continuum validated: Clinical evidence of improved patient outcomes: Part I: Pathophysiology and clinical trial evidence (risk factors through stable coronary artery disease). *Circulation* **2006**, *114*, 2850–2870. [CrossRef]
2. Taquini, A.C., Jr.; Taquini, A.C. The renin-angiotensin system in hypertension. *Am. Heart J.* **1961**, *62*, 558–564. [CrossRef]
3. Genest, J.; Boucher, R.; De Champlain, J.; Veyrat, R.; Chretien, M.; Biron, P.; Tremblay, G.; Roy, P.; Cartier, P. Studies on the renin-angiotensin system in hypertensive patients. *Can. Med. Assoc. J.* **1964**, *90*, 263–268. [PubMed]
4. Vidt, D.G.; Bravo, E.L.; Fouad, F.M. Drug therapy. Captopril. *N. Engl. J. Med.* **1982**, *306*, 214–219. [CrossRef]
5. Staessen, J.A.; Li, Y.; Richart, T. Oral renin inhibitors. *Lancet* **2006**, *368*, 1449–1456. [CrossRef]
6. Iwai, M.; Horiuchi, M. Devil and angel in the renin-angiotensin system: ACE-angiotensin II-AT1 receptor axis vs. ACE2-angiotensin-(1-7)-Mas receptor axis. *Hypertens. Res.* **2009**, *32*, 533–536. [CrossRef]
7. Mancia, G.; Kreutz, R.; Brunström, M.; Burnier, M.; Grassi, G.; Januszewicz, A.; Muiesan, M.L.; Tsioufis, K.; Agabiti-Rosei, E.; Algharably, E.A.E.; et al. 2023 ESH Guidelines for the management of arterial hypertension The Task Force for the management of arterial hypertension of the European Society of Hypertension: Endorsed by the International Society of Hypertension (ISH) and the European Renal Association (ERA). *J. Hypertens.* **2023**, *41*, 1874–2071. [CrossRef]
8. Villela, D.C.; Passos-Silva, D.G.; Santos, R.A. Alamandine: A new member of the angiotensin family. *Curr. Opin. Nephrol. Hypertens.* **2014**, *23*, 130–134. [CrossRef]
9. Tanrıverdi, L.H.; Özhan, O.; Ulu, A.; Yıldız, A.; Ateş, B.; Vardı, N.; Acet, H.A.; Parlakpinar, H. Activation of the Mas receptors by AVE0991 and MrgD receptor using alamandine to limit the deleterious effects of Ang II-induced hypertension. *Fundam. Clin. Pharmacol.* **2023**, *37*, 60–74. [CrossRef]
10. Lau, D.C.; Dhillon, B.; Yan, H.; Szmitko, P.E.; Verma, S. Adipokines: Molecular links between obesity and atheroslcerosis. *Am. J. Physiol. Heart Circ. Physiol.* **2005**, *288*, H2031–H2041. [CrossRef]
11. Schütten, M.T.; Houben, A.J.; de Leeuw, P.W.; Stehouwer, C.D. The Link Between Adipose Tissue Renin-Angiotensin-Aldosterone System Signaling and Obesity-Associated Hypertension. *Physiology* **2017**, *32*, 197–209. [CrossRef]
12. Underwood, P.C.; Adler, G.K. The renin angiotensin aldosterone system and insulin resistance in humans. *Curr. Hypertens. Rep.* **2013**, *15*, 59–70. [CrossRef] [PubMed]
13. Cassandra Mkhize, B.; Mosili, P.; Sethu Ngubane, P.; Khathi, A. The relationship between adipose tissue RAAS activity and the risk factors of prediabetes: A systematic review and meta-analysis. *Adipocyte* **2023**, *12*, 2249763. [CrossRef]
14. Mkhize, B.C.; Mosili, P.; Ngubane, P.S.; Sibiya, N.H.; Khathi, A. Diet-induced prediabetes: Effects on the activity of the renin-angiotensin-aldosterone system in selected organs. *J. Diabetes Investig.* **2022**, *13*, 768–780. [CrossRef] [PubMed]
15. Fagher, K.; Löndahl, M. The combined impact of ankle-brachial index and transcutaneous oxygen pressure on mortality in patients with type 2 diabetes and foot ulcers. *Acta Diabetol.* **2021**, *58*, 1359–1365. [CrossRef] [PubMed]
16. White, A.J.; Cheruvu, S.C.; Sarris, M.; Liyanage, S.S.; Lumbers, E.; Chui, J.; Wakefield, D.; McCluskey, P.J. Expression of classical components of the renin-angiotensin system in the human eye. *J. Renin-Angiotensin-Aldosterone Syst.* **2015**, *16*, 59–66. [CrossRef] [PubMed]
17. Fletcher, E.L.; Phipps, J.A.; Ward, M.M.; Vessey, K.A.; Wilkinson-Berka, J.L. The renin-angiotensin system in retinal health and disease: Its influence on neurons, glia and the vasculature. *Prog. Retin. Eye Res.* **2010**, *29*, 284–311. [CrossRef] [PubMed]

18. Lovshin, J.A.; Lytvyn, Y.; Lovblom, L.E.; Katz, A.; Boulet, G.; Bjornstad, P.; Lai, V.; Cham, L.; Tse, J.; Orszag, A.; et al. Retinopathy and RAAS Activation: Results From the Canadian Study of Longevity in Type 1 Diabetes. *Diabetes Care* **2019**, *42*, 273–280. [CrossRef] [PubMed]
19. Xu, R.; Fang, Z.; Wang, H.; Gu, Y.; Yu, L.; Zhang, B.; Xu, J. Molecular mechanism and intervention measures of microvascular complications in diabetes. *Open Med.* **2024**, *19*, 20230894. [CrossRef]
20. Chaturvedi, N.; Sjolie, A.K.; Stephenson, J.M.; Abrahamian, H.; Keipes, M.; Castellarin, A.; Rogulja-Pepeonik, Z.; Fuller, J.H. Effect of lisinopril on progression of retinopathy in normotensive people with type 1 diabetes. The EUCLID Study Group. EURODIAB Controlled Trial of Lisinopril in Insulin-Dependent Diabetes Mellitus. *Lancet* **1998**, *351*, 28–31. [CrossRef]
21. Tillin, T.; Orchard, T.; Malm, A.; Fuller, J.; Chaturvedi, N. The role of antihypertensive therapy in reducing vascular complications of type 2 diabetes. Findings from the DIabetic REtinopathy Candesartan Trials-Protect 2 study. *J. Hypertens.* **2011**, *29*, 1457–1462. [CrossRef] [PubMed]
22. Singh, B.M.; Mehta, J.L. Interactions between the renin-angiotensin system and dyslipidemia: Relevance in the therapy of hypertension and coronary heart disease. *Arch. Intern. Med.* **2003**, *163*, 1296–1304. [CrossRef] [PubMed]
23. Chen, J.; Mehta, J.L. Interaction of oxidized low-density lipoprotein and the renin-angiotensin system in coronary artery disease. *Curr. Hypertens. Rep.* **2006**, *8*, 139–143. [CrossRef] [PubMed]
24. Kukida, M.; Sawada, H.; Daugherty, A.; Lu, H.S. Megalin: A bridge connecting kidney, the renin-angiotensin system, and atherosclerosis. *Pharmacol. Res.* **2020**, *151*, 104537. [CrossRef] [PubMed]
25. Loh, H.H.; Lim, Q.H.; Chai, C.S.; Goh, S.L.; Lim, L.L.; Yee, A.; Sukor, N. Influence and implications of the renin-angiotensin-aldosterone system in obstructive sleep apnea: An updated systematic review and meta-analysis. *J. Sleep Res.* **2023**, *32*, e13726. [CrossRef] [PubMed]
26. Brewster, U.C.; Perazella, M.A. The renin-angiotensin-aldosterone system and the kidney: Effects on kidney disease. *Am. J. Med.* **2004**, *116*, 263–272. [CrossRef] [PubMed]
27. Benigni, A.; Tomasoni, S.; Gagliardini, E.; Zoja, C.; Grunkemeyer, J.A.; Kalluri, R.; Remuzzi, G. Blocking angiotensin II synthesis/activity preserves glomerular nephrin in rats with severe nephrosis. *J. Am. Soc. Nephrol.* **2001**, *12*, 941–948. [CrossRef]
28. Pop, D.; Zdrenghea, D.; Cucuianu, M.; Zdrenghea, D. *Actualități în Patologia Biochimică a Bolilor Cardiovasculare*; Casa Cărții de Știință: Cluj-Napoca, Romania, 2004; pp. 157–177.
29. Pop, D. *Sistemul Renină-Angiotensina-Aldosteron în Patogeneza Bolilor Cardivasculare*; Clusium: Cluj-Napoca, Romania, 2007; pp. 91–105.
30. Nehme, A.; Zibara, K. Cellular distribution and interaction between extended renin-angiotensin-aldosterone system pathways in atheroma. *Atherosclerosis* **2017**, *263*, 334–342. [CrossRef] [PubMed]
31. Montezano, A.C.; Nguyen Dinh Cat, A.; Rios, F.J.; Touyz, R.M. Angiotensin II and vascular injury. *Curr. Hypertens. Rep.* **2014**, *16*, 431. [CrossRef] [PubMed]
32. Durante, A.; Peretto, G.; Laricchia, A.; Ancona, F.; Spartera, M.; Mangieri, A.; Cianflone, D. Role of the renin-angiotensin-aldosterone system in the pathogenesis of atherosclerosis. *Curr. Pharm. Des.* **2012**, *18*, 981–1004. [CrossRef] [PubMed]
33. Lévy, B.I.; Mourad, J.J. Renin Angiotensin Blockers and Cardiac Protection: From Basis to Clinical Trials. *Am. J. Hypertens.* **2022**, *35*, 293–302. [CrossRef] [PubMed]
34. Heart Outcomes Prevention Evaluation Study Investigators; Yusuf, S.; Sleight, P.; Pogue, J.; Bosch, J.; Davies, R.; Dagenais, G. Effects of an angiotensin-converting-enzyme inhibitor, ramipril, on cardiovascular events in high-risk patients. *N. Engl. J. Med.* **2000**, *342*, 145–153. [CrossRef] [PubMed]
35. Fox, K.M. EURopean trial on reduction of cardiac events with Perindopril in stable coronary Artery disease Investigators. Efficacy of perindopril in reduction of cardiovascular events among patients with stable coronary artery disease: Randomised, double-blind, placebo-controlled, multicentre trial (the EUROPA study). *Lancet* **2003**, *362*, 782–788. [CrossRef] [PubMed]
36. Rodriguez-Granillo, G.A.; Vos, J.; Bruining, N.; Garcia-Garcia, H.M.; de Winter, S.; Ligthart, J.M.; Deckers, J.W.; Bertrand, M.; Simoons, M.L.; Ferrari, R.; et al. Long-term effect of perindopril on coronary atherosclerosis progression (from the perindopril's prospective effect on coronary atherosclerosis by angiography and intravascular ultrasound evaluation [PERSPECTIVE] study). *Am. J. Cardiol.* **2007**, *100*, 159–163. [CrossRef] [PubMed]
37. Magrini, G.; Nicolosi, G.L.; Chiariello, M.; Ferrari, R.; Remme, P.; Tavazzi, L. Razionale, peculiarità e disegno dello studio PREAMI (Perindopril and Remodelling in the Elderly with Acute Myocardial Infarction) [Rationale, characteristics and study design of PREAMI (Perindopril and Remodelling in the Elderly with Acute Myocardial Infraction)]. *Ital. Heart J.* **2005**, *6* (Suppl. S7), 14S–23S. (In Italian) [PubMed]
38. Ceconi, C.; Fox, K.M.; Remme, W.J.; Simoons, M.L.; Bertrand, M.; Parrinello, G.; Kluft, C.; Blann, A.; Cokkinos, D.; Ferrari, R. ACE inhibition with perindopril and endothelial function. Results of a substudy of the EUROPA study: PERTINENT. *Cardiovasc. Res.* **2007**, *73*, 237–246. [CrossRef] [PubMed]
39. Ceconi, C.; Fox, K.M.; Remme, W.J.; Simoons, M.L.; Deckers, J.W.; Bertrand, M.; Parrinello, G.; Kluft, C.; Blann, A.; Cokkinos, D.; et al. ACE inhibition with perindopril and biomarkers of atherosclerosis and thrombosis: Results from the PERTINENT study. *Atherosclerosis* **2009**, *204*, 273–275. [CrossRef]

40. Dahlöf, B.; Sever, P.S.; Poulter, N.R.; Wedel, H.; Beevers, D.G.; Caulfield, M.; Collins, R.; Kjeldsen, S.E.; Kristinsson, A.; McInnes, G.T.; et al. Prevention of cardiovascular events with an antihypertensive regimen of amlodipine adding perindopril as required versus atenolol adding bendroflumethiazide as required, in the Anglo-Scandinavian Cardiac Outcomes Trial-Blood Pressure Lowering Arm (ASCOT-BPLA): A multicentre randomised controlled trial. *Lancet* **2005**, *366*, 895–906.
41. Mancini, G.J.B.; Henry, G.C.; Macaya, C.; O'Neill, B.J.; Pucillo, A.L.; Carere, R.G.; Wargovich, T.J.; Mudra, H.; Lüscher, T.F.; Klibaner, M.I.; et al. ACE inhibition with quinapril improves endothelial vasomotor dysfunction in patients with coronary artery disease. The TREND study. *Circulation* **1994**, *94*, 258–265. [CrossRef]
42. Ambrosioni, E.; Borghi, C.; Magnani, B. The effect of the angiotensin-converting-enzyme inhibitor zofenopril on mortality and morbidity after anterior myocardial infarction. The Survival of Myocardial Infarction Long-Term Evaluation (SMILE) Study Investigators. *N. Engl. J. Med.* **1995**, *332*, 80–85. [CrossRef]
43. Flather, M.D.; Yusuf, S.; Køber, L.; Pfeffer, P.M.; Hall, A.; Murray, P.G.; Torp-Pedersen, C.; Ball, P.S.; Pogue, J.; Moyé, L.; et al. Long-term ACE-inhibitor therapy in patients with heart failure or left-ventricular dysfunction: A systematic overview of data from individual patients. ACE-Inhibitor Myocardial Infarction Collaborative Group. *Lancet* **2000**, *355*, 1575–1581. [CrossRef] [PubMed]
44. Alcocer, L.A.; Bryce, A.; De Padua Brasil, D.; Lara, J.; Cortes, J.M.; Quesada, D.; Rodriguez, P. The Pivotal Role of Angiotensin-Converting Enzyme Inhibitors and Angiotensin II Receptor Blockers in Hypertension Management and Cardiovascular and Renal Protection: A Critical Appraisal and Comparison of International Guidelines. *Am. J. Cardiovasc. Drugs* **2023**, *23*, 663–682. [CrossRef] [PubMed]
45. Benicky, J.; Sanchez-Lemus, E.; Pavel, J.; Saavedra, J.M. Anti-inflammatory efects of angiotensin receptor blockers in the brain and the periphery. *Cell. Mol. Neurobiol.* **2009**, *29*, 781–792. [CrossRef] [PubMed]
46. Kim, M.P.; Zhou, M.; Wahl, L.M. Angiotensin II increases human monocyte matrix metalloproteinase-1 through the AT2 receptor and prostaglandin E2: Implications for atherosclerotic plaque rupture. *J. Leukoc. Biol.* **2005**, *78*, 195–201. [CrossRef] [PubMed]
47. Esteban, V.; Ruperez, M.; Sanchez-Lopez, E.; Rodríguez-Vita, J.; Lorenzo, O.; Demaegdt, H.; Vanderheyden, P.; Egido, J.; Ruiz-Ortega, M. Angiotensin IV activates the nuclear transcription factor-kappaB and related proinfammatory genes in vascular smooth muscle cells. *Circ. Res.* **2005**, *96*, 965–973. [CrossRef] [PubMed]
48. Yang, H.; Zeng, X.J.; Wang, H.X.; Zhang, L.-K.; Dong, X.-L.; Guo, S.; Du, J.; Li, H.-H.; Tang, C.-S. Angiotensin IV protects against angiotensin II-induced cardiac injury via AT4 receptor. *Peptides* **2011**, *32*, 2108–2115. [CrossRef] [PubMed]
49. Telmisartan Randomised AssessmeNt Study in ACE iNtolerant subjects with cardiovascular Disease (TRANSCEND) Investigators; Yusuf, S.; Teo, K.; Anderson, C.; Pogue, J.; Dyal, L.; Copland, I.; Schumacher, H.; Dagenais, G.; Sleight, P. Effects of the angiotensin-receptor blocker telmisartan on cardiovascular events in high-risk patients intolerant to angiotensin-converting enzyme inhibitors: A randomised controlled trial. *Lancet* **2008**, *372*, 1174–1183. [CrossRef] [PubMed]
50. ONTARGET Investigators; Yusuf, S.; Teo, K.K.; Pogue, J.; Dyal, L.; Copland, I.; Schumacher, H.; Dagenais, G.; Sleight, P.; Anderson, C. Telmisartan, ramipril, or both in patients at high risk for vascular events. *N. Engl. J. Med.* **2008**, *358*, 1547–1559. [CrossRef]
51. Dickstein, K.; Kjekshus, J.; OPTIMAAL Steering Committee of the OPTIMAAL Study Group. Effects of losartan and captopril on mortality and morbidity in high-risk patients after acute myocardial infarction: The OPTIMAAL randomised trial. Optimal Trial in Myocardial Infarction with Angiotensin II Antagonist Losartan. *Lancet* **2002**, *360*, 752–760. [CrossRef] [PubMed]
52. Hirohata, A.; Yamamoto, K.; Miyoshi, T.; Hatanaka, K.; Hirohata, S.; Yamawaki, H.; Komatsubara, I.; Murakami, M.; Hirose, E.; Sato, S.; et al. Impact of olmesartan on progression of coronary atherosclerosis a serial volumetric intravascular ultrasound analysis from the OLIVUS (impact of OLmesarten on progression of coronary atherosclerosis: Evaluation by intravascular ultrasound) trial. *J. Am. Coll. Cardiol.* **2010**, *55*, 976–982. [CrossRef]
53. Pfefer, M.A.; McMurray, J.J.; Velazquez, E.J.; Rouleau, J.-L.; Køber, L.; Maggioni, A.P.; Solomon, S.D.; Swedberg, K.; Van de Werf, F.; White, H.; et al. Valsartan, captopril, or both in myocardial infarction complicated by heart failure, left ventricular dysfunction, or both. *N. Engl. J. Med.* **2003**, *349*, 1893–1906. [CrossRef] [PubMed]
54. Visseren, F.L.J.; Mach, F.; Smulders, Y.M.; Carballo, D.; Koskinas, K.C.; Bäck, M.; Benetos, A.; Biffi, A.; Boavida, J.M.; Capodanno, D.; et al. 2021 ESC Guidelines on cardiovascular disease prevention in clinical practice. *Eur. Heart J.* **2021**, *42*, 3227–3337. [CrossRef] [PubMed]
55. Pitt, B.; Remme, W.; Zannad, F.; Neaton, J.; Martinez, F.; Roniker, B.; Bittman, R.; Hurley, S.; Kleiman, J.; Gatlin, M.; et al. Eplerenone, a selective aldosterone blocker, in patients with left ventricular dysfunction after myocardial infarction. *N. Engl. J. Med.* **2003**, *348*, 1309–1321. [CrossRef]
56. Pitt, B.; Zannad, F.; Remme, W.J.; Cody, R.; Castaigne, A.; Perez, A.; Palensky, J.; Wittes, J. The effect of spironolactone on morbidity and mortality in patients with severe heart failure. Randomized Aldactone Evaluation Study Investigators. *N. Engl. J. Med.* **1999**, *341*, 709–717. [CrossRef] [PubMed]
57. Chrysant, S.G. The role of angiotensin II receptors in stroke protection. *Curr. Hypertens. Rep.* **2012**, *14*, 202–208. [CrossRef] [PubMed]
58. Leong, D.P.; McMurray, J.J.V.; Joseph, P.G.; Yusuf, S. From ACE Inhibitors/ARBs to ARNIs in Coronary Artery Disease and Heart Failure (Part 2/5). *J. Am. Coll. Cardiol.* **2019**, *74*, 683–698. [CrossRef] [PubMed]
59. PROGRESS Collaborative Group. Randomised trial of a perindopril-based blood-pressure-lowering regimen among 6,105 individuals with previous stroke or transient ischaemic attack. *Lancet* **2001**, *358*, 1033–1041. [CrossRef] [PubMed]

60. Dahlöf, B.; Devereux, R.; de Faire, U.; Fyhrquist, F.; Hedner, T.; Ibsen, H.; Julius, S.; Kjeldsen, S.; Kristianson, K.; Lederballe-Pedersen, O.; et al. The Losartan Intervention For Endpoint reduction (LIFE) in Hypertension study: Rationale, design, and methods. The LIFE Study Group. *Am. J. Hypertens.* **1997**, *10 Pt 1*, 705–713. [CrossRef] [PubMed]
61. Sever, P. The SCOPE trial. Study on Cognition and Prognosis in the Elderly. *J. Renin-Angiotensin-Aldosterone Syst.* **2002**, *3*, 61–62. [CrossRef]
62. Schrader, J.; Lüders, S.; Kulschewski, A.; Berger, J.; Zidek, W.; Treib, J.; Einhäupl, K.; Diener, H.C.; Dominiak, P.; Acute Candesartan Cilexetil Therapy in Stroke Survivors Study Group. The ACCESS Study: Evaluation of Acute Candesartan Cilexetil Therapy in Stroke Survivors. *Stroke* **2003**, *34*, 1699–1703. [CrossRef]
63. CONSENSUS Trial Study Group. Efects of enalapril on mortality in severe congestive heart failure. Results of the Cooperative North Scandinavian Enalapril Survival Study (CONSENSUS). *N. Engl. J. Med.* **1987**, *316*, 1429–1435. [CrossRef] [PubMed]
64. Yusuf, S.; Pitt, B.; SOLVD Investigators. Effect of enalapril on survival in patients with reduced left ventricular ejection fractions and congestive heart failure. *N. Engl. J. Med.* **1991**, *325*, 293–302. [PubMed]
65. Granger, C.B.; McMurray, J.J.; Yusuf, S.; Held, P.; Michelson, E.L.; Olofsson, B.; Östergren, J.; Pfeffer, M.A.; Swedberg, K.; CHARM Investigators and Committees. Effects of candesartan in patients with chronic heart failure and reduced left-ventricular systolic function intolerant to angiotensin-converting-enzyme inhibitors: The CHARM-Alternative trial. *Lancet* **2003**, *362*, 772–776. [CrossRef] [PubMed]
66. Cohn, J.N.; Tognoni, G.; Valsartan Heart Failure Trial Investigators. A randomized trial of the angiotensin-receptor blocker valsartan in chronic heart failure. *N. Engl. J. Med.* **2001**, *345*, 1667–1675. [CrossRef] [PubMed]
67. Tai, C.; Gan, T.; Zou, L.; Sun, Y.; Zhang, Y.; Chen, W.; Li, J.; Zhang, J.; Xu, Y.; Lu, H.; et al. Effect of angiotensin-converting enzyme inhibitors and angiotensin II receptor blockers on cardiovascular events in patients with heart failure: A meta-analysis of randomized controlled trials. *BMC Cardiovasc. Disord.* **2017**, *17*, 257. [CrossRef] [PubMed]
68. Jong, P.; Demers, C.; McKelvie, R.S.; Liu, P.P. Angiotensin receptor blockers in heart failure: Meta-analysis of randomized controlled trials. *J. Am. Coll. Cardiol.* **2002**, *39*, 463–470. [CrossRef] [PubMed]
69. McDonagh, T.A.; Metra, M.; Adamo, M.; Gardner, R.S.; Baumbach, A.; Böhm, M.; Burri, H.; Butler, J.; Čelutkienė, J.; Chioncel, O.; et al. 2021 ESC Guidelines for the diagnosis and treatment of acute and chronic heart failure. *Eur. Heart J.* **2021**, *42*, 3599–3726. [CrossRef] [PubMed]
70. Matsuda, Y.; Toma, Y.; Matsuzaki, M.; Moritani, K.; Satoh, A.; Shiomi, K.; Ohtani, N.; Kohno, M.; Fujii, T.; Katayama, K. Change of left atrial systolic pressure waveform in relation to left ventricular end-diastolic pressure. *Circulation* **1990**, *82*, 1659–1667. [CrossRef] [PubMed]
71. Li, D.; Pang, L.; Leung, T.K.; Cardin, S.; Wang, Z.; Nattel, S. Effects of angiotensin-converting enzyme inhibition on the development of the atrial fibrillation substrate in dogs with ventricular tachypacinginduced congestive heart failure. *Circulation* **2001**, *104*, 2608–2614. [CrossRef]
72. Goette, A.; Staack, T.; Röcken, C.; Arndt, M.; Geller, J.; Huth, C.; Ansorge, S.; Klein, H.U.; Lendeckel, U. Increased expression of extracellular signal-regulated kinase and angiotensin-converting enzyme in human atria during atrial fibrillation. *J. Am. Coll. Cardiol.* **2000**, *35*, 1669–1677. [CrossRef]
73. Koniari, I.; Artopoulou, E.; Mplani, V.; Mulita, F.; Alexopoulou, E.; Chourdakis, M.; Abo-Elseoud, M.; Tsigkas, G.; Panagiotopoulos, I.; Kounis, N.; et al. Atrial fibrillation in heart failure patients: An update on renin-angiotensin-aldosterone system pathway blockade as a therapeutic and prevention target. *Cardiol. J.* **2023**, *30*, 312–326. [CrossRef] [PubMed]
74. Schmieder, R.E.; Kjeldsen, S.E.; Julius, S.; McInnes, G.T.; Zanchetti, A.; Hua, T.A. Reduced incidence of new-onset atrial fibrillation with angiotensin II receptor blockade: The VALUE trial. *J. Hypertens.* **2008**, *26*, 403–411. [CrossRef] [PubMed]
75. Chaugai, S.; Meng, W.Y.; Ali Sepehry, A. Effects of RAAS Blockers on Atrial Fibrillation Prophylaxis: An Updated Systematic Review and Meta-Analysis of Randomized Controlled Trials. *J. Cardiovasc. Pharmacol. Ther.* **2016**, *21*, 388–404. [CrossRef] [PubMed]
76. Ito, Y.; Yamasaki, H.; Naruse, Y.; Yoshida, K.; Kaneshiro, T.; Murakoshi, N.; Igarashi, M.; Kuroki, K.; Machino, T.; Xu, D.; et al. Effect of eplerenone on maintenance of sinus rhythm after catheter ablation in patients with long-standing persistent atrial fibrillation. *Am. J. Cardiol.* **2013**, *111*, 1012–1018. [CrossRef]
77. Oraii, A.; Healey, J.S.; Kowalik, K.; Pandey, A.K.; Benz, A.P.; Wong, J.A.; Conen, D.; McIntyre, W.F. Mineralocorticoid receptor antagonists and atrial fibrillation: A meta-analysis of clinical trials. *Eur. Heart J.* **2024**, *45*, 756–774. [CrossRef]

Disclaimer/Publisher's Note: The statements, opinions and data contained in all publications are solely those of the individual author(s) and contributor(s) and not of MDPI and/or the editor(s). MDPI and/or the editor(s) disclaim responsibility for any injury to people or property resulting from any ideas, methods, instructions or products referred to in the content.

Review

Role of the Renin-Angiotensin System in Long COVID's Cardiovascular Injuries

Elena Cojocaru [1], Cristian Cojocaru [2,*], Cristiana-Elena Vlad [3,4,†] and Lucian Eva [5,6,†]

1. Morpho-Functional Sciences II Department, Faculty of Medicine, "Grigore T. Popa" University of Medicine and Pharmacy, 700115 Iasi, Romania; elena.cojocaruu@umfiasi.ro
2. Medical III Department, Faculty of Medicine, "Grigore T. Popa" University of Medicine and Pharmacy, 700115 Iasi, Romania
3. Medical II Department, Faculty of Medicine, "Grigore T. Popa" University of Medicine and Pharmacy, 700115 Iasi, Romania; cristiana-elena.vlad@umfiasi.ro
4. "Dr. C. I. Parhon" Clinical Hospital, 700503 Iasi, Romania
5. Faculty of Dental Medicine, "Apollonia" University of Iasi, 700511 Iasi, Romania; lucianeva74@yahoo.com
6. "Prof. Dr. Nicolae Oblu" Clinic Emergency Hospital, 700309 Iasi, Romania
* Correspondence: cristian.cojocaru@umfiasi.ro; Tel.: +40-744-423-240
† These authors contributed equally to this work.

Abstract: The renin-angiotensin system (RAS) is one of the biggest challenges of cardiovascular medicine. The significance of the RAS in the chronic progression of SARS-CoV-2 infection and its consequences is one of the topics that are currently being mostly discussed. SARS-CoV-2 undermines the balance between beneficial and harmful RAS pathways. The level of soluble ACE2 and membrane-bound ACE2 are both upregulated by the endocytosis of the SARS-CoV-2/ACE2 complex and the tumor necrosis factor (TNF)-α-converting enzyme (ADAM17)-induced cleavage. Through the link between RAS and the processes of proliferation, the processes of fibrous remodelling of the myocardium are initiated from the acute phase of the disease, continuing into the long COVID stage. In the long term, RAS dysfunction may cause an impairment of its beneficial effects leading to thromboembolic processes and a reduction in perfusion of target organs. The main aspects of ACE2—a key pathogenic role in COVID-19 as well as the mechanisms of RAS involvement in COVID cardiovascular injuries are studied. Therapeutic directions that can be currently anticipated in relation to the various pathogenic pathways of progression of cardiovascular damage in patients with longCOVID have also been outlined.

Keywords: renin-angiotensin system; ACE2; angiotensin II; signal transduction; cardiovascular system; long COVID

1. Introduction

The first cases of COVID-19, the disease caused by the novel coronavirus SARS-CoV-2, were reported in December 2019 in the city of Wuhan, Hubei Province, China. However, the initial cluster of cases was reported to the World Health Organization (WHO) on 31 December 2019. This marked the beginning of global recognition and response to the COVID-19 pandemic. Due to this, on 30 January 2020, the WHO declared a Public Health Emergency of International Concern and on 11 March 2020 classified the outbreak as a pandemic [1]. Over 767 million confirmed cases and more than 6.9 million deaths had been reported globally as of 4 June 2023 [2]. According to WHO, no matter the age or the severity of the first symptoms, the individuals exposed to SARS-CoV-2 are susceptible to the post-COVID-19 disease, also known as long COVID [3,4]. Informal patient reports of previously healthy people enduring persistent symptoms and not fully healing following infection with SARS-CoV-2 began to appear in April 2020, not long after the pandemic's onset. These patients began to identify as "Long Haulers," and thus the term "Long COVID" was invented [5].

The Department of Health and Human Services, in collaboration with the Centers for Disease Control and Prevention and other partners, issued the Long COVID definition: "Long COVID is broadly defined as signs, symptoms, and conditions that continue or develop after initial COVID-19 or SARS-CoV-2 infection. The signs, symptoms, and conditions are present four weeks or more after the initial phase of infection; may be multisystemic; and may present with a relapsing–remitting pattern and progression or worsening over time, with the possibility of severe and life-threatening events even months or years after infection. Long COVID is not one condition. It represents many potentially overlapping entities, likely with different biological causes and different sets of risk factors and outcomes" [6].

Based on an estimation of 10% of infected individuals and more than 651 million COVID-19 cases worldwide, at least 65 million people worldwide have long-term COVID [7]. Long COVID can be divided into two phases based on how long the symptoms last. Post-acute COVID refers to symptoms that last longer than 3 weeks and up to 12 weeks, whereas chronic COVID refers to symptoms that last longer than 12 weeks [8,9].

Age and sex gender are related to long-COVID progression. In fact, women and/or young people have a higher probability of developing long COVID than men, but the risk level levels out around the age of 60 [10]. Regardless of whether individuals were hospitalized, a study found that 45% of COVID-19 survivors had a variety of unresolved symptoms after less than four months [11].

There is disagreement among experts on the type of symptoms that COVID-19 may be responsible for, which further complicates issues. Another study revealed that 12.7% of COVID-19 patients in the general population will experience persistent somatic symptoms after COVID-19 by considering the symptoms that worsened and could be linked to COVID-19 and while correcting for seasonal variations and non-infectious health aspects of the pandemic on symptom dynamics [7]. According to a study based on 9764 participants, the 12 main symptoms that characterize the most of the long COVID disease are: loss of smell or taste, post-exertional malaise, chronic cough, brain fog, thirst, heart palpitations, chest pain, fatigue, dizziness, gastrointestinal symptoms, issues with sexual desire or capacity, and abnormal movements. The prevalence of long COVID across the various study groups ranged from 10 to 23% based on the 12 defining symptoms, depending on when they got the disease and whether they were already diagnosed with long COVID when they enrolled in the study. In addition, this study revealed that long COVID was more prevalent in those who were unvaccinated and those who experienced it more than once [4].

It has been observed that in patients with a history of cardiovascular disease, higher serum concentrations of ACE2 may be responsible for the increased severity of COVID-19 disease [12]. Furthermore, the use of ACE inhibitor therapies may increase the risk of SARS-CoV-2 infection. However, in patients who have recovered, Ang II levels normalized [13]. Patients who have required a prolonged hospital stay or have died have associated elevated serum troponin levels as an expression of ischemic and non-ischemic cardiac lesions [14].

In addition to hemodynamic actions, local RAS has multiple functions, including regulation of cell growth, differentiation, proliferation, apoptosis, reactive oxygen species generation, tissue inflammation and fibrosis, and hormone secretion. In this regard, local RAS has alternative pathways regulated by biologically active peptides (e.g., Ang IV, Ang A, alamandine, and angioprotectin), additional specific receptors (e.g., pro-renin receptor), and alternative pathways for Ang II generation (e.g., renin-independent mechanisms of Ang-peptide generation from Ang (1–12)) [15]. The precise mechanisms of myocardial injury in patients with COVID-19 are still unclear. It is also not known whether the myocardial injury is a direct effect of the virus or a response to systemic inflammation or both.

Among the factors of poor prognosis in patients with COVID-19, older age, presence of comorbidities, and smoking were cited. A study published by Cheng et al. found that older age is associated with reduced ACE2 and consequently reduced susceptibility to infection [16].

2. Methods

We have conducted a systematic review of the literature to provide updated evidence on the role of RAS during the long COVID-19. The following databases have been searched: Medline, Google Scholar, and WHO Global Research Database on COVID-19 from 1 January 2020 to 10 June 2023. We screened original articles and reviews in English, and the search terms included "long COVID", "renin-angiotensin system", "ACE2", "cardiovascular effects", "WHO long COVID", "SARS-COV-2", as well as "persistent symptoms".

3. The Alteration of RAS Components—A Key Role in SARS-CoV-2 Infection

The RAS is composed of a series of regulatory peptides that play important roles in the body's physiological functions, such as maintaining fluid and electrolyte balance, controlling blood pressure, controlling vascular permeability, and promoting tissue growth [17].

RAS has two opposite axes [18] (Figure 1):

1. Harmful arm-ACE/Ang II/angiotensin II receptor type 1 (AT1R) pathway; because it separates Ang II from Ang I and interacts with the AT1R to cause vasoconstriction, inflammation, hypertrophy, blood coagulation, cell growth and proliferation, extracellular matrix remodelling, and stimulation of oxidative stress and fibrosis.
2. Protective arm-ACE2-Ang (1–7)-MasR receptor (MasR); Ang (1–7) interacts with MasR to promote vasodilation, anti-inflammatory effects, antifibrotic effects, antiproliferative effects, and vascular protection and mediates endothelial nitric oxide synthase activation and suppress apoptosis, negatively regulating RAS. The Ang (1–7) is produced either by the cleavage of Ang II by ACE2 or through the metabolism of inactive Ang (1–9) (cleaved from Ang I by ACE2). The cardiac antihypertrophic actions of Ang II and Ang (1–7) are mediated through the activation of Ang type 2 receptors (AT2R) and MasR, respectively [19]. Since only AT1R is blocked, using an ARB (angiotensin receptor blockers) for the therapy of COVID-19 patients may have favorable outcomes because any formed Ang II may have anti-inflammatory effects through its interaction with AT2R or through conversion by residual ACE2 to Ang (1–7) acting through AT2R, MasR, and Mas-related G protein-coupled receptor D) [20]. Thus, by preventing or reversing Ang II-induced cardiac hypertrophy, AT1R inhibition or activation of AT2R and Mas receptors as well as decreased oxidative stress may have a cardioprotective role.

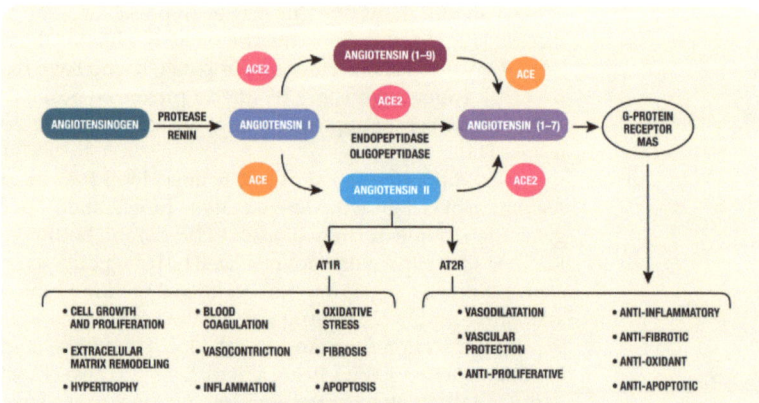

Figure 1. The two axes of the RAS.

The balance between the two axes defines health status. Conditions including lung disease, hypertension, and heart failure have all been related to an imbalance between these two systems [21].

The key receptor at the SARS-CoV-2 entry, ACE2, is considerably reduced in COVID-19 infection, which inhibits the protective effect of the ACE2-Ang (1–7)-MasR arm. Dysregulation of the RAS, direct viral toxicity, endothelial cell destruction, and thrombo-inflammation, as well as dysregulation of the immune system, are among the mechanisms allowing multi-organ damage caused by SARS-CoV-2 infection. The involvement of these mechanisms in the pathophysiology of COVID-19 is currently not fully understood [18]. RAS dysregulation and ACE-2 mediated viral entry, as well as tissue damage, may be also secondary to sepsis, even though some of these mechanisms, such as systemic cytokine release and microcirculatory dysfunction, may be important factors to the pathogenesis of COVID-19.

Due to the involvement of ACE2, a key part of the RAS, this system has been linked to COVID-19. The RAS is stated to be involved in COVID-19 in various ways.

3.1. SARS-CoV-2 Cell Entrance

The SARS-CoV-2 virus can enter human cells (Figure 2), particularly those in the respiratory tract, through the ACE2 receptor. The virus's spike protein links to ACE2, allowing the virus to enter host cells more easily. An essential phase in the infection process is this interaction between the virus and ACE2.

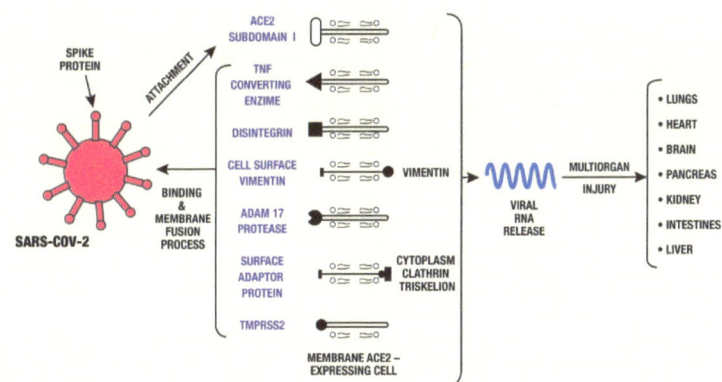

Figure 2. SARS-CoV-2 cell entrance.

A summary of the steps involved in the direct viral entrance is provided as follows:

3.1.1. The ACE2 receptor, which is widely expressed on the surface of a variety of human cells, including those in the respiratory tract, lungs, heart, and other organs, is recognized and bound to the spike protein on the surface of SARS-CoV-2 [18].

3.1.2. After initial attachment, the spike protein undergoes a process termed priming. Some transmembrane proteinases and proteins, including vimentin and clathrin, may be involved in the binding and membrane fusion processes. Examples include transmembrane protease serine 2 (TMPRSS2), disintegrin and ADAM17, and TNF-converting enzyme. For instance, both TMPRSS2 and ADAM17 have the ability to cleave ACE2 in order to increase viral uptake and ectodomain shedding, respectively [22–25]. However, additional proteases may also be involved in this process. The spike protein's structural modifications brought on by priming enable it to merge with the host cell membrane.

3.1.3. Membrane Fusion: In order to merge with the host cell membrane, the primed spike protein undergoes a conformational shift. This fusion enables the virus to release its genetic material and enter the cytoplasm of the host cell.

3.1.4. Genetic Material Release: The viral genetic material, which takes the form of ribonucleic acid (RNA), is released once inside the host cell. The instructions for generating new viral components are encoded by the viral RNA.

3.1.5. Replication and Assembly: The mechanisms of the host cell use the viral RNA to create viral proteins and reproduce the viral genome. In the host cell, new virus particles are assembled.

3.1.6. Release of New Viral Particles: the newly formed viral particles are released from the infected host cell, either by cell lysis (cell apoptosis) or by budding from the host cell membrane. These released viral particles can continue to infect other cells and continue the cycle of infection.

Because of this, SARS-CoV-2 can infect different systems in the body, leading to a range of COVID-19-related symptoms and complications.

3.2. ACE2—A Key Pathogenic Role in COVID-19

The components of the RAS and, in particular, ACE2—the central element of this system—play a particular role in the pathological states of COVID-19 via the following mechanisms:

3.2.1. ACE2 Downregulation

SARS-CoV-2 infection has been associated with the downregulation of ACE2 expression. The binding of the virus to ACE2 and its internalization during the infection process can lead to a decrease in the number of ACE2 receptors on the cell surface. This downregulation of ACE2 could have several implications, as ACE2 is involved in various physiological processes, including the regulation of blood pressure and the balance of the RAS.

3.2.2. Imbalance in the RAS

ACE2 is responsible for converting Ang II to Ang (1–7), which has vasodilatory and anti-inflammatory effects. The downregulation of ACE2 during SARS-CoV-2 infection can result in an imbalance between Ang II and Ang (1–7) levels. This imbalance may contribute to increased levels of Ang II, which is known to promote vasoconstriction, inflammation, and fibrosis.

3.2.3. Inflammatory Response and Lung Injury

Dysregulation of the RAS system, along with the release of pro-inflammatory cytokines and chemokines, can contribute to an excessive and uncontrolled inflammatory response in the lungs. This inflammatory response can lead to lung injury, acute respiratory distress syndrome (ARDS), and other complications associated with severe COVID-19.

3.2.4. The Kallikrein-Kinin System (KKS)

KKS, the degradation pathways of which are regulated by ACE and ACE2, is a system linked to RAS. Through the cleavage of kininogens by kallikrein, the KKS increases in inflammation and causes an increase in bradykinin [26]. An increased risk of thrombotic events such as pulmonary embolism, myocardial infarction, and stroke is reported after COVID-19 infection [27]. Considering that the kallikrein–kinin system can activate the coagulation system through factor XII, therapeutical approaches that rebalance the RAS and the closely linked KKS may be able to reduce the hyperinflammatory and procoagulant state experienced by long-COVID patients [28]. Since blocking the AT1R restores RAS equilibrium as a result of a reduction in ADAM17, an increase in ACE2 subsequently, and a downregulation of the B1 bradykinin receptor, targeting RAS in long-COVID might be favourable. Studies have shown that AT1R antagonists/blockade (AT1RB) are beneficial for long-COVID patients because they prevent the ACE-dependent degradation of bradykinin [29,30]. Using computational modelling, it was recently discovered that AT1RB could be changed to have a stronger affinity for the ACE2-receptor, competing with or scavenging for the spike protein of the SARS-CoV-2 [29]. Targeting RAS has not been found to benefit outcomes for adult patients hospitalized for acute COVID-19, according to three recent studies. Based on an assumption that overactivation of the RAS pathway would result in worse COVID-19 results, the REMAP-CAP investigators have studied the use of an ACE inhibitor or an ARB against no RAS inhibition [31]. In two trials, TXA-127 and TRV-027, respectively, which have both been evaluated versus placebo, researchers from the ACTIV-4 Host Tissue platform got a novel approach to correct imbalances in Ang (1–7) and Ang II caused by RAS dysregulation [32,33]. However, it should be noted that the US and European heart associations have recommended changing RAS-inhibitor administrations. The BRACE CORONA and REPLACE COVID randomized trials contributed to support the theory that the need for the medications was a marker for patients who have a greater underlying burden of chronic health conditions and cardio-vascular risk rather than the fact that the medications themselves were probably not causing any harm [34,35]. However, the REMAP-CAP study has found that the administration of RAS inhibitors could attenuate Ang II levels and thus improve outcomes.

If ACE2 is bound to SARS-CoV-2, the normal conversion of Ang II to Ang (1–7) would be inhibited, and there would be a relative increase in Ang II activity. The addition of TXA-127, a synthetic Ang (1–7), to COVID therapy in the hope of achieving RAS balance, would limit the effects of Ang II. Another solution is TRV-027, an agonist of the Ang II type 1 receptor [34,35]. However, these two drugs have not worked as expected. The REMAP-CAP results provide conclusive proof that this approach is not appropriate for acute treatment [31]. Despite this, there is still a possibility that RAS inhibitors could be used in a randomized trial to treat long COVID, whereas RAS dysregulation has been considered to be a cause.

The treatment of corticosteroids has no impact on persistent symptoms in hospitalized patients because long COVID-19 displays multiorgan symptoms that are likely brought on by a variety of mechanisms [36].

4. RAS Involvement in COVID Cardiovascular Injuries

In the cardiovascular system, RAS is involved in the pathogenesis of COVID-19 through the following pathways:

4.1. Inflammatory Response

Infection with the SARS-CoV-2 virus has introduced a new paradigm in the understanding of the immune response through the extent and diversity of the body disorders. Considered from a broad point of view, cellular infection with SARS-CoV-2 causes excessive production of virions and secondary endoplasmic reticulum stress [37]. To interrupt the massive production of virions, the infected cell dies as part of the defense mechanism [38]. ACE2 plays a role in mediating the interaction between the virus and host cells, being the entry receptor for the SARS-CoV-2 virus [39]. From a physiological perspective, ACE2 regulates the RAS, which plays a role in the homeostasis of the human body. ACE2 expression is more pronounced in alveolar epithelial cells, renal tubular cells, Leydig cells as well as in cells in the gastrointestinal tract or heart [40]. Furthermore, cardiomyocytes and type II alveolar cells have the ability to positively influence the transmission and generation of viruses thus increasing the vulnerability of both organs [41]. It should be noted that the effects of RAS, whether determined by local or systemic components, are hormonally influenced, and thus there is a gender-related variability.

Following the pathogenic processes, after infection, viral replication, and migration, multiple inflammatory pathways are triggered, through which macrophages and dendritic cells are activated and proinflammatory cytokines and chemokines are secreted; there are multiple inflammatory mechanisms that induce deregulation of the RAS. These processes are often excessive and self-induced and can be responsible for tissue destruction and host death, as follows:

4.1.1. Secretion of inflammatory factors by macrophages is mediated by SARS-CoV-2 S, N, and E proteins [42]. Protein N is responsible for enhancing the secretion of inflammatory factors such as TNF-alpha, interleukin (IL) 6, and IL-10 [43]. It has been suggested that the S protein may bind to Toll-like receptor (TLR) 4, which in turn increases ACE2 expression, and the S1 subunit induces activation of nuclear factor-κB (NF-κb) and mitogen-activated protein kinase (MAPK) pathways [42,44].

4.1.2. Binding of Ang II to target cells activates the Janus kinase 2/signal transducer and activator of transcription 3 (JAK2/STAT) (STAT1/2/3) pathway, resulting in increased production of proinflammatory cytokines [45]. Activation of this pathogenic pathway leads to the loss of the ACE2 receptor and RAS dysfunction [46].

4.1.3. Through the p38 MAPK pathway, the immune response and ACE2 endocytosis are impaired, leading to ACE2 down-regulation, RAS disruption, and Ang II accumulation [47]. As an exacerbation of inflammatory processes occurs, the extracellular domain of ACE2 breaks down, altering RAS balance, which ultimately induces myocarditis, cardiovascular complications, and ARDS [48]. Ang II plays a role in increasing cardiomyocyte contractility and the transformation of fibroblasts in the heart into myofibroblasts with profibrotic action [49]. As inflammatory processes are sustained over time and exacerbated by Ang II and oxidative stress, cardiac remodelling, and destructive effects on the vasculature occur [40].

An important factor in the pathogenesis of long COVID is oxidative stress. The main pathogenic processes of long COVID include a systemic hyperinflammatory state and coagulopathy, which are exacerbated by inflammation and oxidative stress. There is a vicious cycle of oxidative stress, inflammation, and long-COVID disease progression because thrombosis and inflammation lead to the reactivation of reactive oxygen species (ROS). Inflammation is triggered by ROS, which also affect the endothelium, cause microthrombi and neuroinflammation, stimulate the production of autoantibodies, and interfere with the synthesis of neurotransmitters [50]. Through activating NADPH Oxidase, ACE2 stimulates oxidative stress, an effect that is mainly mediated by Ang II [51]. An increase in Ang II and ROS set by the SARS-CoV-2 infection triggers oxidative stress and multisystemic injuries [52]. According to a study, the degree of severity of the disease and poor outcomes are associated with higher levels of oxidative stress markers, lower levels of antioxidant indicators, and lower levels of ACE2 expression in hospitalized COVID-19 patients [53].

Once these imbalances and fibrotic remodelling processes are initiated, RAS dysfunction and serum ACE2 activity appear to be hardly and incompletely reversible. As a result, the disturbances described by long-COVID syndrome, produced not only after severe infections but also after persistent, subclinical infections, are associated with RAS dysfunction [54].

The range of cells that can be infected by the virus is also notable, with viral genes being detected in the cardiac conduction system, which explains the arrhythmias observed in these patients [55]. A study published by Tavazzi et al. reveals, by analysis of biopsy specimens, that myocardial interstitial cells undergo viral invasion and inflammation [56]. Studies on the presence of viral particles in cells of the cardiovascular system are not entirely clear. A study that followed the morphological analysis of 276 patients who died after infection with different variants of concern of SARS-CoV-2 virus revealed the presence of spike protein, but no viral RNA has been found in the myocardium. Electron microscopy observed virion-like particles on the surface of the vascular endothelium [57].

Based on single-cell RNA sequencing and histology studies, human cardiomyocytes have been found to express the SARS-CoV-2 receptor ACE2, especially in patients with cardiovascular disorders, which suggests that SARS-CoV-2 may target cardiomyocytes [58]. In the cytoplasm and nucleus, cardiomyocytes present a low level of TMPRSS2 [59], but significant levels of cathepsin B and L, indicating that endocytosis is the primary route for SARS-CoV-2 entry by cathepsin-dependent manner [58,60]. These results indicating the role of cathepsins in cardiomyocyte infection support the consideration of cathepsin L inhibitors in anti-COVID therapy [61]. Cardiomyocytes are permissive for SARS-CoV-2 infection, according to in vitro studies on human induced pluripotent stem cells (hiPS) and isolated adult cardiomyocytes as well as in vivo animal models [58]. Thus, it has been shown that cardiomyocytes produce viral spike glycoproteins and have intracellular double-stranded viral RNAs. SARS-CoV-2 infectivity was confirmed in 3D cardiospheric tissue, and increased viral RNA concentrations have been found in supernatants from infected cardiomyocytes. The expression of the viral spike protein was effectively suppressed by neutralizing antibodies, recombinant ACE2, or inhibition of the RNA polymerase with remdesivir.

A significantly greater interaction between SARS-CoV-2 and the receptor can result from the microvascular pericytes expressing a significant amount of ACE2. As a result, endothelial dysfunction is caused by the lesion of the pericytes set by the dissemination of the viral infection [62]. The microvasculature's lining pericytes, as well as vascular smooth muscle cells, fibroblasts, and cardiomyocytes, exhibit significant cardiac expression of ACE2 [63].

In addition to the exacerbated inflammatory response, the pathogenic pathways of which in relation to RAS have been described above, the association between long COVID syndrome and cardiovascular disorders can be linked to the viral genes being incorporated into the deoxyribonucleic acid of infected cells, which results in the maintenance of procoagulant processes and other cardiovascular complications [64]. From these perspectives, at least long-term RAS imbalance and cellular remodelling are a continuous process, and the final stage of evolution is fibrosis of endothelial structures, cardiomyocytes, or the conduction system.

In addition to inflammatory mechanisms, myocardial damage may also occur as a result of vascular endothelial disorders, cardiomyocyte apoptosis, increased mechanical stress, thrombus formation, or inadequate perfusion [14].

4.2. Endothelial Dysfunction

A dynamic connection between the circulating blood and different tissues is achieved by the vascular endothelium. The regulation of vascular tone and maintenance of vascular homeostasis depend entirely on the vascular endothelium, an active paracrine, endocrine, and autocrine organ [65].

As a multisystem disease, COVID-19 is partly caused by damage to the vascular endothelium. After infection, there may be residual repercussions and long-term sequelae, which may be caused by chronic endothelial dysfunction [66]. These COVID-19 long-term effects have been considered the next upcoming public health global emergency, and scientific data on the size and magnitude of the situation are urgently needed to assist in the planning of a proper healthcare response. The ACE2 receptors expressed by endothelial cells are used by SARS-CoV-2 to infect the host and cause clear and distinct systemic endotheliitis [67].

Endotheliitis and infection-mediated endothelial injury can cause excessive thrombin production, inhibit fibrinolysis, and activate complement pathways which may result in microvascular dysfunction and the deposition of micro thrombi [68].

As mentioned above, SARS-CoV-2 enters in cells via the SARS-CoV ACE2 receptor, primes the S protein via the serine protease TMPRSS2, and thus, directly infects the endothelium. In addition, an authorized TMPRSS2 inhibitor limits this entry and could be used as a therapy option [23].

Endothelial damage may be a major contributing factor to the persistence of long-COVID symptoms, together with platelet pathology and the presence of micro clots in the circulation [69].

A study has revealed higher autoantibody levels in long-COVID patients [70]. These antibodies approach antiphospholipid antibodies, which are the main cause of antiphospholipid syndrome in the general population. The autoantibodies attach to cell surfaces, where they stimulate neutrophils, platelets, and endothelial cells, leading to thrombosis at the blood vessel wall barrier [71]. There are individuals with a history of SARS-CoV-2 infection, which have anti-ACE2 antibodies. These patients have reduced plasma levels of soluble ACE2. Additionally, exogenous ACE2 activity is inhibited by plasma from these patients. Thus, following SARS-CoV-2 infection, the ACE2 antibodies are produced and reduce ACE2 activity. This might result in more Ang II circulating, which promotes the proinflammatory status that underlies the long-COVID symptoms. Post-Acute Sequelae after SARS-CoV-2 infection may be treated with recombinant soluble ACE2 protein [70].

Endotheliitis present in COVID-19 patients may be the cause of the systemic impairment of microcirculatory function in various arterial beds and its clinical consequences. This statement supports the use of ACE inhibitors, anti-cytokine, anti-inflammatory, and statin drugs, to stop viral replication [67].

4.3. Microvascular Abnormalities: Hypercoagulability and Thrombus Formation

Patients with COVID-19 present infection-mediated endothelial injury, marked by increased von Willebrand factor levels and endotheliitis, characterized by the presence of activated neutrophils and macrophages in a number of vascular beds, including the lungs, kidney, heart, small intestine, and liver. These changes can trigger excessive thrombin production, inhibit fibrinolysis, and activate complement pathways, initiating thrombo-inflammation and ultimately leading to micro thrombus deposition and microvascular dysfunction.

The current literature reveals that SARS-CoV-2 infection increases a patient's risk for thromboembolic diseases. A recent study showed that platelet hyperactivity is associated with detectable virus RNA in blood and that COVID-19 patients display higher platelet activation in vivo, as shown by enhanced integrin $\alpha IIb\beta 3$ activation and P-selectin expression. In addition, platelets had high levels of ACE2 and TMPRSS2 expression, two essential cellular elements involved in SARS-CoV-2 cell entry, and SARS-CoV-2 and its spike protein induce platelet activity and thrombus formation via the MAPK pathway, which is downstream of ACE2. These findings imply that treatment with anti-spike monoclonal antibody and recombinant human ACE2 protein can prevent platelet activation and thrombus formation caused by SARS-CoV-2 [72].

The occurrence of microvascular dysfunction in long-COVID patients may be due to peripheral activation of ACE2 receptors or exacerbation of proinflammatory cytokines that may remain in circulation even after the infection subside [73]. ACE2 causes endothelial injury that results in endothelial dysfunction, microvascular inflammation, and thrombosis, as mechanisms of microvascular abnormalities in COVID-19 [74]. This observational study revealed that the underlying cause of angina-like chronic chest pain in people who have recovered from COVID-19 is coronary microvascular ischemia [75].

4.4. Fibrosis

Fibrosis is the outcome of almost all chronic inflammatory diseases. Fibrosis can be considered a consequence of a disordered wound-healing process and can be directly related to the severity of the acute event [76]. Different mechanisms of lung injury have been described in COVID-19, with both viral and immune-mediated mechanisms being involved [77]. Studies have shown that SARS-CoV-2 infection can lead to damage to epithelial structures and is associated with the presence of fibrotic tissue through excess collagen (fibrosis). Among the multiple effects of RAS that include the regulation of cell growth, apoptosis, and inflammation, tissue fibrosis is controlled. The practical existence of a tissue RAS that may be independent of the circulating RAS causes confusion in the overall interpretation of the role of this system, given the existence of alternative pathways related to active peptides or Ang II generation [15].

Given the short time since the outbreak of the COVID-19 pandemic and with efforts largely focused on acute case management, there are insufficient data to assess the long-term impact of infection, spontaneous reversibility, or potential benefits of initiating antifibrotic therapy in patients with post-infection sequelae.

RAS does not act by itself but is closely related to other inflammatory mechanisms, oxidative stress, and endothelial dysfunction, which contribute to the development and maintenance of cardiac lesions. Some effects of RAS such as synaptic remodelling, improved cell survival, cellular signal transmission, and antioxidant effects should not be omitted [78].

Fibrosis as a fibro-proliferative disease also recognizes a genetic predisposition and is correlated with chronic inflammation. In relation to post-COVID fibrosis, the transforming growth factor-β signalling pathway, the Wingless/int signalling pathway, and the Yes-associated protein/transcriptional coactivator PDZ-binding motif signalling pathway have been proposed as mechanisms of production [79]. Another pathway of fibrotic disease production is based on the release of fibroblast growth factors, transforming the growth factor released from megakaryocytes [80]. Due to intense inflammatory processes, there is a continuous release of these mediators, which can lead to inadequate remodelling and fibrosis [81], although current data suggest the influence of the thrombotic process as a precursor of pulmonary and hepatic fibrosis [82]. There are still no possibilities of therapeutic intervention due to the difficulty of separation from physiological healing processes.

Thille et al. described in a cohort of 159 autopsies of patients with ARDS that 4% of patients with a disease duration of less than 1 week, 24% of patients with a disease duration between 1 and 3 weeks, and 61% of patients with a disease duration greater than 3 weeks, developed fibrosis [83]. The repair process involves regeneration by native stem cells and deposition of connective tissue to replace areas of defect [84]. Alveolar macrophages play a central role in this process by phagocytosing alveolar debris and producing cytokines and growth factors involved in the repair pathway [85].

In SARS-CoV-2 infection, the main cause of aggravation of the disease course is considered to be cytokine storm with excessive release of metalloproteinases producing epithelial and endothelial damage, vascular endothelial cell growth factor (VEGF), and cytokines such as IL-6 and TNFα involved in the fibrotic process [86]. VEGF and fibroblast growth factors stimulate the migration and proliferation of intact endothelial cells leading to pulmonary capillary angiogenesis [87]. The detection of fibrotic changes at the beginning of the disease suggests an attempt to repair the lesions that appeared in the acute phase. However, it is too early in the disease progression to determine whether this finding resolves in time or results in permanent fibrosis.

5. Conclusions

We have attempted to summarize the pathophysiological mechanisms resulting from an unbalanced RAS system that may contribute to long COVID's symptoms.

However, further research on the underlying mechanisms is needed. Recent data suggest an intricate immune-inflammatory signaling network involving the imbalance of the RAS. To date, there is not yet a biomarker for the long-COVID prediction. Due to the wide range of symptoms, it is important to thoroughly evaluate the patients in order to exclude any other causes or comorbidities, notably persistent inflammation, microvascular cardiac dysfunction, microclots, persistent endothelial dysfunction, and fibrosis, which have been all described in long COVID.

The variety of cardiovascular symptoms in long COVID requires extensive evaluation and assessment, which challenges clinicians and determines the design of future clinical trials. Applying a variety of approaches to reset the RAS may be a possibility for managing patients with long COVID.

Association between long-COVID syndrome and cardiovascular disorders can be linked to the viral genes being incorporated into the deoxyribonucleic acid of infected cells, which results in the maintenance of a vicious cycle of oxidative stress, inflammation, as well as procoagulant processes. Microvascular dysfunction in long-COVID patients may be due to peripheral activation of ACE2 receptors or exacerbation of proinflammatory cytokines that may remain in circulation even after the infection subsides. Long-term RAS imbalance and cellular remodelling is a continuous process, and the final stage of evolution is fibrosis of endothelial structures, cardiomyocytes, or the conduction system. Therapeutical approaches that rebalance the RAS may be able to reduce the hyperinflammatory and procoagulant state experienced by long-COVID patients.

Long COVID continues to concern the medical world and compels us to expand the diagnostic methods and therapeutic guidelines to protect patients from the harmful effects of SARS-CoV-2 proteins. Multi-organ damage and cardiovascular complications in particular require detailed genetic and molecular studies, as well as translational and randomized trials.

6. Future Perspectives

Long-term longitudinal follow-up of cardiovascular dysfunction in patients with long COVID is needed to better understand the mechanism behind the persistence of symptoms and to target therapy.

The connection between endothelial dysfunction and long-COVID-19 symptoms is becoming clearer. The medical care of long COVID patients may benefit from dividing risks into categories.

Because of the particular nature of long-COVID cardiovascular symptoms, it is likely that several methods of therapy will be necessary to manage care for the patients.

Studies on the biological factors that could lead to the onset of long-COVID symptoms are required. These factors may include the presence of SARS-CoV-2 persistent reservoirs in specific tissues, the reactivation of other pathogens under immune dysregulation, the interaction of viruses with hosts, the host microbiome with the human virome, coagulation disorders, and autoimmunity triggered by peptides from the pathogen and the host that are structurally similar.

Thus, teams with a variety of expertise, such as pathology, virology, immunometabolism, physical therapy, and rehabilitation, must collaborate on studies on long-COVID patients.

Author Contributions: Conceptualization, C.C., C.-E.V., L.E. and E.C.; methodology, E.C.; software, C.C.; validation, C.-E.V., C.C., L.E. and E.C.; formal analysis, E.C.; investigation, C.-E.V.; resources, C.C., E.C., L.E. and C.-E.V.; data curation, L.E.; writing—original draft preparation, C.C. and E.C.; writing—review and editing, L.E. and C.-E.V.; visualization, L.E.; supervision, C.C.; project administration, C.-E.V. All authors have read and agreed to the published version of the manuscript.

Funding: This research received no external funding.

Institutional Review Board Statement: Not applicable.

Informed Consent Statement: Not applicable.

Data Availability Statement: Additional data are available from the corresponding author upon reasonable request.

Acknowledgments: Thanks to Florin Pinzariu, Graphic Art Lecturer at "George Enescu" National University of Arts, Iasi, Romania, for the graphic design of the figures and the graphical abstract of this manuscript.

Conflicts of Interest: The authors declare no conflict of interest.

References

1. World Health Organization. Coronavirus Disease (COVID-19) Pandemic. Available online: https://www.who.int/europe/emergencies/situations/covid-19 (accessed on 6 June 2023).
2. World Health Organization. Weekly Epidemiological Update on COVID-19—8 June 2023. Available online: https://www.who.int/publications/m/item/weekly-epidemiological-update-on-covid-19{-}{-}-8-june-2023 (accessed on 9 June 2023).
3. World Health Organization. Post COVID-19 Condition (Long COVID). Available online: https://www.who.int/europe/newsroom/fact-sheets/item/post-covid-19-condition (accessed on 10 June 2023).
4. Thaweethai, T.; Jolley, S.E.; Karlson, E.W.; Levitan, E.B.; Levy, B.; McComsey, G.A.; McCorkell, L.; Nadkarni, G.N.; Parthasarathy, S.; Singhe, U.; et al. Development of a Definition of Postacute Sequelae of SARS-CoV-2 Infection. *JAMA* **2023**, *329*, 1934–1946. [CrossRef]
5. Rubin, R. As Their Numbers Grow, COVID-19 "Long Haulers" Stump Experts. *JAMA* **2020**, *324*, 1381–1383. [CrossRef] [PubMed]
6. COVID.gov—What Is Long COVID. Available online: https://www.covid.gov/longcovid/definitions (accessed on 10 June 2023).
7. Ballering, A.V.; van Zon, S.K.R.; Olde Hartman, T.C.; Rosmalen, J.G.M. Persistence of somatic symptoms after COVID-19 in the Netherlands: An observational cohort study. *Lancet* **2022**, *400*, 452–461. [CrossRef]
8. Raveendran, A.V. Long COVID-19, Challenges in the diagnosis and proposed diagnostic criteria. *Diabetes Metab. Syndr.* **2021**, *15*, 145–146. [CrossRef] [PubMed]

9. Raveendran, A.V.; Jayadevan, R.; Sashidharan, S. Long COVID: An overview. *Diabetes Metab. Syndr.* **2021**, *15*, 869–875. [CrossRef] [PubMed]
10. Sudre, C.H.; Murray, B.; Varsavsky, T.; Graham, M.S.; Penfold, R.S.; Bowyer, R.C.; Pujol, J.C.; Klaser, K.; Antonelli, M.; Canas, L.S.; et al. Attributes and predictors of long COVID. *Nat. Med.* **2021**, *27*, 626–631. [CrossRef]
11. O'Mahoney, L.L.; Routen, A.; Gillies, C.; Ekezie, W.; Welford, A.; Zhang, A.; Karamchandani, U.; Simms-Williams, N.; Cassambai, S.; Ardavani, A.; et al. The prevalence and long-term health effects of Long COVID among hospitalised and non-hospitalised populations: A systematic review and meta-analysis. *EClinicalMedicine* **2023**, *55*, 101762. [CrossRef]
12. Liu, Z.; Xiao, X.; Wei, X.; Li, J.; Yang, J.; Tan, H.; Zhu, J.; Zhang, Q.; Wu, J.; Liu, L.; et al. Composition and divergence of coronavirus spike proteins and host ACE2 receptors predict potential intermediate hosts of SARS-CoV-2. *J. Med. Virol.* **2020**, *92*, 595–601. [CrossRef]
13. Driggin, E.; Madhavan, M.V.; Bikdeli, B.; Chuich, T.; Laracy, J.; Biondi-Zoccai, G.; Brown, T.S.; Der Nigoghossian, C.; Zidar, D.A.; Haythe, J.; et al. Cardiovascular Considerations for Patients, Health Care Workers, and Health Systems during the COVID-19 Pandemic. *J. Am. Coll. Cardiol.* **2020**, *75*, 2352–2371. [CrossRef]
14. Bavishi, C.; Bonow, R.O.; Trivedi, V.; Abbott, J.D.; Messerli, F.H.; Bhatt, D.L. Special Article—Acute myocardial injury in patients hospitalized with COVID-19 infection: A review. *Prog. Cardiovasc. Dis.* **2020**, *63*, 682–689. [CrossRef]
15. Campbell, D.J. Clinical relevance of local Renin Angiotensin systems. *Front. Endocrinol.* **2014**, *14*, 113. [CrossRef]
16. Cheng, W.A.; Turner, L.; Marentes Ruiz, C.J.; Tanaka, M.L.; Congrave-Wilson, Z.; Lee, Y.; Jumarang, J.; Perez, S.; Peralta, A.; Pannaraj, P.S. Clinical manifestations of COVID-19 differ by age and obesity status. *Influenza Other Respir. Viruses* **2022**, *16*, 255–264. [CrossRef]
17. Vaduganathan, M.; Vardeny, O.; Michel, T.; McMurray, J.J.V.; Pfeffer, M.A.; Solomon, S.D. Renin–Angiotensin–Aldosterone System Inhibitors in Patients with COVID-19. *N. Eng. J. Med.* **2020**, *382*, 1653–1659. [CrossRef]
18. Costa, L.B.; Perez, L.G.; Palmeira, V.A.; Macedo e Cordeiro, T.; Teatini Ribeiro, V.; Lanza, K.; Simões e Silva, A.C. Insights on SARS-CoV-2 Molecular Interactions With the Renin-Angiotensin System. *Front. Cell Dev. Biol.* **2020**, *8*, 559841. [CrossRef] [PubMed]
19. Bhullar, S.K.; Dhalla, N.S. Angiotensin II-Induced Signal Transduction Mechanisms for Cardiac Hypertrophy. *Cells* **2022**, *11*, 3336. [CrossRef] [PubMed]
20. Lumbers, E.R.; Head, R.; Smith, G.R.; Delforce, S.J.; Jarrott, B.; Martin, J.H.; Pringle, K.G. The interacting physiology of COVID-19 and the renin-angiotensin-aldosterone system: Key agents for treatment. *Pharmacol. Res. Perspect.* **2022**, *10*, e00917. [CrossRef] [PubMed]
21. Simko, F.; Hrenak, J.; Adamcova, M.; Paulis, L. Renin-Angiotensin-Aldosterone System: Friend or Foe-The Matter of Balance. Insight on History, Therapeutic Implications and COVID-19 Interactions. *Int. J. Mol. Sci.* **2021**, *22*, 3217. [CrossRef]
22. Ni, W.; Yang, X.; Yang, D.; Bao, J.; Li, R.; Xiao, Y.; Hou, C.; Wang, H.; Liu, J.; Yang, D.; et al. Role of angiotensin-converting enzyme 2 (ACE2) in COVID-19. *Crit. Care* **2020**, *24*, 422. [CrossRef]
23. Hoffmann, M.; Kleine-Weber, H.; Schroeder, S.; Krüger, N.; Herrler, T.; Erichsen, S.; Schiergens, T.S.; Herrler, G.; Wu, N.H.; Nitsche, A.; et al. SARS-CoV-2 Cell Entry Depends on ACE2 and TMPRSS2 and Is Blocked by a Clinically Proven Protease Inhibitor. *Cell* **2020**, *181*, 271–280.e8. [CrossRef]
24. Heurich, A.; Hofmann-Winkler, H.; Gierer, S.; Liepold, T.; Jahn, O.; Pöhlmann, S. TMPRSS2 and ADAM17 cleave ACE2 differentially and only proteolysis by TMPRSS2 augments entry driven by the severe acute respiratory syndrome coronavirus spike protein. *J. Virol.* **2014**, *88*, 1293–1307. [CrossRef]
25. Li, F. Structure, Function, and Evolution of Coronavirus Spike Proteins. *Annu. Rev. Virol.* **2016**, *3*, 237–261. [CrossRef] [PubMed]
26. Tabassum, A.; Iqbal, M.S.; Sultan, S.; Alhuthali, R.A.; Alshubaili, D.I.; Sayyam, R.S.; Abyad, L.M.; Qasem, A.H.; Arbaeen, A. Dysregulated Bradykinin: Mystery in the Pathogenesis of COVID 19. *Mediat. Inflamm.* **2022**, *2022*, 7423537. [CrossRef] [PubMed]
27. Raman, B.; Bluemke, D.A.; Lüscher, T.F.; Neubauer, S. Long COVID: Post-acute sequelae of COVID-19 with a cardiovascular focus. *Eur. Heart J.* **2022**, *43*, 1157–1172. [CrossRef] [PubMed]
28. Maas, C.; Renne, T. Coagulation factor XII in thrombosis and inflammation. *Blood* **2018**, *131*, 1903–1909. [CrossRef] [PubMed]
29. Schieffer, E.; Schieffer, B. The rationale for the treatment of long-COVID symptoms—A cardiologist's view. *Front. Cardiovasc. Med.* **2022**, *9*, 992686. [CrossRef]
30. Schmidt, B.; Schieffer, B. Angiotensin II AT1 receptor antagonists. Clinical implications of active metabolites. *J. Med. Chem.* **2003**, *46*, 2261–2270. [CrossRef]
31. Writing Committee for the REMAP-CAP Investigators. Effect of Angiotensin-Converting Enzyme Inhibitor and Angiotensin Receptor Blocker Initiation on Organ Support–Free Days in Patients Hospitalized with COVID-19, A Randomized Clinical Trial. *JAMA* **2023**, *329*, 1183–1196. [CrossRef]
32. Self, W.H.; Shotwell, M.S.; Gibbs, K.W.; de Wit, M.; Files, C.; Harkins, M.; Hudock, K.M.; Merck, L.H.; Moskowitz, A.; Apodaca, K.D.; et al. Renin-Angiotensin System Modulation with Synthetic Angiotensin (1–7) and Angiotensin II Type 1 Receptor–Biased Ligand in Adults with COVID-19, Two Randomized Clinical Trials. *JAMA* **2023**, *329*, 1170–1182. [CrossRef]
33. Lee, M.M.Y.; McMurray, J.J.V. Lack of Benefit of Renin-Angiotensin System Inhibitors in COVID-19. *JAMA* **2023**, *329*, 1155–1156. [CrossRef] [PubMed]

34. Lopes, R.D.; Macedo, A.V.S.; de Barros E Silva, P.G.M.; Moll-Bernardes, R.J.; Feldman, A.; D'Andréa Saba Arruda, G.; de Souza, A.S.; de Albuquerque, D.C.; Mazza, L.; Santos, M.F.; et al. Continuing versus suspending angiotensin-converting enzyme inhibitors and angiotensin receptor blockers: Impact on adverse outcomes in hospitalized patients with severe acute respiratory syndrome coronavirus 2 (SARS-CoV-2)–The BRACE CORONA Trial. *Am. Heart J.* **2020**, *226*, 49–59. [CrossRef]
35. Cohen, J.B.; Hanff, T.C.; William, P.; Sweitzer, N.; Rosado-Santander, N.R.; Medina, C.; Rodriguez-Mori, J.R.; Renna, N.; Chang, T.I.; Corrales-Medina, V.; et al. Continuation versus discontinuation of renin–angiotensin system inhibitors in patients admitted to hospital with COVID-19, A prospective, randomised, open-label trial. *Lancet Respir. Med.* **2021**, *9*, 275–284. [CrossRef]
36. Chan Sui Ko, A.; Candellier, A.; Mercier, M.; Joseph, C.; Carette, H.; Basille, D.; Lion-Daolio, S.; Devaux, S.; Schmit, J.L.; Lanoix, J.P.; et al. No Impact of Corticosteroid Use During the Acute Phase on Persistent Symptoms Post-COVID-19. *Int. J. Gen. Med.* **2022**, *18*, 6645–6651. [CrossRef] [PubMed]
37. Fung, T.S.; Liu, D.X. Coronavirus infection, ER stress, apoptosis and innate immunity. *Front. Microbiol.* **2014**, *5*, 296. [CrossRef] [PubMed]
38. Inde, Z.; Croker, B.A.; Yapp, C.; Joshi, G.N.; Spetz, J.; Fraser, C.; Qin, X.; Xu, L.; Deskin, B.; Ghelf, E.; et al. Age-dependent regulation of SARS-CoV-2 cell entry genes and cell death programs correlates with COVID-19 severity. *Sci. Adv.* **2021**, *7*, 8609–8627. [CrossRef]
39. Aleksova, A.; Ferro, F.; Gagno, G.; Cappelletto, C.; Santon, D.; Rossi, M.; Ippolito, G.; Zumla, A.; Beltrami, A.P.; Sinagra, G. COVID-19 and renin-angiotensin system inhibition: Role of angiotensin converting enzyme 2 (ACE2)—Is there any scientific evidence for controversy? *J. Intern. Med.* **2020**, *288*, 410–421. [CrossRef] [PubMed]
40. Ashraf, U.M.; Abokor, A.A.; Edwards, J.M.; Edwards, J.M.; Waigi, E.W.; Royfman, R.S.; Hasan, S.A.M.; Smedlund, K.B.; Gregio Hardy, A.M.; Chakravarti, R.; et al. SARS-CoV-2, ACE2 expression, and systemic organ invasion. *Physiol. Genom.* **2021**, *53*, 51–60. [CrossRef] [PubMed]
41. Zhang, H.; Penninger, J.M.; Li, Y.; Zhong, N.; Slutsky, A.S. Angiotensin-converting enzyme 2 (ACE2) as a SARS-CoV-2 receptor: Molecular mechanisms and potential therapeutic target. *Intensive Care Med.* **2020**, *46*, 586–590. [CrossRef]
42. Choudhury, A.; Mukherjee, S. In silico studies on the comparative characterization of the interactions of SARS-CoV-2 spike glycoprotein with ACE-2 receptor homologs and human TLRs. *J. Med. Virol.* **2020**, *92*, 2105–2113. [CrossRef]
43. Polidoro, R.B.; Hagan, R.S.; de Santis Santiago, R.; Schmidt, N.W. Overview: Systemic Inflammatory Response Derived From Lung Injury Caused by SARS-CoV-2 Infection Explains Severe Outcomes in COVID-19. *Front. Immunol.* **2020**, *11*, 1626. [CrossRef]
44. Aboudounya, M.M.; Heads, R.J. COVID-19 and Toll-Like Receptor 4 (TLR4): SARS-CoV-2 May Bind and Activate TLR4 to Increase ACE2 Expression, Facilitating Entry and Causing Hyperinflammation. *Mediat. Inflamm.* **2021**, *2021*, 8874339. [CrossRef]
45. Recinos, A.; LeJeune, W.S.; Sun, H.; Lee, C.Y.; Tieu, B.C.; Lu, M.; Hou, T.; Boldogh, I.; Tilton, R.G.; Brasier, A.R. Angiotensin II induces IL-6 expression and the Jak-STAT3 pathway in aortic adventitia of LDL receptor-deficient mice. *Atherosclerosis* **2007**, *194*, 125–133. [CrossRef] [PubMed]
46. Jiang, Y.; Zhao, T.; Zhou, X.; Xiang, Y.; Gutierrez-Castrellon, P.; Ma, X. Inflammatory pathways in COVID-19, Mechanism and therapeutic interventions. *MedComm* **2022**, *3*, e154. [CrossRef] [PubMed]
47. Cuadrado, A.; Nebreda, A.R. Mechanisms and functions of p38 MAPK signalling. *Biochem. J.* **2010**, *429*, 403–417. [CrossRef] [PubMed]
48. Grimes, J.M.; Grimes, K.V. p38 MAPK inhibition: A promising therapeutic approach for COVID-19. *J. Mol. Cell. Cardiol.* **2020**, *144*, 63–65. [CrossRef]
49. Sriram, K.; Insel, P.A. A hypothesis for pathobiology and treatment of COVID-19, The centrality of ACE1/ACE2 imbalance. *Br. J. Pharmacol.* **2020**, *177*, 4825–4844. [CrossRef] [PubMed]
50. Vollbracht, C.; Kraft, K. Oxidative Stress and HyperInflammation as Major Drivers of Severe COVID-19 and Long COVID: Implications for the Benefit of HighDose Intravenous Vitamin C. *Front. Pharmacol.* **2022**, *13*, 899198. [CrossRef]
51. de las Heras, N.; Martín Giménez, V.M.; Ferder, L.; Manucha, W.; Lahera, V. Implications of Oxidative Stress and Potential Role of Mitochondrial Dysfunction in COVID-19, Therapeutic Effects of Vitamin D. *Antioxidants* **2020**, *9*, 897. [CrossRef]
52. Zhu, Z.; Zheng, Z.; Liu, J. Comparison of COVID-19 and Lung Cancer via Reactive Oxygen Species Signaling. *Front. Oncol.* **2021**, *11*, 708263. [CrossRef]
53. Alobaidy, A.S.H.; Elhelaly, M.; Amer, M.E.; Shemies, R.A.; Othman, A.I.; El-Missiry, M.A. Angiotensin converting enzyme 2 gene expression and markers of oxidative stress are correlated with disease severity in patients with COVID-19. *Mol. Biol. Rep.* **2023**, *50*, 5827–5836. [CrossRef]
54. Khazaal, S.; Harb, J.; Rima, M.; Annweiler, C.; Wu, Y.; Cao, Z.; Khattar, Z.A.; Legros, C.; Kovacic, H.; Fajloun, Z.; et al. The Pathophysiology of Long COVID throughout the Renin-Angiotensin System. *Molecules* **2022**, *27*, 2903. [CrossRef]
55. Jakovac, H.; Ferenčić, A.; Stemberger, C.; Mohar Vitezić, B.; Cuculić, D. Detection of SARS-CoV-2 Antigens in the AV-Node of a Cardiac Conduction System—A Case Report. *Trop. Med. Infect. Dis.* **2022**, *7*, 43. [CrossRef] [PubMed]
56. Tavazzi, G.; Pellegrini, C.; Maurelli, M.; Belliato, M.; Sciutti, F.; Bottazzi, A.; Sepe, P.A.; Resasco, T.; Camporotondo, R.; Bruno, R.; et al. Myocardial localization of coronavirus in COVID-19 cardiogenic shock. *Eur. J. Heart Fail.* **2020**, *22*, 911–915. [CrossRef] [PubMed]
57. Mitrofanova, L.B.; Makarov, I.A.; Gorshkov, A.N.; Runov, A.L.; Vonsky, M.S.; Pisareva, M.M.; Komissarov, A.B.; Makarova, T.A.; Li, Q.; Karonova, T.L.; et al. Comparative Study of the Myocardium of Patients from Four COVID-19 Waves. *Diagnostics* **2023**, *13*, 1645. [CrossRef] [PubMed]

58. Bojkova, D.; Wagner, J.U.G.; Shumliakivska, M.; Aslan, G.S.; Saleem, U.; Hansen, A.; Luxán, G.; Günther, S.; Duc Pham, M.; Krishnan, J.; et al. SARS-CoV-2 infects and induces cytotoxic effects in human cardiomyocytes. *Cardiovasc. Res.* 2020, *116*, 2207–2215. [CrossRef]
59. Sakamoto, A.; Kawakami, R.; Kawai, K.; Gianatti, A.; Pellegrini, D.; Kutys, R.; Guo, L.; Mori, M.; Cornelissen, A.; Sato, Y.; et al. ACE2 (Angiotensin-Converting Enzyme 2) and TMPRSS2 (Transmembrane Serine Protease 2) Expression and Localization of SARS-CoV-2 Infection in the Human Heart. *Arterioscler. Thromb. Vasc. Biol.* 2021, *41*, 542–544. [CrossRef]
60. Yang, J.; Chen, T.; Zhou, Y. Mediators of SARS-CoV-2 entry are preferentially enriched in cardiomyocytes. *Hereditas* 2021, *158*, 4. [CrossRef]
61. Liu, T.; Luo, S.; Libby, P.; Shi, G.P. Cathepsin L-selective inhibitors: A potentially promising treatment for COVID-19 patients. *Pharmacol. Ther.* 2020, *213*, 107587. [CrossRef]
62. Nappi, F.; Avtaar Singh, S.S. Endothelial Dysfunction in SARS-CoV-2 Infection. *Biomedicines* 2022, *10*, 654. [CrossRef]
63. Tucker, N.R.; Chaffin, M.; Bedi, K.C., Jr.; Papangeli, I.; Akkad, A.D.; Arduini, A.; Hayat, S.; Eraslan, G.; Muus, C.; Bhattacharyya, R.P.; et al. Myocyte-Specific Upregulation of ACE2 in Cardiovascular Disease. *Circulation* 2020, *142*, 708–710.
64. Di Toro, A.; Bozzani, A.; Tavazzi, G.; Urtis, M.; Giuliani, L.; Pizzoccheri, R.; Aliberti, F.; Fergnani, V.; Arbustini, E. Long COVID: Long-term effects? *Eur. Hear J. Suppl.* 2021, *23*, E1–E5. [CrossRef]
65. Flammer, A.J.; Anderson, T.; Celermajer, D.S.; Creager, M.A.; Deanfield, J.; Ganz, P.; Hamburg, N.M.; Lüscher, T.F.; Shechter, M.; Taddei, S.; et al. The assessment of endothelial function: From research into clinical practice. *Circulation* 2012, *126*, 753–767. [CrossRef]
66. Del Turco, S.; Vianello, A.; Ragusa, R.; Caselli, C.; Basta, G. COVID-19 and cardiovascular consequences: Is the endothelial dysfunction the hardest challenge? *Thromb. Res.* 2020, *196*, 143–151. [CrossRef] [PubMed]
67. Varga, Z.; Flammer, A.J.; Steiger, P.; Haberecker, M.; Andermatt, R.; Zinkernagel, A.S.; Mehra, M.R.; Schuepbach, R.A.; Ruschitzka, F.; Moch, H. Endothelial cell infection and endotheliitis in COVID-19. *Lancet* 2020, *395*, 1417–1418. [CrossRef] [PubMed]
68. Proal, A.D.; VanElzakker, M.B. Long COVID or Post-acute Sequelae of COVID-19 (PASC): An Overview of Biological Factors That May Contribute to Persistent Symptoms. *Front. Microbiol.* 2021, *12*, 698169. [CrossRef] [PubMed]
69. Pretorius, E.; Vlok, M.; Venter, C.; Bezuidenhout, J.A.; Laubscher, G.J.; Steenkamp, J.; Kell, D.B. Persistent clotting protein pathology in Long COVID/Post-Acute Sequelae of COVID-19 (PASC) is accompanied by increased levels of antiplasmin. *Cardiovasc. Diabetol.* 2021, *20*, 172. [CrossRef]
70. Arthur, J.M.; Forrest, J.C.; Boehme, K.W.; Kennedy, J.L.; Owens, S.; Herzog, C.; Liu, J.; Harville, T.O. Development of ACE2 autoantibodies after SARS-CoV-2 infection. *PLoS ONE* 2021, *16*, e0257016. [CrossRef]
71. Shi, H.; Zuo, Y.; Navaz, S.; Harbaugh, A.; Hoy, C.K.; Gandhi, A.A.; Sule, G.; Yalavarthi, S.; Gockman, K.; Madisonet, J.A.; et al. Endothelial Cell–Activating Antibodies in COVID-19. *Arthritis Rheumatol.* 2022, *74*, 1132–1138. [CrossRef]
72. Zhang, S.; Liu, Y.; Wang, X.; Yang, L.; Li, H.; Wang, Y.; Liu, M.; Zhao, X.; Xie, Y.; Yang, Y.; et al. SARS-CoV-2 binds platelet ACE2 to enhance thrombosis in COVID-19. *J. Hematol. Oncol.* 2020, *13*, 120. [CrossRef]
73. Team, R.A. The microvascular hypothesis underlying neurologic manifestations of long COVID-19 and possible therapeutic strategies. *Cardiovasc. Endocrinol. Metab.* 2021, *10*, 193–203. [CrossRef]
74. Vallejo Camazón, N.; Teis, A.; Martínez Membrive, M.J.; Llibre, C.; Bayés-Genís, A.; Mateu, L. Long COVID-19 and microvascular disease-related angina. *Rev. Esp. Cardiol.* 2022, *75*, 444–446. [CrossRef]
75. Davis, H.E.; McCorkell, L.; Vogel, J.M.; Topol, E.J. Long COVID: Major findings, mechanisms and recommendations. *Nat. Rev. Microbiol.* 2023, *21*, 133–146. [CrossRef] [PubMed]
76. Strieter, R.M.; Mehrad, B. New mechanisms of pulmonary fibrosis. *Chest* 2009, *136*, 1364–1370. [CrossRef] [PubMed]
77. Liu, J.; Zheng, X.; Tong, Q.; Li, W.; Wang, B.; Sutter, K.; Trilling, M.; Dittmer, M.U.; Yang, D. Overlapping and discrete aspects of the pathology and pathogenesis of the emerging human pathogenic coronaviruses SARS-CoV, MERS-CoV, and 2019-nCoV. *J. Med. Virol.* 2020, *92*, 491–494. [CrossRef]
78. Méndez-García, L.A.; Escobedo, G.; Minguer-Uribe, A.G.; Viurcos-Sanabria, R.; Aguayo-Guerrero, J.A.; Carrillo-Ruiz, J.D.; Solleiro-Villavicencio, H. Role of the renin-angiotensin system in the development of COVID-19-associated neurological manifestations. *Front. Cell Neurosci.* 2022, *16*, 977039. [CrossRef]
79. Piersma, B.; Bank, R.A.; Boersema, M. Signaling in Fibrosis: TGF-β, WNT, and YAP/TAZ Converge. *Front. Med.* 2015, *2*, 59. [CrossRef]
80. Castro-Malaspina, H.; Rabellino, E.M.; Yen, A.; Nachman, R.L.; Moore, M.A. Human megakaryocyte stimulation of proliferation of bone marrow fibroblasts. *Blood* 1981, *57*, 781–787. [CrossRef]
81. George, P.M.; Wells, A.U.; Jenkins, R.G. Pulmonary fibrosis and COVID-19, The potential role for antifibrotic therapy. *Lancet Respir. Med.* 2020, *8*, 807–815. [CrossRef]
82. Bitto, N.; Liguori, E.; La Mura, V. Coagulation, Microenvironment and Liver Fibrosis. *Cells* 2018, *7*, 85. [CrossRef]
83. Thille, A.W.; Esteban, A.; Fernández-Segoviano, P.; Rodriguez, J.M.; Aramburu, J.A.; Vargas-Errazuriz, P.; Martin-Pellicer, A.; Lorente, J.A.; Frutos-Vivar, F. Chronology of histological lesions in acute respiratory distress syndrome with diffuse alveolar damage: A prospective cohort study of clinical autopsies. *Lancet Respir. Med.* 2013, *1*, 395–401. [CrossRef]
84. Huang, W.T.; Akhter, H.; Jiang, C.; MacEwen, M.; Ding, Q.; Antony, V.; Thannickal, V.J.; Liu, R.M. Plasminogen activator inhibitor 1, fibroblast apoptosis resistance, and aging-related susceptibility to lung fibrosis. *Exp. Gerontol.* 2015, *61*, 62–75. [CrossRef] [PubMed]

85. Venkataraman, T.; Frieman, M.B. The role of epidermal growth factor receptor (EGFR) signaling in SARS coronavirus-induced pulmonary fibrosis. *Antivir. Res.* **2017**, *143*, 142. [CrossRef] [PubMed]
86. Raj, V.S.; Mou, H.; Smits, S.L.; Dekkers, D.H.W.; Muller, M.A.; Dijkman, R.; Muth, D.; Demmers, J.A.A.; Zaki, A.; Fouchier, R.A.M.; et al. Dipeptidyl peptidase 4 is a functional receptor for the emerging human coronavirus-EMC. *Nature* **2013**, *495*, 251–254. [CrossRef] [PubMed]
87. Kobayashi, T.; Tanaka, K.; Fujita, T.; Umezawa, H.; Amano, H.; Yoshioka, K.; Naito, Y.; Hatano, M.; Kimura, S.; Tatsumi, K.; et al. Bidirectional role of IL-6 signal in pathogenesis of lung fibrosis. *Respir. Res.* **2015**, *16*, 99. [CrossRef] [PubMed]

Disclaimer/Publisher's Note: The statements, opinions and data contained in all publications are solely those of the individual author(s) and contributor(s) and not of MDPI and/or the editor(s). MDPI and/or the editor(s) disclaim responsibility for any injury to people or property resulting from any ideas, methods, instructions or products referred to in the content.

Article

Angiotensin II Exposure In Vitro Reduces High Salt-Induced Reactive Oxygen Species Production and Modulates Cell Adhesion Molecules' Expression in Human Aortic Endothelial Cell Line

Nikolina Kolobarić [†], Nataša Kozina [†], Zrinka Mihaljević and Ines Drenjančević *

Department of Physiology and Immunology, Faculty of Medicine Osijek, J. J. Strossmayer University of Osijek, J. Huttlera 4, 31000 Osijek, Croatia; nbdujmusic@mefos.hr (N.K.); nkozina@mefos.hr (N.K.); zmihaljevic@mefos.hr (Z.M.)

* Correspondence: ines.drenjancevic@mefos.hr; Tel.: +385-912-241-406
[†] These authors contributed equally to this work.

Abstract: Background/Objectives: Increased sodium chloride (NaCl) intake led to leukocyte activation and impaired vasodilatation via increased oxidative stress in human/animal models. Interestingly, subpressor doses of angiotensin II (AngII) restored endothelium-dependent vascular reactivity, which was impaired in a high-salt (HS) diet in animal models. Therefore, the present study aimed to assess the effects of AngII exposure following high salt (HS) loading on endothelial cells' (ECs') viability, activation, and reactive oxygen species (ROS) production. **Methods**: The fifth passage of human aortic endothelial cells (HAECs) was cultured for 24, 48, and 72 h with NaCl, namely, the control (270 mOsmol/kg), HS320 (320 mOsmol/kg), and HS350 (350 mOsmol/kg). AngII was administered at the half-time of the NaCl incubation (10^{-4}–10^{-7} mol/L). **Results**: The cell viability was significantly reduced after 24 h in the HS350 group and in all groups after longer incubation. AngII partly preserved the viability in the HAECs with shorter exposure and lower concentrations of NaCl. Intracellular hydrogen peroxide (H_2O_2) and peroxynitrite ($ONOO^-$) significantly increased in the HS320 group following AngII exposure compared to the control, while it decreased in the HS350 group compared to the HS control. A significant decrease in superoxide anion ($O_2^{·-}$) formation was observed following AngII exposure at 10^{-5}, 10^{-6}, and 10^{-7} mol/L for both HS groups. There was a significant decrease in intracellular adhesion molecule 1 (ICAM-1) and endoglin expression in both groups following treatment with 10^{-4} and 10^{-5} mol/L of AngII. **Conclusions**: The results demonstrated that AngII significantly reduced ROS production at HS350 concentrations and modulated the viability, proliferation, and activation states in ECs.

Keywords: angiotensin II (AngII); cell adhesion molecules (CAMs); endothelium; reactive oxygen species (ROS); sodium chloride (NaCl)

1. Introduction

Complex interactions between vascular smooth muscle cells (VSMCs), endothelial cells (ECs), and immune cells contribute to vascular health and its physiological function [1,2]. Endothelial dysfunction (ED), an early indicator of vascular dysfunction (VD), is characterized by the loss of the endothelium's anti-inflammatory, anticoagulant, and vasodilatory properties, marking one of the initial stages in cardiovascular disease (CVDs) [3–6], such as hypertension, atherosclerosis, stroke, obesity, diabetes, and thrombosis [7–11].

A powerful stimulator of vascular inflammation and oxidative stress is a high intake of sodium chloride (NaCl) [12–14], which leads to ED by promoting vascular low-grade inflammation and impaired vasodilation even in healthy, normotensive individuals [13–16]. For example, in young, healthy individuals, a seven-day high salt (HS) intake significantly impairs endothelium-dependent vasodilation in macro- and microcirculation without affecting arterial blood pressure (BP), fluid retention, or body composition [17]. The HS diet is known to increase the risk of heart failure and hypertension [18–22]. Excessive salt intake is a significant, yet preventable, contributor to CVD-related deaths globally. Implementing appropriate dietary restrictions, lifestyle modifications, and effective pharmacological interventions, all of which emphasize sodium reduction to help regulate blood pressure, improve endothelial function [23], and mitigate these negative trends [19,24–28]. Pharmacological treatments for hypertension include antihypertensive medications, such as angiotensin-converting enzyme (ACE) inhibitors [29–31], angiotensin II (AngII) receptor blockers (ARBs) [32–36], calcium channel blockers (CCBs) [37–39], and diuretics.

The effects of an HS diet are associated with alterations in the redox equilibrium between reactive oxygen species (ROS) generation and antioxidant mechanisms [40]. The bioavailability and production of endothelium-derived vasodilator nitric oxide (NO), along with other endothelium-derived vasodilator factors, are heavily influenced by increased ROS and overall vascular oxidative stress [13,41,42].

Notably, HS intake suppresses the renin–angiotensin system and AngII levels, which is the main regulatory system for the maintenance of blood volume and arterial blood pressure by vasoconstriction, increased aldosterone synthesis, and stimulation of the sympathetic nervous system [43–45]. In the vasculature, under normal conditions, higher levels of AngII induce ROS generation in vascular smooth muscle and ECs [46]. However, paradoxically, low levels of AngII also lead to increased ROS generation. For example, Ćosić et al. (2016) reported a significant increase in intracellular ROS levels and down-regulation of antioxidant enzyme expression, accompanied by impaired flow-induced dilation (FID) of middle cerebral arteries (MCAs) in Sprague-Dawley (SD) rats after a seven-day HS dietary intake [42]. On the other hand, subpressor low-dose exposure to AngII has shown beneficial effects on redox balance and restoration of flow-induced dilation after salt-induced impairment of endothelium-dependent vasodilation [47,48]. Interestingly, we have demonstrated that increased NaCl dietary intake induced low-grade systemic inflammation, which could be related to suppressed AngII levels and involved changes in Th17 and Treg cell distribution, a shift in lipid and arachidonic acid metabolism, and vascular wall remodeling in animal models [49] and in humans on an HS diet [13]. Since HS interferes with cellular response through the alteration of key molecules involved in the inflammatory response [13,50,51], using supraphysiological doses of sodium in vitro helps to model pathological conditions, amplify cellular responses, and reveal mechanisms that are not observable under normal baseline conditions. Specifically, it was shown that excess salt leads to cellular senescence, promoting inflammatory/fibrotic response and reducing NO generation in human and animal cells [52–54]. A previous study by Dmitrieva and Burg (2015) [55] showed that elevated extracellular NaCl affects adhesion molecules' gene expression in human umbilical vein endothelial cells (HUVECs), directly involving it in the facilitation of vascular changes. Nevertheless, existing research does not appear to address the effects of AngII supplementation in a cell culture model of high NaCl intake, nor its implications for regulating redox balance in ECs. Thus, the present study hypothesizes that increasing concentrations of NaCl in the cell medium will affect the viability and ROS generation in ECs and will alter the activation state of the cells. At the same time, AngII addition in physiological doses will prevent these changes and exhibit a protective effect in the EC culture model.

The aim of this study was to assess the NaCl-dose response accompanied by AngII exposure in vitro on human aortic endothelial cells (HAECs) viability, intracellular ROS production, and cell adhesion molecules' (CAMs') expression (intercellular adhesion molecule 1, ICAM-1; vascular cell adhesion molecule 1, VCAM-1; E-selectin; and endoglin) following treatment, to further elucidate underlying mechanisms and effectors in salt-induced endothelial damage.

2. Materials and Methods

2.1. Materials and Chemical Reagents

HAECs were purchased from Innoprot (Barcelona, Spain). Cell culture flasks and well plates were purchased from TPP Techno Plastic Products AG (Trasagiden, Switzerland). Human Large Vessel Endothelial Cell Basal Medium, low serum growth supplement (LSGS), and Trypsin-EDTA (0.25%) were purchased from Gibco (Thermo Fisher Scientific, Waltham, MA, USA). AngII was purchased from Merck (Darmstadt, Germany). NaCl was purchased from Gram-Mol d.o.o. (Zagreb, Croatia). The 3-(4, 5-dimethylthiazolyl-2)-2 and 5-diphenyltetrazolium bromide (MTT) were purchased from Invitrogen (Thermo Fisher Scientific, Waltham, MA, USA).

2.2. HAECs

The HAECs were carefully thawed, seeded in tissue-appropriate cell culture flasks for adherent cells (T-25 cm^2), and placed in an incubator (Shel Lab, CO_2 Series, Sheldon Manufacturing Inc., Cornelius, OR, USA) under the following conditions: ~37 °C, 5% CO_2, and >80% humidity level. Human Large Vessel Endothelial Cell Basal Medium supplemented with LSGS was used throughout our experiment. HAECs were monitored daily, and basal media was changed every other day until reaching confluence. For experimental purposes, the fifth passage (P5) of cells was used at approximately 80% confluence. Trypsinization was used as a method for the detachment of the HAECs from the flask/plate surface.

2.3. Cell Culture Treatment: NaCl and AngII

Cell culture treatment was performed in 24 and 96-well plates, depending on the following protocols. The osmolality of the control medium was 270 mOsmol/kg (133 mmol/L) (CTRL group). The HS medium was prepared by adding NaCl to the total osmolality of (1) 320 mOsmol/kg (158 mmol/L; HS320 group) and (2) 350 mOsmol/kg (173 mmol/L; HS350 group) [55–58]. During the experiment, the control medium was replaced by the high NaCl medium for 24, 48, and 72 h. The cell medium used for experimental purposes was not supplemented with LSGS, or in other words, it was serum-free.

Different concentrations of AngII were administered at half of the NaCl incubation (for 24 h after 12 h, for 48 h after 24 h, and for 73 h after 36 h). AngII was added in the following concentrations: 10^{-4}, 10^{-5}, 10^{-6}, and 10^{-7} mol/L.

For experiments including only HS exposure and flow cytometry analysis of CAMs' expression, 5 biological replicates (n = 5) were performed, with each replicate carried out over a 6-month period, using newly thawed vials of cells and conducted at the appropriate passage. However, for oxidative stress analysis, which included AngII exposure, 3 biological replicates (n = 3) were performed due to technical challenges encountered during the study period. All experiments were performed in 5 technical replicates.

2.4. Cell Viability and Metabolic Activity

MTT assay was performed as described by Shiwakoti et al. (2020) [59]. Cells were seeded in 96-well plates and treated with appropriate NaCl concentrations and 10^{-6} mol/L of AngII, as described above. After 24, 48, and 72 h, 10 µL of MTT stock solution was added to each well, and the plate was placed in an incubator for 4 h to produce formazan crystals. After incubation, 100 µL of MTT solvent was added to dissolve the created formazan crystals. The intensity of staining was proportional to the number of living cells. Plates were read spectrophotometrically at 595 nm using a microplate reader (BioRad PR 3100 TSC, Bio-Rad Laboratories, Hercules, CA, USA).

Cell viability was calculated as a percentage of the untreated control by subtracting the absorbance of treated cells from the absorbance of untreated cells, dividing by the absorbance of untreated cells, and multiplying by 100.

2.5. Flow Cytometry

The samples were collected after the HAECs were exposed for 72 h to different concentrations of NaCl and AngII (36 h), which were administered to the cell cultures in 24-well plates. The cells were washed in 1× phosphate-buffered saline (PBS) and prepared for the appropriate staining protocol (approximately 10^5 cells/mL of medium).

The FACS Canto II flow cytometer (BD Bioscience, Franklin Lakes, NJ, USA; 488 excitation laser and 530/30 BP analysis filter) was used for the assessment of intracellular ROS production, as well as the CAMs' expression. Data analysis and visualization were performed using Flow Logic software v.8 (Inivai Technologies, Mentone, Australia).

2.5.1. Intracellular ROS Production

A dichlorofluorescein diacetate (DCF-DA) assay was used to determine the levels of hydrogen peroxide (H_2O_2) and peroxynitrite ($ONOO^-$). DCF-DA is a non-fluorescent, cell-permeable dye that is deacetylated inside the cell by cellular esterases to DCFH (2′,7′-dichlorodihydrofluorescein). DCFH is oxidized in the presence of ROS in a fluorescent DCF (2′,7′-dichlorofluorescein), which can be detected as an indicator of oxidative stress. The dihydroethidium (DHE) assay was used to determine the level of superoxide anion ($O_2^{·-}$) in the HAECs. This is another cell-permeable dye, which, upon entering the cell, is oxidized by superoxide to form ethidium, which emits red fluorescence that can be quantified to assess the levels of superoxide production within the cell.

Cells were resuspended in 100 µL of 1× PBS and incubated with DCF-DA or DHE (10 µM final concentration) for 30 min at +4 °C. Following the incubation period, the samples stained with DCF-DA were re-suspended in PBS and immediately analyzed, while samples stained with DHE required additional rinsing with PBS before resuspension in PBS and cytometer reading. Following initial readings, phorbol 12-myristate 13-acetate (PMA) was added to each sample to stimulate ROS production. Data are expressed as geomean fluorescence intensity (GMFI) in FLH-1 and FLH-2 channels. Assays were performed according to previously reported protocols that were optimized for this purpose [60,61].

2.5.2. CAMs' Expression

Cells were resuspended and washed twice before staining with appropriate antibodies. The following antibodies were used for staining the HAECs: ICAM-1 (CD54, clone: 15.2, Proteintech Group, Inc., Rosemont, IL, USA), VCAM-1 (CD106, clone: 51-10C9; BD Pharmigen Inc., Franklin Lakes, NJ, USA), E-selectin (CD62E, clone: 68-5H11; BD Pharmigen Inc., Franklin Lakes, NJ, USA), and endoglin (CD105, clone: 266; BD Pharmigen Inc., Franklin Lakes, NJ, USA).

2.6. Statistical Analysis

The normality of the residuals, where appropriate, was assessed using the Shapiro–Wilk test, while the homogeneity of variances was determined using the Brown–Forsythe test. Significant differences between treatment groups were determined using one-way ANOVA and the following post hoc analyses: Dunnett's test, which compared each treatment group to the control, and Tukey's HSD test, which compared all treatment groups against each other ($p < 0.05$).

3. Results

3.1. Assessment of Cell Metabolic Activity (MTT Assay)

Changes in cellular metabolic activity (i.e., an indicator of cell viability) of HAECs following incubation with different NaCl concentrations before and after the addition of AngII in the cell media are shown in Figure 1. Cell metabolic activity was significantly reduced after 24, 48, and 72 h at 350 mOsmol/kg NaCl concentration compared to the CTRL. The addition of AngII to the HS320 incubated cells did not alter cell metabolic activity at 24 and 72 h time points, except for 48 h, where AngII significantly decreased metabolic activity compared to the CTRL and HS320 conditions. The addition of AngII to the HS350 incubated cells resulted in preserved metabolic activity of cells after 24 and 72 h, while it significantly decreased metabolic activity after 48 h of incubation.

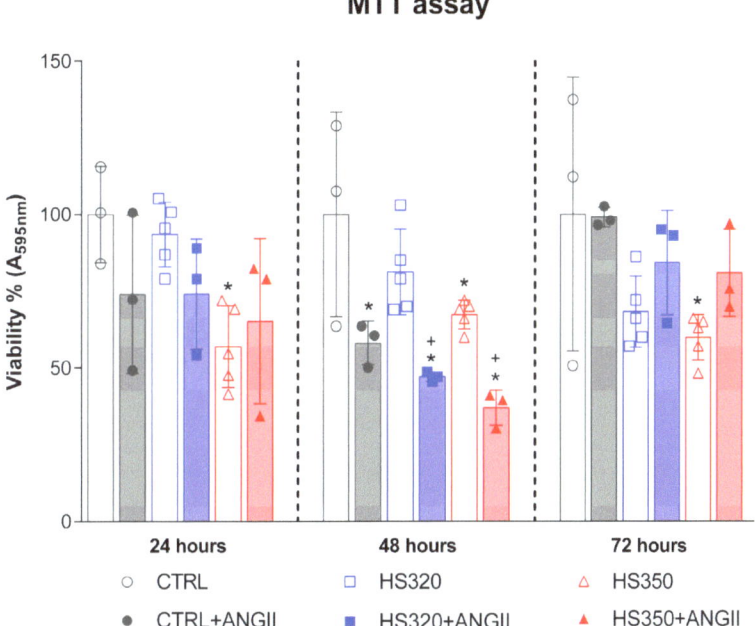

*One-way ANOVA, p<0.05: 24 hours p=0.027; 48 hours p=0.0005; 72 hours p=0.05

Figure 1. Cellular metabolic activity of HAECs after treatment with different concentrations of NaCl and AngII assessed via MTT assay. A—absorbance; nm—nanometers; CTRL—control; HS—high salt; AngII—angiotensin II; One-way ANOVA; *,+ significance level $p < 0.05$ (* compared to control group; + compared to HS group before AngII).

3.2. Assessment of Intracellular ROS Production: Analysis of DCF-DA and DHE Fluorescence Signals

In the HS320 group, treatment with NaCl did not have any significant effect on the formation of H_2O_2 and $ONOO^-$ in the HAECs. Following 10^{-4} and 10^{-5} mol/L of AngII exposure, H_2O_2 and $ONOO^-$ formation was not different from the levels observed in the HS320 condition without AngII but increased significantly with AngII 10^{-6} and 10^{-7} mol/L, suggesting that higher doses of AngII suppressed H_2O_2 and $ONOO^-$ production. Formation of H_2O_2 and $ONOO^-$ was significantly increased in the HS350 group compared to the CTRL group (Figure 2). In the HS350 group, a significant decrease was detected after administering AngII in concentrations from 10^{-4} to 10^{-6} mol/L.

Production of $O_2^{\cdot-}$ in HAECs was significantly decreased in HS320 compared to CTRL, while there were no significant changes in the formation of $O_2^{\cdot-}$ in the HS350 group compared to the CTRL group (Figure 3). $O_2^{\cdot-}$ level was significantly decreased in the HS320 group supplemented with AngII compared to the CTRL but increased compared to the HS320 group without AngII. In the HS350 group, $O_2^{\cdot-}$ production was significantly decreased after the addition of AngII compared to the CTRL and HS350 group without AngII.

Additionally, PMA stimulation that was performed following the initial baseline readings, specifically under HS conditions (HS320 and HS350), is presented as Supplementary Figures for DCF-DA (Figure S1) and DHE assays (Figure S2).

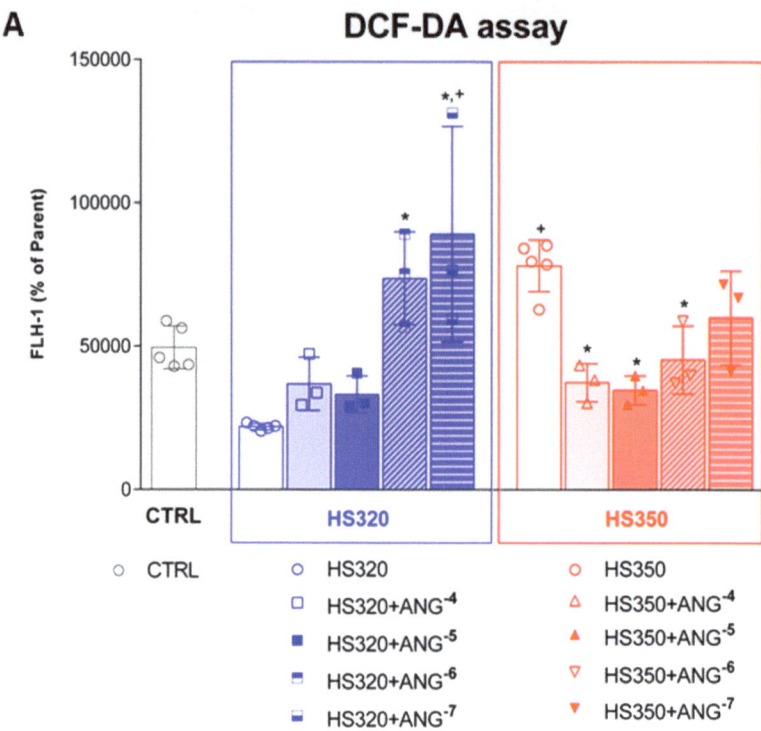

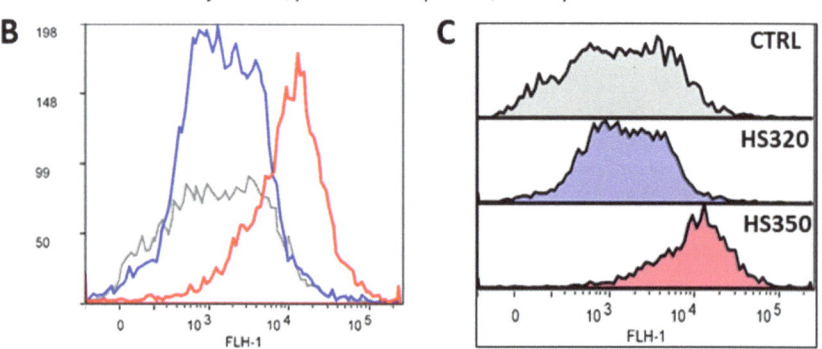

Figure 2. Formation of hydrogen peroxide and peroxynitrite in HAECs following high-salt treatment accompanied by AngII exposure (**A**). Representative histogram overlay (**B**) and stacked histograms (**C**). Grey color representing CTRL group, blue color representing HS320 group, red color representing HS350 group. Results are expressed as geometric mean fluorescence intensity (GMFI). DCF-DA—dichlorofluorescein diacetate; CTRL—control; HS—high salt; One-way ANOVA; *,+ significance level $p < 0.05$ (* compared to HS group before AngII; + compared to control group).

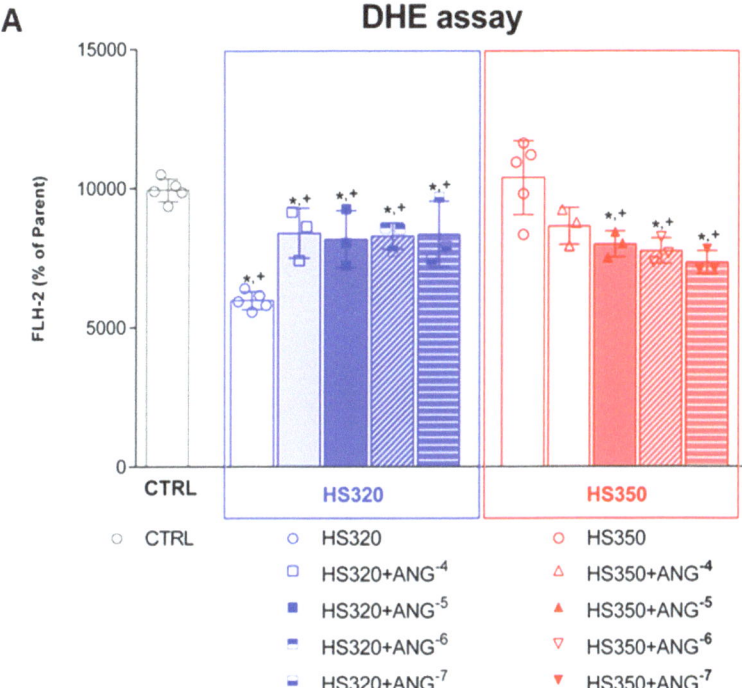

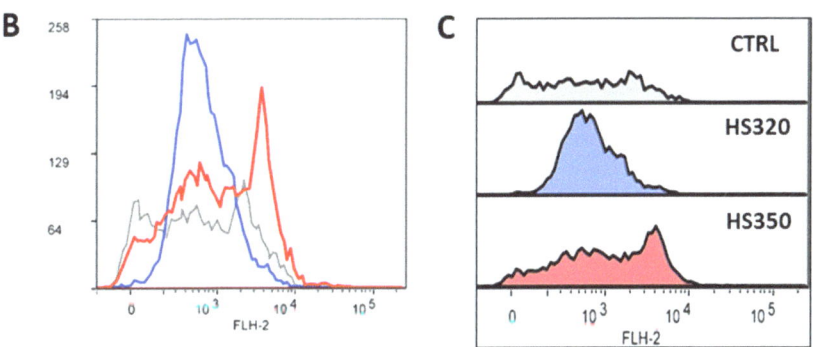

Figure 3. Formation of superoxide in HAECs following high-salt treatment accompanied by AngII exposure (**A**). Representative histogram overlay (**B**) and stacked histograms (**C**). Grey color representing CTRL group, blue color representing HS320 group, red color representing HS350 group. Results are expressed as geometric mean fluorescence intensity (GMFI). DHE—dihydroethidium; CTRL—control; HS—high salt; One-way ANOVA; *,+ significance level $p < 0.05$ (* compared to HS group before AngII; + compared to control group).

3.3. Quantitative Analysis of CAM Expression: Flow Cytometry Analysis

Changes in CAM expressions following treatment with NaCl and different concentrations of AngII are shown in Figure 4. HS did not affect VCAM-1 expression (Figure 4A). However, VCAM-1 was significantly increased in the HS320 group following the addition of AngII at a 10^{-4} mol/L dose compared to the CTRL and HS groups, and in the HS350 group following AngII at 10^{-5} mol/L compared to the HS group. ICAM-1 expression was significantly decreased in the HS320 and HS350 groups following the addition of 10^{-4} mol/L of AngII compared to the CTRL (Figure 4B).

A significant decrease in endoglin expression occurred in both HS groups supplemented with 10^{-5} mol/L of AngII compared to the HS and CTRL groups (Figure 4C). There were no significant changes in the expression of E-selectin in all study groups, with or without AngII exposure (Figure 4D).

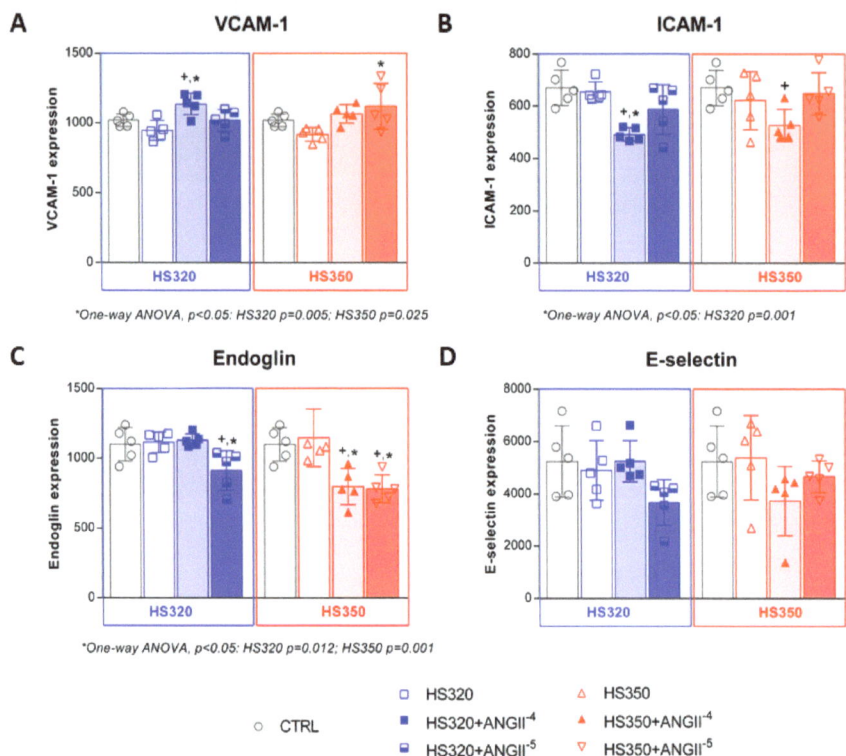

Figure 4. Changes in CAMs' expression: VCAM-1 (**A**), ICAM-1 (**B**), Endoglin (**C**), and E-selectin (**D**) in HAECs following high-salt treatment accompanied by AngII exposure. Grey color representing CTRL group, blue color representing HS320 group, red color represent-ing HS350 group. Results are expressed as geometric mean fluorescence intensity (GMFI). CAMs—cell adhesion molecules; HAECs—human aortic endothelial cells; VCAM-1—vascular cell adhesion molecule 1; ICAM-1—intracellular adhesion molecule 1; HS—high salt; CTRL—control; AngII—angiotensin II; One-way ANOVA; *,+ significance level $p < 0.05$ (* compared to HS group before AngII; + compared to control group).

3.4. Protein Expression of AT1 and AT2 Receptors

Results that are presented in the Supplementary Materials (Figure S3) showed that protein expression of both AT receptors was significantly decreased in animals on the HS diet compared to control animals. AngII partially restored the AT1 receptor expression compared to the HS group but was still lower than in the CTRL group. The AT2 receptor was not affected by the AngII treatment.

4. Discussion

Other than regulating vascular tone, blood flow, and coagulation, ECs also secrete various cytokines, chemokines, growth factors, and adhesion molecules. This makes them important players in inflammation and leukocyte-EC adhesion, which underlies all major cardio-metabolic diseases [4,7,62,63].

Some of the noteworthy findings in the present study are the following: (1) higher doses and prolonged exposure to NaCl decrease the ECs' metabolic activity and increase the ECs' ROS production, thus making ECs an important source of ROS leading to endothelial dysfunction; (2) AngII exposure significantly reduces ROS production of ECs and can protect their metabolic activity to some extent; (3) AngII can modulate the expression of CAMs (as demonstrated by a decrease in ICAM-1 and endoglin expression with certain utilized doses of AngII). Taken together, the present study demonstrates the direct effects of NaCl on ECs' activation status and redox balance, which could be altered by AngII.

As previously discussed, HS intake is a significant source of ROS, leading to systemic oxidative stress, decreased antioxidant activity, and impaired endothelial function [64,65]. NaCl, being the universal stressor, provokes an adaptive response in cells to mitigate the damage and maintain viability under hyperosmotic conditions. Generally, the elevation of osmolality above 400 mOsmol/kg H_2O progressively impairs cell proliferation, and while cell cycle arrest provides short-term adaptation, it ultimately leads to cell death (500 mOsmol/kg), while lower concentration for shorter periods induces reversible changes in cells [66]. In our current study, we aimed to evaluate the NaCl-dose and time-dependent response in conjunction with AngII exposure in vitro, focusing on the HAECs' metabolic activity, intracellular ROS production, and the expression of CAMs (ICAM-1, VCAM-1, E-selectin, and endoglin) following treatment. The results suggest that a NaCl concentration of 320 mOsmol/kg in the HAECs culture has a hormetic effect (bi-phasic dose–response curve), yielding a lower level of ROS, especially superoxide anion [67–70]. There are several potential explanations for this occurrence. HS alters the redox state of cells. A low dose of this stressor might produce mild stress-altering gene expression and protein synthesis. As demonstrated earlier, this further affects enzymes involved in ROS production, such as NADPH oxidase and mitochondrial respiratory complex, activating protective mechanisms like antioxidant pathways and consequently leading to reduced ROS [64,66,71,72], suggesting that cellular tolerance occurs when cells are exposed to NaCl concentrations close to physiological levels, such as 320 mOsmol/kg. However, Dmitrieva et al. (2004) and others warned that these surviving, adapted cells may be damaged beyond repair at the DNA level despite rapid proliferation and minimal apoptosis [66,72–74]. Furthermore, Xu et al. (2009) co-cultured ECs with VSMCs and reported that ECs upregulate antioxidants, more accurately, thioredoxin, consequently decreasing ROS production by the VSMCs in a state of stress [75]. These findings are supportive of our present results but also present a challenging perspective for future investigations.

On the other hand, a concentration of 350 mOsmol/kg appears to be a critical point for NaCl-induced oxidative stress in our cultured HAECs, significantly increasing H_2O_2 and $ONOO^-$ production (DCF-DA assay). A significant increase in $O_2^{\cdot-}$ was observed at a NaCl concentration of 350 mOsmol/kg following PMA stimulation (DHE assay), presented in Figure S1 (Supplementary Materials). These differences in ROS production at different NaCl concentrations might result from different metabolic pathways activated by NaCl, ranging from oxidase activation to structural/mitochondrial damage [76–78].

The endothelium, a metabolically active layer of cells, acts as a selective barrier separating the vascular wall and circulating blood [79]. Its intricate role includes functioning as an active endocrine, paracrine, and autocrine organ necessary for maintaining vascular homeostasis [80–82]. It is well documented that higher supraphysiological concentrations of AngII increase ROS production via NADPH-oxidase [83,84]. Even at low concentrations, such as 10^{-7} mol/L, AngII induces protein nitration in endothelial cells, with the extent of nitration increasing in a concentration-dependent manner as the concentration of AngII rises [85]. However, too-low levels of AngII have been shown to increase ROS in the vasculature, too [48,86,87], rendering the conclusion that a physiological range of AngII is needed to balance oxidative stress. Previously, we have shown that three days of a subpressor dose of AngII supplementation in HS-fed rats restored the FID of MCAs and significantly increased GPx4 and extracellular SOD antioxidative enzyme expression [48]. Furthermore, the HS diet significantly reduced the expression of Cu/Zn SOD and Mn SOD in the cerebral resistance arteries of the HS-fed rats, while an infusion of AngII restored protein Cu/Zn SOD (but not Mn SOD) expression [88]. A study involving the two-week infusion of AngII in sham rats and rats with myocardial infarction suggested the activation of counter-regulatory mechanisms by AngII, as blood pressure changes and vascular remodeling were minimal in the animals with myocardial infarction leading to compensation of hypertensive and growth stimulatory effects of AngII observed in sham rats. Namely, rats with myocardial infarction exhibited concomitant increases in plasma ANP and NO synthase activity following AngII infusion, demonstrating AngII-activated protective mechanisms related to blood vessel structure and function [89]. In the present study, AngII exposure in our cell cultures significantly altered metabolic activity following HS treatment after 48 h. Viability was reduced after 48 h but seemingly restored after 72 h of incubation. It has been reported that AngII exposure increases the viability of human mammary epithelial cells (184A1) through mitochondrial metabolism, as well as changes in cell behavior via AT1R overexpression [90]. Interestingly, AngII stimulates the Na/K pump in the proximal tubular cell culture and acutely stimulates the transcellular sodium transport [91], thus preventing an increase in cellular osmolality. If such an effect of AngII is present in the ECs, it remains to be investigated and demonstrates that it could also contribute a piece of the puzzle to the beneficial effect of AngII in the endothelium.

One possible explanation for an observed beneficial effect of AngII on the metabolic activity of the NaCl-stressed HAECs is the upregulation of antioxidative systems. This has been demonstrated in animal models of rats on an HS diet and in Dahl salt-sensitive rats [46,48]. In the vasculature, subpressor doses of AngII have led to restored microvascular reactivity and increased expression of antioxidative enzymes, such as SOD. Interestingly, it was previously reported that AngII, as a stress inducer, increases the expression of Nrf2-related genes in rat and mice neuronal cell lines, consequently leading to an increase in cell viability [92]. Consequently, future research should not only focus on the functions and activation pathways of Nrf2 but also explore its broader implications and applications.

The results showed that elevated levels of NaCl lead to an increase in oxidative stress and cause changes in the expression of certain CAMs. However, these changes do not occur uniformly in the same direction or at the same concentration of NaCl. This variability is likely due to the involvement of different signaling pathways, which may respond differently to NaCl, resulting in diverse effects on oxidative stress and the regulation of CAMs [13,93]. Endothelial pro-inflammatory phenotype is generated by increased pro-inflammatory cytokines production (IL-1β, IL-6, TNF-α) alongside C-reactive protein (CRP). This phenotype is characterized by an increase in CAMs, such as E-selectin, ICAM-1, and VCAM-1 [94–96]. Animal studies have shown upregulation in CAM expression following the HS diet [97–99]. Leukocyte migration and vascular adhesion mediated by ICAM-1 and VCAM-1 regulate inflammation and homeostasis in various diseases [98,100]. HS intake also stimulates the expression of E-selectin, monocyte chemotactic protein-1 (MCP-1), and endothelin 1 (ET1) [101,102]. ICAM-1 plays an important role in vascular inflammation and represents an attractive target for future treatment due to its increased expression in response to stressors [103,104]. The results of the present study showed decreased ICAM-1 expression following AngII exposure in a moderate NaCl concentration. Furthermore, endoglin is an angiogenesis/vascular remodeling participator with its soluble form that tends to be increased in inflammatory/vascular pathological conditions (atherosclerosis, hypertension, type 2 diabetes, and dyslipidaemia) and endothelial injury events [105–107]. In our study, endoglin expression was significantly decreased following AngII exposure in all three HS groups. This is in line with the notion that oxidative stress upregulated endoglin, e.g., in the placenta [108]. Since AngII can decrease oxidative stress, as demonstrated in this and previous studies, the reduction in endoglin levels is warranted.

While our study lays a foundation for future research on the mechanisms of AngII in relation to cell viability and activation, it has certain limitations. Subsequent studies should incorporate pharmacological experiments and explore the effects of angiotensin receptor blockers on ROS production and the expression of adhesion molecules in HAECs following AngII treatment. Indeed, in our previous work on rat animal models, we have demonstrated that the inhibition of AT1R leads to impaired endothelium-mediated vasodilation in response to changes in flow in isolated, pressurized, middle cerebral arteries and increases oxidative stress [93]. Further, results on protein expression of the AT1 and AT2 receptors in brain blood vessels (BBV) provided in Supplementary Figure (Figure S3) suggest that permissive effects of AngII on vasodilation and oxidative balance include and are mediated mainly via AT1 receptors.

One limitation of this study that may arise is the osmolarity control of the experiments that could have been performed by, e.g., mannitol or sorbitol. However, in this study, we opted not to perform this experiment due to growing evidence that increasing the osmolarity of the culture medium by the addition of sorbitol or mannitol (100 mM) did not show alteration in cellular response, as shown in neutrophils [56] or rat glial cell culture [109], but results suggested the sole effect of high sodium and not of increased osmolarity. Furthermore, the mechanisms of action and consequences of mannitol vs. NaCl are different [110], and mannitol also induces apoptosis in ECs [111]. Importantly, the effect of increased osmolarity is reversible [112], as shown in the experiments on mIMCD3 mouse renal collecting duct cells, where, with up to 600 mosmol/kg, the effect was only transient, and by 12 h at 550 mosmol/kg, the effect was reversible to normal. Thus, our control condition is the physiological concentration of NaCl (i.e., 270 mOsmol/L).

While our study focused on measuring intracellular ROS (baseline and following PMA stimulation), we acknowledge that calcium signaling, which can modulate PKC activity, was not directly assessed. This represents a limitation of our study, particularly in the context of HS conditions where calcium-dependent PKC-signaling pathways are relevant [113]. Future research should explore this aspect, including the measurement of intracellular calcium levels and their relationship to PKC activation and ROS generation.

5. Conclusions

The present study demonstrated that a longer duration and higher concentration of NaCl have detrimental effects on ECs' metabolic activity and induce them as a source of ROS. Importantly, the present study demonstrated that AngII can decrease oxidative stress and alter the activation state of ECs, thus providing beneficial conditions for cell survival. The potential common denominator of observed effects of AngII is the upregulation of antioxidative systems. Thus, future research should focus on further elucidating the mechanisms underlying AngII-induced antioxidative defense modulation and its potential therapeutic implications.

Supplementary Materials: The following supporting information can be downloaded at: https://www.mdpi.com/article/10.3390/biomedicines12122741/s1, Figure S1: Intracellular production of hydrogen peroxide and peroxynitrite (DCF-DA) in HAECs following high-salt treatment without and upon stimulation with PMA. Results are expressed as geometric mean fluorescence intensity (GMFI). DCF-DA–dichlorofluorescein diacetate; HAECs–human aortic endothelial cells; CTRL–control; HS–high salt; PMA–phorbol 12-myristate 13-acetate; One-way ANOVA; * significance level $p < 0.05$; Figure S2: Intracellular production of superoxide anion (DHE) in HAECs following high-salt treatment without and upon stimulation with PMA. Results are expressed as geometric mean fluorescence intensity (GMFI). DHE–dihydroethidium; HAECs–human aortic endothelial cells; CTRL–control; HS–high salt; PMA–phorbol 12-myristate 13-acetate; One-way ANOVA; * significance level $p < 0.05$; Figure S3: The relative protein expression (normalized to β-actin as a reference protein and loading control) and representative blots of the AT1 and AT2 receptors in brain surface vessels in the CTRL, HS, and HS and ANGII groups of Sprague-Dawley rats determined by Western blot method. CTRL–control; HS–high salt; ANGII–angiotensin II; AT1–angiotensin II type 1; AT2–angiotensin II type 2; One-way ANOVA; * significance level $p < 0.05$.

Author Contributions: Conceptualization, N.K. (Nikolina Kolobarić), N.K. (Nataša Kozina) and I.D.; Data curation, N.K. (Nikolina Kolobarić), N.K. (Nataša Kozina), Z.M. and I.D.; Formal analysis, N.K. (Nikolina Kolobarić), N.K. (Nataša Kozina), Z.M. and I.D.; Funding acquisition, I.D.; Investigation, N.K. (Nikolina Kolobarić), N.K. (Nataša Kozina), Z.M. and I.D.; Methodology, N.K. (Nikolina Kolobarić), N.K. (Nataša Kozina), Z.M. and I.D.; Project administration, I.D.; Resources, I.D.; Software, N.K. (Nikolina Kolobarić), N.K. (Nataša Kozina) and I.D.; Supervision, I.D.; Validation, I.D.; Visualization, N.K. (Nikolina Kolobarić), N.K. (Nataša Kozina), Z.M. and I.D.; Writing—original draft, N.K. (Nikolina Kolobarić), N.K. (Nataša Kozina), Z.M. and I.D.; Writing—review and editing, N.K. (Nikolina Kolobarić), N.K. (Nataša Kozina), Z.M. and I.D. All authors have read and agreed to the published version of the manuscript.

Funding: This research was funded by the Faculty of Medicine Osijek Institutional grant IP1-MEFOS-2019 (PI I.D.), IP1-MEFOS-2020 (PI I.D.), IP3-MEFOS-2021 project (PI I.D.), IP1-MEFOS-2022 project (PI I.D.), and IP18-MEFOS-2023 (PI I.D.), and by European Structural and Investment Funds through a grant from the Croatian National Science Center of Excellence for Personalized Health Care, Josip Juraj Strossmayer University of Osijek (# KK.01.1.1.01.0010.).

Institutional Review Board Statement: Not applicable.

Informed Consent Statement: Not applicable.

Data Availability Statement: The original contributions presented in this study are included in the article/Supplementary Materials. Further inquiries can be directed to the corresponding author.

Conflicts of Interest: The authors declare no conflicts of interest.

Abbreviations

ACE	angiotensin-converting enzyme
AngII	angiotensin II
ANOVA	analysis of variance
ARBs	angiotensin II receptor blockers
AT1R	angiotensin type 1 receptor
CAMs	cellular adhesion molecules
CCBs	calcium channel blockers
CO_2	carbon dioxide
CRP	C-reactive protein
Cu	copper
DCFDA	dichlorofluorescein diacetate
DCFH	$2',7'$-dichlorofluorescein
DHE	dihydroethidium
ET-1	endothelin 1
FID	flow-induced dilation
FLH	fluorescence height
GMFI	geometric mean fluorescence intensity
GPx	glutathione peroxidase
H_2O_2	hydrogen peroxide
HAECs	human aortic endothelial cells
HS	high salt
HUVECs	human umbilical vein endothelial cells
ICAM-1	intercellular adhesion molecule 1
IL	interleukin
MCAs	middle cerebral arteries
MCP-1	monocyte chemotactic protein-1
Mn	manganese
MTT	3-(4, 5-dimethylthiazolyl-2)-2, 5-diphenyltetrazolium bromide
NaCl	sodium chloride
NADPH	nicotinamide adenine dinucleotide phosphate
Na/K	sodium/potassium pump
NO	nitric oxide
$O_2{}^{\cdot-}$	superoxide anion

ONOO⁻	peroxynitrite
PBS	phosphate-buffered saline
PKC	protein kinase C
PMA	phorbol 12-myristate 13-acetate
Prx	peroxiredoxin
ROS	reactive oxygen species
SD	Sprague-Dawley
SOD	superoxide dismutase
Th17	T helper 17 cells
TNF-α	tumor necrosis factor alpha
Treg	T regulatory cells
VCAM-1	vascular adhesion molecule 1
Zn	zinc

References

1. Méndez-Barbero, N.; Gutiérrez-Muñoz, C.; Colio, L.M.B. Cellular Crosstalk between Endothelial and Smooth Muscle Cells in Vascular Wall Remodeling. *Int. J. Mol. Sci.* **2021**, *22*, 7284. [CrossRef] [PubMed]
2. Beck-Joseph, J.; Lehoux, S. Molecular Interactions Between Vascular Smooth Muscle Cells and Macrophages in Atherosclerosis. *Front. Cardiovasc. Med.* **2021**, *8*, 737934. [CrossRef]
3. Versari, D.; Daghini, E.; Virdis, A.; Ghiadoni, L.; Taddei, S. Endothelial dysfunction as a target for prevention of cardiovascular disease. *Diabetes Care* **2009**, *32* (Suppl. S2), S314–S321. [CrossRef]
4. Widlansky, M.E.; Gokce, N.; Keaney, J.F.; Vita, J.A. The clinical implications of endothelial dysfunction. *J. Am. Coll. Cardiol.* **2003**, *42*, 1149–1160. [CrossRef]
5. Fisher, M. Injuries to the vascular endothelium: Vascular wall and endothelial dysfunction. *Rev. Neurol. Dis.* **2008**, *5* (Suppl. S1), S4–S11. [PubMed]
6. Huang, Y.; Song, C.; He, J.; Li, M. Research progress in endothelial cell injury and repair. *Front. Pharmacol.* **2022**, *13*, 997272. [CrossRef]
7. Sun, H.-J.; Wu, Z.-Y.; Nie, X.-W.; Bian, J.-S. Role of Endothelial Dysfunction in Cardiovascular Diseases: The Link Between Inflammation and Hydrogen Sulfide. *Front. Pharmacol.* **2020**, *10*, 1568. [CrossRef] [PubMed]
8. Widmer, R.J.; Lerman, A. Endothelial dysfunction and cardiovascular disease. *Glob. Cardiol. Sci. Pract.* **2014**, *2014*, 291–308. [CrossRef]
9. Little, P.J.; Askew, C.D.; Xu, S.; Kamato, D. Endothelial Dysfunction and Cardiovascular Disease: History and Analysis of the Clinical Utility of the Relationship. *Biomedicines* **2021**, *9*, 699. [CrossRef]
10. Cai, H.; Harrison, D.G. Endothelial Dysfunction in Cardiovascular Diseases: The Role of Oxidant Stress. *Circ. Res.* **2000**, *87*, 840–844. [CrossRef]
11. Calila, H.; Bălășescu, E.; Nedelcu, R.I.; Ion, D.A. Endothelial Dysfunction as a Key Link between Cardiovascular Disease and Frailty: A Systematic Review. *J. Clin. Med.* **2024**, *13*, 2686. [CrossRef] [PubMed]
12. Ertuglu, L.A.; Mutchler, A.P.; Yu, J.; Kirabo, A. Inflammation and oxidative stress in salt sensitive hypertension; The role of the NLRP3 inflammasome. *Front. Physiol.* **2022**, *13*, 1096296. [CrossRef] [PubMed]
13. Knezović, A.; Kolobarić, N.; Drenjančević, I.; Mihaljević, Z.; Šušnjara, P.; Jukić, I.; Stupin, M.; Kibel, A.; Marczi, S.; Mihalj, M.; et al. Role of Oxidative Stress in Vascular Low-Grade Inflammation Initiation Due to Acute Salt Loading in Young Healthy Individuals. *Antioxidants* **2022**, *11*, 444. [CrossRef] [PubMed]
14. Mihalj, M.; Matić, A.; Mihaljević, Z.; Barić, L.; Stupin, A.; Drenjančević, I. Short-Term High-NaCl Dietary Intake Changes Leukocyte Expression of VLA-4, LFA-1, and Mac-1 Integrins in Both Healthy Humans and Sprague-Dawley Rats: A Comparative Study. *Mediat. Inflamm.* **2019**, *2019*, e6715275. [CrossRef]
15. Tolj, I.; Stupin, A.; Drenjančević, I.; Šušnjara, P.; Perić, L.; Stupin, M. The Role of Nitric Oxide in the Micro- and Macrovascular Response to a 7-Day High-Salt Diet in Healthy Individuals. *Int. J. Mol. Sci.* **2023**, *24*, 7159. [CrossRef]
16. Wu, Q.; Burley, G.; Li, L.; Lin, S.; Shi, Y. The role of dietary salt in metabolism and energy balance: Insights beyond cardiovascular disease. *Diabetes Obes. Metab.* **2023**, *25*, 1147–1161. [CrossRef]
17. Barić, L.; Drenjančević, I.; Matić, A.; Stupin, M.; Kolar, L.; Mihaljević, Z.; Lenasi, H.; Šerić, V.; Stupin, A. Seven-Day Salt Loading Impairs Microvascular Endothelium-Dependent Vasodilation without Changes in Blood Pressure, Body Composition and Fluid Status in Healthy Young Humans. *Kidney Blood Press. Res.* **2019**, *44*, 835–847. [CrossRef] [PubMed]
18. Yuan, M.; Yan, D.; Wang, Y.; Qi, M.; Li, K.; Lv, Z.; Gao, D.; Ning, N. Sodium intake and the risk of heart failure and hypertension: Epidemiological and Mendelian randomization analysis. *Front. Nutr.* **2024**, *10*, 1263554. [CrossRef]
19. Stamler, J. The INTERSALT Study: Background, methods, findings, and implications. *Am. J. Clin. Nutr.* **1997**, *65*, 626S–642S. [CrossRef]
20. He, J.; Whelton, P.K. Commentary: Salt intake, hypertension and risk of cardiovascular disease: An important public health challenge. *Int. J. Epidemiol.* **2002**, *31*, 327–331. [CrossRef]

21. Strazzullo, P.; D'Elia, L.; Kandala, N.-B.; Cappuccio, F.P. Salt intake, stroke, and cardiovascular disease: Meta-analysis of prospective studies. *Br. Med. J. BMJ* **2009**, *339*, b4567. [CrossRef]
22. Wang, Y.-J.; Yeh, T.-L.; Shih, M.-C.; Tu, Y.-K.; Chien, K.-L. Dietary Sodium Intake and Risk of Cardiovascular Disease: A Systematic Review and Dose-Response Meta-Analysis. *Nutrients* **2020**, *12*, 2934. [CrossRef] [PubMed]
23. Qi, H.; Liu, Z.; Cao, H.; Sun, W.-P.; Peng, W.-J.; Liu, B.; Dong, S.-J.; Xiang, Y.-T.; Zhang, L. Comparative Efficacy of Antihypertensive Agents in Salt-Sensitive Hypertensive Patients: A Network Meta-Analysis. *Am. J. Hypertens.* **2018**, *31*, 835–846. [CrossRef]
24. Vargas-Meza, J.; Nilson, E.A.F.; Nieto, C.; Khandpur, N.; Denova-Gutiérrez, E.; Valero-Morales, I.; Barquera, S.; Campos-Nonato, I. Modelling the impact of sodium intake on cardiovascular disease mortality in Mexico. *BMC Public Health* **2023**, *23*, 983. [CrossRef] [PubMed]
25. Williams, B.; Mancia, G.; Spiering, W.; Agabiti Rosei, E.; Azizi, M.; Burnier, M.; Clement, D.L.; Coca, A.; de Simone, G.; Dominiczak, A.; et al. 2018 ESC/ESH Guidelines for the management of arterial hypertension: The Task Force for the management of arterial hypertension of the European Society of Cardiology and the European Society of Hypertension. *J. Hypertens.* **2018**, *36*, 1953. [CrossRef] [PubMed]
26. Riegel, B.; Lee, S.; Hill, J.; Daus, M.; Baah, F.O.; Wald, J.W.; Knafl, G.J. Patterns of adherence to diuretics, dietary sodium and fluid intake recommendations in adults with heart failure. *Heart Lung J. Crit. Care* **2019**, *48*, 179–185. [CrossRef]
27. He, F.J.; Li, J.; Macgregor, G.A. Effect of longer term modest salt reduction on blood pressure: Cochrane systematic review and meta-analysis of randomised trials. *BMJ Clin. Res. Ed.* **2013**, *346*, f1325. [CrossRef]
28. Jablonski, K.L.; Racine, M.L.; Geolfos, C.J.; Gates, P.E.; Chonchol, M.; McQueen, M.B.; Seals, D.R. Dietary sodium restriction reverses vascular endothelial dysfunction in middle-aged/older adults with moderately elevated systolic blood pressure. *J. Am. Coll. Cardiol.* **2013**, *61*, 335–343. [CrossRef]
29. Heart Outcomes Prevention Evaluation Study Investigators; Yusuf, S.; Sleight, P.; Pogue, J.; Bosch, J.; Davies, R.; Dagenais, G. Effects of an angiotensin-converting-enzyme inhibitor, ramipril, on cardiovascular events in high-risk patients. *N. Engl. J. Med.* **2000**, *342*, 145–153. [CrossRef]
30. Danchin, N.; Cucherat, M.; Thuillez, C.; Durand, E.; Kadri, Z.; Steg, P.G. Angiotensin-Converting Enzyme Inhibitors in Patients With Coronary Artery Disease and Absence of Heart Failure or Left Ventricular Systolic Dysfunction: An Overview of Long-term Randomized Controlled Trials. *Arch. Intern. Med.* **2006**, *166*, 787–796. [CrossRef]
31. Solomon, S.D.; Rice, M.M.; Jablonski, A.K.; Jose, P.; Domanski, M.; Sabatine, M.; Gersh, B.J.; Rouleau, J.; Pfeffer, M.A.; Braunwald, E. Renal Function and Effectiveness of Angiotensin-Converting Enzyme Inhibitor Therapy in Patients With Chronic Stable Coronary Disease in the Prevention of Events with ACE inhibition (PEACE) Trial. *Circulation* **2006**, *114*, 26–31. [CrossRef] [PubMed]
32. Yusuf, S.; Pfeffer, M.A.; Swedberg, K.; Granger, C.B.; Held, P.; McMurray, J.J.V.; Michelson, E.L.; Olofsson, B.; Ostergren, J.; CHARM Investigators and Committees. Effects of candesartan in patients with chronic heart failure and preserved left-ventricular ejection fraction: The CHARM-Preserved Trial. *Lancet* **2003**, *362*, 777–781. [CrossRef] [PubMed]
33. Lithell, H.; Hansson, L.; Skoog, I.; Elmfeldt, D.; Hofman, A.; Olofsson, B.; Trenkwalder, P.; Zanchetti, A. SCOPE Study Group The Study on Cognition and Prognosis in the Elderly (SCOPE): Principal results of a randomized double-blind intervention trial. *J. Hypertens.* **2003**, *21*, 875–886. [CrossRef]
34. Dézsi, C.A. A review of clinical studies on angiotensin II receptor blockers and risk of cancer. *Int. J. Cardiol.* **2014**, *177*, 748–753. [CrossRef]
35. Munger, M.A. Use of Angiotensin Receptor Blockers In Cardiovascular Protection: Current Evidence and Future Directions. *Pharm. Ther.* **2011**, *36*, 22.
36. Suzuki, H.; Kanno, Y.; Sugahara, S.; Ikeda, N.; Shoda, J.; Takenaka, T.; Inoue, T.; Araki, R. Effect of angiotensin receptor blockers on cardiovascular events in patients undergoing hemodialysis: An open-label randomized controlled trial. *Am. J. Kidney Dis. Off. J. Natl. Kidney Found.* **2008**, *52*, 501–506. [CrossRef]
37. Cummings, D.M.; Amadio, P.; Nelson, L.; Fitzgerald, J.M. The role of calcium channel blockers in the treatment of essential hypertension. *Arch. Intern. Med.* **1991**, *151*, 250–259. [CrossRef]
38. Jones, K.E.; Hayden, S.L.; Meyer, H.R.; Sandoz, J.L.; Arata, W.H.; Dufrene, K.; Ballaera, C.; Lopez Torres, Y.; Griffin, P.; Kaye, A.M.; et al. The Evolving Role of Calcium Channel Blockers in Hypertension Management: Pharmacological and Clinical Considerations. *Curr. Issues Mol. Biol.* **2024**, *46*, 6315–6327. [CrossRef]
39. Lee, E.M. Calcium channel blockers for hypertension: Old, but still useful. *Cardiovasc. Prev. Pharmacother.* **2023**, *5*, 113–125. [CrossRef]
40. Ogita, H.; Liao, J.K. Endothelial Function and Oxidative Stress. *Endothel. J. Endothel. Cell Res.* **2004**, *11*, 123–132. [CrossRef]
41. Barić, L.; Drenjančević, I.; Mihalj, M.; Matić, A.; Stupin, M.; Kolar, L.; Mihaljević, Z.; Mrakovčić-Šutić, I.; Šerić, V.; Stupin, A. Enhanced Antioxidative Defense by Vitamins C and E Consumption Prevents 7-Day High-Salt Diet-Induced Microvascular Endothelial Function Impairment in Young Healthy Individuals. *J. Clin. Med.* **2020**, *9*, 843. [CrossRef] [PubMed]
42. Cosic, A.; Jukic, I.; Stupin, A.; Mihalj, M.; Mihaljevic, Z.; Novak, S.; Vukovic, R.; Drenjancevic, I. Attenuated flow-induced dilatation of middle cerebral arteries is related to increased vascular oxidative stress in rats on a short-term high salt diet. *J. Physiol.* **2016**, *594*, 4917–4931. [CrossRef]
43. Fyhrquist, F.; Metsärinne, K.; Tikkanen, I. Role of angiotensin II in blood pressure regulation and in the pathophysiology of cardiovascular disorders. *J. Hum. Hypertens.* **1995**, *9* (Suppl. S5), S19–S24.

44. Patel, P.; Sanghavi, D.K.; Morris, D.L.; Kahwaji, C.I. Angiotensin II. In *StatPearls*; StatPearls Publishing: Treasure Island, FL, USA, 2024.
45. Timmermans, P.B.M.W.M.; Benfield, P.; Chiu, A.T.; Herblin, W.F.; Wong, P.C.; Smith, R.D. Angiotensin II Receptors and Functional Correlates. *Am. J. Hypertens.* **1992**, *5*, 221S–235S. [CrossRef]
46. Zhu, J.; Drenjancevic-Peric, I.; McEwen, S.; Friesema, J.; Schulta, D.; Yu, M.; Roman, R.J.; Lombard, J.H. Role of superoxide and angiotensin II suppression in salt-induced changes in endothelial Ca^{2+} signaling and NO production in rat aorta. *Am. J. Physiol. Heart Circ. Physiol.* **2006**, *291*, H929–H938. [CrossRef]
47. Drenjančević-Perić, I.; Jelaković, B.; Lombard, J.H.; Kunert, M.P.; Kibel, A.; Gros, M. High-Salt Diet and Hypertension: Focus on the Renin-Angiotensin System. *Kidney Blood Press. Res.* **2011**, *34*, 1–11. [CrossRef] [PubMed]
48. Matic, A.; Jukic, I.; Mihaljevic, Z.; Kolobaric, N.; Stupin, A.; Kozina, N.; Bujak, I.T.; Kibel, A.; Lombard, J.H.; Drenjancevic, I. Low-dose angiotensin II supplementation restores flow-induced dilation mechanisms in cerebral arteries of Sprague-Dawley rats on a high salt diet. *J. Hypertens.* **2022**, *40*, 441. [CrossRef] [PubMed]
49. Mihalj, M.; Štefanić, M.; Mihaljević, Z.; Kolobarić, N.; Jukić, I.; Stupin, A.; Matić, A.; Frkanec, R.; Tavčar, B.; Horvatić, A.; et al. Early Low-Grade Inflammation Induced by High-Salt Diet in Sprague Dawley Rats Involves Th17/Treg Axis Dysregulation, Vascular Wall Remodeling, and a Shift in the Fatty Acid Profile. *Cell. Physiol. Biochem.* **2024**, *58*, 83–103. [PubMed]
50. Miyauchi, H.; Geisberger, S.; Luft, F.C.; Wilck, N.; Stegbauer, J.; Wiig, H.; Dechend, R.; Jantsch, J.; Kleinewietfeld, M.; Kempa, S.; et al. Sodium as an Important Regulator of Immunometabolism. *Hypertens. Dallas Tex 1979* **2024**, *81*, 426–435. [CrossRef]
51. Hunter, R.W.; Dhaun, N.; Bailey, M.A. The impact of excessive salt intake on human health. *Nat. Rev. Nephrol.* **2022**, *18*, 321–335. [CrossRef]
52. Li, J.; White, J.; Guo, L.; Zhao, X.; Wang, J.; Smart, E.J.; Li, X.-A. Salt inactivates endothelial nitric oxide synthase in endothelial cells. *J. Nutr.* **2009**, *139*, 447–451. [CrossRef] [PubMed]
53. Pap, D.; Pajtók, C.; Veres-Székely, A.; Szebeni, B.; Szász, C.; Bokrossy, P.; Zrufkó, R.; Vannay, Á.; Tulassay, T.; Szabó, A.J. High Salt Promotes Inflammatory and Fibrotic Response in Peritoneal Cells. *Int. J. Mol. Sci.* **2023**, *24*, 13765. [CrossRef] [PubMed]
54. Yamakami, Y.; Yonekura, R.; Matsumoto, Y.; Takauji, Y.; Miki, K.; Fujii, M.; Ayusawa, D. High concentrations of NaCl induce cell swelling leading to senescence in human cells. *Mol. Cell. Biochem.* **2016**, *411*, 117–125. [CrossRef] [PubMed]
55. Dmitrieva, N.I.; Burg, M.B. Elevated Sodium and Dehydration Stimulate Inflammatory Signaling in Endothelial Cells and Promote Atherosclerosis. *PLoS ONE* **2015**, *10*, e0128870. [CrossRef]
56. Mazzitelli, I.; Bleichmar, L.; Melucci, C.; Gerber, P.P.; Toscanini, A.; Cuestas, M.L.; Diaz, F.E.; Geffner, J. High Salt Induces a Delayed Activation of Human Neutrophils. *Front. Immunol.* **2022**, *13*, 831844. [CrossRef]
57. Ju, H.K.; Hwang, S.-J.; Jeon, C.-J.; Lee, G.M.; Yoon, S.K. Use of NaCl prevents aggregation of recombinant COMP–Angiopoietin-1 in Chinese hamster ovary cells. *J. Biotechnol.* **2009**, *143*, 145–150. [CrossRef]
58. Jakic, B.; Buszko, M.; Cappellano, G.; Wick, G. Elevated sodium leads to the increased expression of HSP60 and induces apoptosis in HUVECs. *PLoS ONE* **2017**, *12*, e0179383. [CrossRef]
59. Shiwakoti, S.; Adhikari, D.; Lee, J.P.; Kang, K.-W.; Lee, I.-S.; Kim, H.J.; Oak, M.-H. Prevention of Fine Dust-Induced Vascular Senescence by Humulus lupulus Extract and Its Major Bioactive Compounds. *Antioxidants* **2020**, *9*, 1243. [CrossRef]
60. Kolar, L.; Šušnjara, P.; Stupin, M.; Stupin, A.; Jukić, I.; Mihaljević, Z.; Kolobarić, N.; Bebek, I.; Nejašmić, D.; Lovrić, M.; et al. Enhanced Microvascular Adaptation to Acute Physical Stress and Reduced Oxidative Stress in Male Athletes Who Consumed Chicken Eggs Enriched with n-3 Polyunsaturated Fatty Acids and Antioxidants-Randomized Clinical Trial. *Life* **2023**, *13*, 2140. [CrossRef]
61. Kong, J.; Hu, X.-M.; Cai, W.-W.; Wang, Y.-M.; Chi, C.-F.; Wang, B. Bioactive Peptides from Skipjack Tuna Cardiac Arterial Bulbs (II): Protective Function on UVB-Irradiated HaCaT Cells through Antioxidant and Anti-Apoptotic Mechanisms. *Mar. Drugs* **2023**, *21*, 105. [CrossRef]
62. Barthelmes, J.; Nägele, M.P.; Ludovici, V.; Ruschitzka, F.; Sudano, I.; Flammer, A.J. Endothelial dysfunction in cardiovascular disease and Flammer syndrome—Similarities and differences. *EPMA J.* **2017**, *8*, 99–109. [CrossRef] [PubMed]
63. Lippi, M.; Stadiotti, I.; Pompilio, G.; Sommariva, E. Human Cell Modeling for Cardiovascular Diseases. *Int. J. Mol. Sci.* **2020**, *21*, 6388. [CrossRef] [PubMed]
64. Ristow, M.; Schmeisser, K. Mitohormesis: Promoting Health and Lifespan by Increased Levels of Reactive Oxygen Species (ROS). *Dose-Response* **2014**, *12*, 288–341. [CrossRef] [PubMed]
65. Schulz, E.; Gori, T.; Münzel, T. Oxidative stress and endothelial dysfunction in hypertension. *Hypertens. Res.* **2011**, *34*, 665–673. [CrossRef]
66. Burg, M.B.; Ferraris, J.D.; Dmitrieva, N.I. Cellular Response to Hyperosmotic Stresses. *Physiol. Rev.* **2007**, *87*, 1441–1474. [CrossRef]
67. Calabrese, E.J. Hormesis: A fundamental concept in biology. *Microb. Cell* **2014**, *1*, 145–149. [CrossRef]
68. Calabrese, E.J.; Baldwin, L.A. The Hormetic Dose-Response Model Is More Common than the Threshold Model in Toxicology. *Toxicol. Sci.* **2003**, *71*, 246–250. [CrossRef] [PubMed]
69. Calabrese, E.J.; Mattson, M.P. How does hormesis impact biology, toxicology, and medicine? *NPJ Aging Mech. Dis.* **2017**, *3*, 13. [CrossRef]
70. Calabrese, V.; Cornelius, C.; Dinkova-Kostova, A.T.; Calabrese, E.J.; Mattson, M.P. Cellular Stress Responses, The Hormesis Paradigm, and Vitagenes: Novel Targets for Therapeutic Intervention in Neurodegenerative Disorders. *Antioxid. Redox Signal.* **2010**, *13*, 1763–1811. [CrossRef]

71. Kunsch, C.; Medford, R.M. Oxidative Stress as a Regulator of Gene Expression in the Vasculature. *Circ. Res.* **1999**, *85*, 753–766. [CrossRef]
72. Kültz, D.; Chakravarty, D. Hyperosmolality in the form of elevated NaCl but not urea causes DNA damage in murine kidney cells. *Proc. Natl. Acad. Sci. USA* **2001**, *98*, 1999–2004. [CrossRef] [PubMed]
73. Dmitrieva, N.I.; Cai, Q.; Burg, M.B. Cells adapted to high NaCl have many DNA breaks and impaired DNA repair both in cell culture and in vivo. *Proc. Natl. Acad. Sci. USA* **2004**, *101*, 2317–2322. [CrossRef] [PubMed]
74. Mukherjee, S.; Dutta, A.; Chakraborty, A. External modulators and redox homeostasis: Scenario in radiation-induced bystander cells. *Mutat. Res. Mutat. Res.* **2021**, *787*, 108368. [CrossRef] [PubMed]
75. Xu, S.; He, Y.; Vokurkova, M.; Touyz, R.M. Endothelial Cells Negatively Modulate Reactive Oxygen Species Generation in Vascular Smooth Muscle Cells. *Hypertension* **2009**, *54*, 427–433. [CrossRef]
76. Ali, T.; Li, D.; Ponnamperumage, T.N.F.; Peterson, A.K.; Pandey, J.; Fatima, K.; Brzezinski, J.; Jakusz, J.A.R.; Gao, H.; Koelsch, G.E.; et al. Generation of Hydrogen Peroxide in Cancer Cells: Advancing Therapeutic Approaches for Cancer Treatment. *Cancers* **2024**, *16*, 2171. [CrossRef]
77. Andrés, C.M.C.; Pérez de la Lastra, J.M.; Juan, C.A.; Plou, F.J.; Pérez-Lebeña, E. Chemistry of Hydrogen Peroxide Formation and Elimination in Mammalian Cells, and Its Role in Various Pathologies. *Stresses* **2022**, *2*, 256–274. [CrossRef]
78. Nauseef, W. Detection of superoxide anion and hydrogen peroxide production by cellular NADPH oxidases. *Biochim. Biophys. Acta* **2013**, *1840*, 757–767. [CrossRef]
79. Pahwa, R.; Goyal, A.; Bansal, P.; Jialal, I. Chronic Inflammation. In *StatPearls*; StatPearls Publishing: Treasure Island, FL, USA, 2020.
80. Bonetti, P.O.; Lerman, L.O.; Lerman, A. Endothelial Dysfunction. *Arterioscler. Thromb. Vasc. Biol.* **2003**, *23*, 168–175. [CrossRef] [PubMed]
81. Hadi, H.A.; Carr, C.S.; Al Suwaidi, J. Endothelial Dysfunction: Cardiovascular Risk Factors, Therapy, and Outcome. *Vasc. Health Risk Manag.* **2005**, *1*, 183–198.
82. Park, K.-H.; Park, W.J. Endothelial Dysfunction: Clinical Implications in Cardiovascular Disease and Therapeutic Approaches. *J. Korean Med. Sci.* **2015**, *30*, 1213–1225. [CrossRef]
83. Nguyen Dinh Cat, A.; Montezano, A.C.; Burger, D.; Touyz, R.M. Angiotensin II, NADPH Oxidase, and Redox Signaling in the Vasculature. *Antioxid. Redox Signal.* **2013**, *19*, 1110–1120. [CrossRef] [PubMed]
84. Frey, R.S.; Ushio-Fukai, M.; Malik, A.B. NADPH Oxidase-Dependent Signaling in Endothelial Cells: Role in Physiology and Pathophysiology. *Antioxid. Redox Signal.* **2009**, *11*, 791–810. [CrossRef] [PubMed]
85. Mihm, M.J.; Wattanapitayakul, S.K.; Piao, S.-F.; Hoyt, D.G.; Bauer, J.A. Effects of angiotensin II on vascular endothelial cells: Formation of receptor-mediated reactive nitrogen species. *Biochem. Pharmacol.* **2003**, *65*, 1189–1197. [CrossRef]
86. Drenjancevic-Peric, I.; Lombard, J.H. Reduced Angiotensin II and Oxidative Stress Contribute to Impaired Vasodilation in Dahl Salt-Sensitive Rats on Low-Salt Diet. *Hypertension* **2005**, *45*, 687–691. [CrossRef]
87. Drenjancevic-Peric, I.; Frisbee, J.C.; Lombard, J.H. Skeletal Muscle Arteriolar Reactivity in SS.BN13 Consomic Rats and Dahl Salt-Sensitive Rats. *Hypertension* **2003**, *41*, 1012–1015. [CrossRef] [PubMed]
88. Durand, M.J.; Lombard, J.H. Low-Dose Angiotensin II Infusion Restores Vascular Function in Cerebral Arteries of High Salt–Fed Rats by Increasing Copper/Zinc Superoxide Dimutase Expression. *Am. J. Hypertens.* **2013**, *26*, 739. [CrossRef]
89. Heeneman, S.; Smits, J.F.M.; Leenders, P.J.A.; Schiffers, P.M.H.; Daemen, M.J.A.P. Effects of Angiotensin II on Cardiac Function and Peripheral Vascular Structure During Compensated Heart Failure in the Rat. *Arterioscler. Thromb. Vasc. Biol.* **1997**, *17*, 1985–1994. [CrossRef]
90. Piastowska-Ciesielska, A.W.; Domińska, K.; Nowakowska, M.; Gajewska, M.; Gajos-Michniewicz, A.; Ochędalski, T. Angiotensin modulates human mammary epithelial cell motility. *J. Renin Angiotensin Aldosterone Syst.* **2014**, *15*, 419–429. [CrossRef]
91. Hanna, F.S.; Alkhouri, S.; Rajagopalan, C.; Ji, K.; Mattingly, R.R.; Yingst, D.R. Ang II acutely stimulates Na,K-pump in cells from proximal tubules by increasing its phosphorylation at S938 via a PI3K/AKT pathway. *Physiol. Rep.* **2022**, *10*, e15508. [CrossRef]
92. Parga, J.A.; Rodriguez-Perez, A.I.; Garcia-Garrote, M.; Rodriguez-Pallares, J.; Labandeira-Garcia, J.L. Angiotensin II induces oxidative stress and upregulates neuroprotective signaling from the NRF2 and KLF9 pathway in dopaminergic cells. *Free Radic. Biol. Med.* **2018**, *129*, 394–406. [CrossRef]
93. Jukic, I.; Mihaljevic, Z.; Matic, A.; Mihalj, M.; Kozina, N.; Selthofer-Relatic, K.; Mihaljevic, D.; Koller, A.; Tartaro Bujak, I.; Drenjancevic, I. Angiotensin II type 1 receptor is involved in flow-induced vasomotor responses of isolated middle cerebral arteries: Role of oxidative stress. *Am. J. Physiol. Heart Circ. Physiol.* **2021**, *320*, H1609–H1624. [CrossRef] [PubMed]
94. Medina-Leyte, D.J.; Zepeda-García, O.; Domínguez-Pérez, M.; González-Garrido, A.; Villarreal-Molina, T.; Jacobo-Albavera, L. Endothelial Dysfunction, Inflammation and Coronary Artery Disease: Potential Biomarkers and Promising Therapeutical Approaches. *Int. J. Mol. Sci.* **2021**, *22*, 3850. [CrossRef]
95. Theofilis, P.; Sagris, M.; Oikonomou, E.; Antonopoulos, A.S.; Siasos, G.; Tsioufis, C.; Tousoulis, D. Inflammatory Mechanisms Contributing to Endothelial Dysfunction. *Biomedicines* **2021**, *9*, 781. [CrossRef] [PubMed]
96. Nafisa, A.; Gray, S.G.; Cao, Y.; Wang, T.; Xu, S.; Wattoo, F.H.; Barras, M.; Cohen, N.; Kamato, D.; Little, P.J. Endothelial function and dysfunction: Impact of metformin. *Pharmacol. Ther.* **2018**, *192*, 150–162. [CrossRef] [PubMed]

97. Takahashi, H.; Nakagawa, S.; Wu, Y.; Kawabata, Y.; Numabe, A.; Yanagi, Y.; Tamaki, Y.; Uehara, Y.; Araie, M. A high-salt diet enhances leukocyte adhesion in association with kidney injury in young Dahl salt-sensitive rats. *Hypertens. Res. Off. J. Jpn. Soc. Hypertens.* **2017**, *40*, 912–920. [CrossRef]
98. Kaur, R.; Singh, V.; Kumari, P.; Singh, R.; Chopra, H.; Emran, T.B. Novel insights on the role of VCAM-1 and ICAM-1: Potential biomarkers for cardiovascular diseases. *Ann. Med. Surg.* **2022**, *84*, 104802. [CrossRef]
99. Granger, D.N.; Senchenkova, E. Leukocyte–Endothelial Cell Adhesion. In *Inflammation and the Microcirculation*; Morgan & Claypool Life Sciences: Willston, VT, USA, 2010.
100. Singh, V.; Kaur, R.; Kumari, P.; Pasricha, C.; Singh, R. ICAM-1 and VCAM-1: Gatekeepers in various inflammatory and cardiovascular disorders. *Clin. Chim. Acta Int. J. Clin. Chem.* **2023**, *548*, 117487. [CrossRef]
101. Afsar, B.; Kuwabara, M.; Ortiz, A.; Yerlikaya, A.; Siriopol, D.; Covic, A.; Rodriguez-Iturbe, B.; Johnson, R.J.; Kanbay, M. Salt Intake and Immunity. *Hypertension* **2018**, *72*, 19–23. [CrossRef]
102. Ma, P.; Li, G.; Jiang, X.; Shen, X.; Li, H.; Yang, L.; Liu, W. NFAT5 directs hyperosmotic stress-induced fibrin deposition and macrophage infiltration via PAI-1 in endothelium. *Aging* **2020**, *13*, 3661–3679. [CrossRef]
103. Bui, T.M.; Wiesolek, H.L.; Sumagin, R. ICAM-1: A master regulator of cellular responses in inflammation, injury resolution, and tumorigenesis. *J. Leukoc. Biol.* **2020**, *108*, 787–799. [CrossRef]
104. Muro, S.; Gajewski, C.; Koval, M.; Muzykantov, V.R. ICAM-1 recycling in endothelial cells: A novel pathway for sustained intracellular delivery and prolonged effects of drugs. *Blood* **2005**, *105*, 650–658. [CrossRef] [PubMed]
105. Gallardo-Vara, E.; Gamella-Pozuelo, L.; Perez-Roque, L.; Bartha, J.L.; Garcia-Palmero, I.; Casal, J.I.; López-Novoa, J.M.; Pericacho, M.; Bernabeu, C. Potential Role of Circulating Endoglin in Hypertension via the Upregulated Expression of BMP4. *Cells* **2020**, *9*, 988. [CrossRef] [PubMed]
106. Rossi, E.; Bernabeu, C.; Smadja, D.M. Endoglin as an Adhesion Molecule in Mature and Progenitor Endothelial Cells: A Function Beyond TGF-β. *Front. Med.* **2019**, *6*, 10. [CrossRef] [PubMed]
107. Schoonderwoerd, M.J.A.; Goumans, M.-J.T.H.; Hawinkels, L.J.A.C. Endoglin: Beyond the Endothelium. *Biomolecules* **2020**, *10*, 289. [CrossRef]
108. Jena, M.K.; Sharma, N.R.; Petitt, M.; Maulik, D.; Nayak, N.R. Pathogenesis of Preeclampsia and Therapeutic Approaches Targeting the Placenta. *Biomolecules* **2020**, *10*, 953. [CrossRef]
109. Arimochi, H.; Morita, K. High salt culture conditions suppress proliferation of rat C6 glioma cell by arresting cell-cycle progression at S-phase. *J. Mol. Neurosci.* **2005**, *27*, 293–301. [CrossRef]
110. Kinsman, B.J.; Browning, K.N.; Stocker, S.D. NaCl and osmolarity produce different responses in organum vasculosum of the lamina terminalis neurons, sympathetic nerve activity and blood pressure. *J. Physiol.* **2017**, *595*, 6187–6201. [CrossRef] [PubMed]
111. Malek, A.M.; Goss, G.G.; Jiang, L.; Izumo, S.; Alper, S.L. Mannitol at Clinical Concentrations Activates Multiple Signaling Pathways and Induces Apoptosis in Endothelial Cells. *Stroke* **1998**, *29*, 2631–2640. [CrossRef]
112. Michea, L.; Ferguson, D.R.; Peters, E.M.; Andrews, P.M.; Kirby, M.R.; Burg, M.B. Cell cycle delay and apoptosis are induced by high salt and urea in renal medullary cells. *Am. J. Physiol. Ren. Physiol.* **2000**, *278*, F209–F218. [CrossRef]
113. Li, W.; Lv, J.; Wu, J.; Zhou, X.; Jiang, L.; Zhu, X.; Tu, Q.; Tang, J.; Liu, Y.; He, A.; et al. Maternal high-salt diet altered PKC/MLC20 pathway and increased ANG II receptor-mediated vasoconstriction in adult male rat offspring. *Mol. Nutr. Food Res.* **2016**, *60*, 1684–1694. [CrossRef]

Disclaimer/Publisher's Note: The statements, opinions and data contained in all publications are solely those of the individual author(s) and contributor(s) and not of MDPI and/or the editor(s). MDPI and/or the editor(s) disclaim responsibility for any injury to people or property resulting from any ideas, methods, instructions or products referred to in the content.

Article

Cardiovascular and Renal Effects Induced by Alpha-Lipoic Acid Treatment in Two-Kidney-One-Clip Hypertensive Rats

Déborah Victória Gomes Nascimento [1], Darlyson Ferreira Alencar [2], Matheus Vinicius Barbosa da Silva [1], Danilo Galvão Rocha [3], Camila Ferreira Roncari [3], Roberta Jeane Bezerra Jorge [2,3], Renata de Sousa Alves [2], Richard Boarato David [3], Wylla Tatiana Ferreira e Silva [1], Lígia Cristina Monteiro Galindo [1] and Thyago Moreira de Queiroz [1,*]

[1] Laboratory of Nutrition, Physical Activity and Phenotypic Plasticity, Federal University of Pernambuco—UFPE, Vitória de Santo Antão 55608-680, Brazil; deborah.gnascimento@ufpe.br (D.V.G.N.); matheus.viniciusbarbosa@ufpe.br (M.V.B.d.S.); wylla.silva@ufpe.br (W.T.F.e.S.); ligia.mgalindo@ufpe.br (L.C.M.G.)

[2] Department of Morphology, School of Medicine, Federal University of Ceará—UFC, Fortaleza 60430-160, Brazil; darlysonferreira_alencar@hotmail.com (D.F.A.); robertajeane@ufc.br (R.J.B.J.); renata.alves@ufc.br (R.d.S.A.)

[3] Department of Physiology and Pharmacology, School of Medicine, Federal University of Ceará—UFC, Fortaleza 60430-160, Brazil; d.galvaorocha@gmail.com (D.G.R.); camilafroncari@ufc.br (C.F.R.); rdavid@ufc.br (R.B.D.)

* Correspondence: thyago.queiroz@ufpe.br

Abstract: α-Lipoic acid (LA) is an antioxidant of endogenous production, also obtained exogenously. Oxidative stress is closely associated with hypertension, which causes kidney injury and endothelial dysfunction. Here, we evaluated the cardiovascular and renal effects of LA in the two-kidney-one-clip (2K1C) hypertension model. The rats were divided into four groups: Sham surgery (Sham), the two-kidneys-one-clip (2K1C) group, and groups treated with LA for 14 days (Sham-LA and 2K1C-LA). No changes were observed in the pattern of food, water intake, and urinary volume. The left/right kidney weight LKw/RKw ratio was significantly higher in 2K1C animals. LA treatment did not reverse the increase in cardiac mass. In relation to vascular reactivity, there was an increase in the potency of phenylephrine (PHE) curve in the hypertensive animals treated with LA compared to the 2K1C group and also compared to the Sham group. Vasorelaxation induced by acetylcholine (Ach) and sodium nitroprusside (SNP) were not improved by treatment with LA. Urea and creatinine levels were not altered by the LA treatment. In conclusion, the morphological changes in the aorta and heart were not reversed; however, the treatment with LA mitigated the contraction increase induced by the 2K1C hypertension.

Keywords: antioxidant; lipoic acid; hypertension; 2K1C; oxidative stress; renal function; renin–angiotensin system; vascular reactivity

1. Introduction

Hypertension is the main risk factor for cardiovascular and kidney diseases and has been associated with a high risk of cardiovascular morbidity and mortality [1]. A decrease in blood pressure (BP) induces a reduction in cardiovascular risk [2]. Hypertension is a highly complex disease and several mechanisms underlie its pathophysiology, involving central nervous system imbalances, endothelial dysfunction, and renal defects [3].

Renovascular hypertension (RVH) has been widely studied as a prototype of angiotensin-dependent hypertension [4]. 2K1C is a classic animal model used in hypertension studies, as it significantly resembles renal hypertension as observed in humans [5]. The process of reducing renal flow induced by placing a clip on the renal artery promotes an increase in BP and vasoconstriction via the angiotensin-II (Ang-II) type 1 receptor (AT1R) [6].

Almost two decades ago, Griendling et al. (1994) [7] reported, for the first time, that Ang-II, a product of RAS cascade, activates vascular smooth muscle (VSM) nicotinamide adenine dinucleotide phosphate [NAD(P)H] oxidase, an important cellular source of reactive oxygen species (ROS), with a crucial role in inducing the state of oxidative stress [8].

It is possible to suggest that the use of experimental antioxidant therapy attenuates or prevents the development of hypertension through actions that include direct scavenging of ROS, mimicking antioxidants such as superoxide dismutase (SOD), and inhibiting the NAD(P)H oxidase enzyme [9–13], in addition to reducing the overexpression of AT1R [12,14] and attenuating cardiac and renal dysfunctions generated by hypertension [12,14].

Evidence has demonstrated that the use of antioxidants in the treatment of hypertension is promising. The therapeutic use of LA in hypertension is justified by its ability to restore the level of endogenous antioxidants and prevent the deleterious modification of the sulfhydryl group in Ca^{2+} channels [15]. Studies in experimental models of hypertension, including RVH, showed that treatment with LA was able to generate a variety of beneficial effects, involving reduced expression of NADPH oxidase subunits, in addition to attenuating sympathetic hyperexcitation [16], improving baroreflex sensitivity [10,11] and increasing SOD and glutathione (GSH) levels [17].

Furthermore, LA has been associated with an improvement in the endothelial synthesis of nitric oxide (NO), the leading vasodilator agent of the cardiovascular system [18–20], in addition to attenuating the increase in blood pressure by reducing the expression and activity of a disintegrin and metalloprotease 17 (ADAM17), an enzyme that promotes hypertensive effects through the release of a variety of inflammatory cytokines and is also responsible for the cleavage of angiotensin-converting enzyme type 2 (ACE2) [6,11].

Considering the biological importance of LA, the current study aimed to investigate the cardiovascular and renal effects induced by oral treatment with LA in 2K1C hypertensive rats.

2. Materials and Methods

2.1. Animals and Ethical Approval

Adult male Wistar rats (150–180 g) were housed in conditions of controlled temperature (21 ± 1 °C) and exposed to a 12 h light–dark cycle with free access to food (standard rodent pellets Nuvilab CR-1, Quimtia, Colombo, Brazil) and tap water. All procedures described in the present study are in accordance with the Institutional Animal Care and Use Committee of the Federal University of Ceará CEUA/UFC (protocol #2867020519). The animals were randomly allocated into four experimental groups, according to the surgical protocol performed and treatment: Sham surgery (Sham), n = 5; 2K1C (2K1C), n = 6; Sham surgery + lipoic acid (Sham-LA), n = 5; and 2K1C + lipoic acid (2K1C-LA), n = 5.

2.2. Induction of 2K1C Renovascular Hypertension

Rats underwent surgical procedures to develop renovascular hypertension (2K1C Goldblatt model), as described by Queiroz et al. (2012) [10]. Briefly, under combined ketamine and xylazine anesthesia (75 and 10 mg/kg, i.p., respectively), a midline abdominal incision was made. The left renal artery was exposed and isolated over a short segment by blunt dissection. A U-shaped silver clip (0.2 mm internal diameter) was placed over the vessel at a site proximal to the abdominal aorta and the wound closed and sutured. A Sham procedure, which entailed the entire surgery except for renal artery clipping, served as control. After the procedures, animals received an intramuscular injection of antibiotic (penicillin—24,000 IUs plus streptomycin—10 mg; Pentabiótico Veterinário—Zoetis, Campinas, São Paulo, Brazil) and an s.c. injection of analgesic/anti-inflammatory (ketoprofen 5 mg/kg; Agener União, Embu-Guaçu, Brazil). Rats were returned to their home cages and were observed for six weeks to develop hypertension.

2.3. Treatment with α-Lipoic Acid

LA was dissolved in a solution of NaOH (5N) and then with isotonic saline. A total of 28 days (4 weeks) after surgery to implant the clip in the renal artery (2K1C) or Sham surgery, the animals began treatment with oral administration, via gavage, of LA (60 mg/kg) or vehicle once a day for 14 days. During the second week (8th to 14th day) of treatment, animals were placed in metabolic cages. During this period, water, food intake and urine were measured daily.

2.4. Analysis of Water Intake, Food Intake, and Diuresis

From the 8th day of treatment, the animals were housed in metabolic cages; urine was collected through a container attached to the bottom of the cage, quantified using a graduated cylinder, and a sample was stored at $-12\ °C$ for biochemical analysis. Water and food intake were recorded daily. The daily food intake was measured from the difference between each final and initial chow weight remaining in the cage food container after a period of 24 h. The daily water intake was measured from the difference between each final and initial volume measured from a polypropylene bottle (100 mL capacity with divisions to the nearest mL) with a stainless-steel spout attached to the cage, after a period of 24 h.

2.5. Assessment of Blood Pressure

Blood pressure was assessed by the non-invasive method of tail plethysmography. Initially, the rats underwent an adaptation period of 3 days in a cylindrical container to minimize eventual bias in recordings due to stress. Initially, the rats were heated for 10 min in a heating box containing a 150-watt ceramic heat bulb to promote caudal artery dilation. Next, they were placed in the containment cylinder. An occluder and a sensor were fitted to the proximal portion of each rat's tail. At the time of testing, they were coupled to an electric sphygmomanometer connected to a signal transduction system (MRBP System, IITC Life Science, Woodland Hills, CA, USA) and to a computer containing suitable software for continuous recording; then, the mean arterial pressure was calculated properly.

2.6. Histomorphometric Analyses

After removal, the kidneys (right and left) were sectioned in the sagittal plane; one-half was placed in histological cassettes for analysis and immersed in a container subjected to chemical fixation with 10% aqueous formalin solution, followed by dehydration with ethanol and soaking in paraffin. The paraffin blocks were submitted to microtomy with serial section, with a thickness of five micrometers. The other half was placed in centrifuge tubes for the study of oxidative stress.

The heart was sectioned transversely in the apex region, before being placed in a centrifuge tube for the same investigation as mentioned above, and the remainder was placed in Falcon tubes containing 10% formalin together with an abdominal aorta ring, sectioned transversely. The organs in the tubes were kept at $-12\ °C$. Samples were stained with hematoxylin–eosin (HE).

The histological evaluation was carried out at the Center for Studies in Microscopy and Image Processing (NEMPI) of the Federal University of Ceará, and the slides were read by a single examiner.

Subsequently, the photomicrographs were analyzed using ImageJ software, version 1.44 (Research Services Branch, U.S. National Institutes of Health, Bethesda, MD, USA). All data were placed in an Excel spreadsheet for further statistical analysis. Images were captured using a light microscope coupled to a camera with an LAZ 3.5 acquisition system (Leica DM1000, Wetzlar, Germany).

2.6.1. Aortic Lumen Analysis

To perform the analysis of the aortic lumen, ImageJ software was used. One (1) photomicrograph of each animal was taken, and the inner region was used for the study.

This analysis is important because in hypertensive rats, the aortic lumen tends to be reduced in response to the increase in vascular resistance because of vasoconstriction.

2.6.2. Measurement of the Tunica Adventitia/Media Ratio

The assessment of the proportion of collagen fiber deposition around the animals' aortas was measured by the adventitial/mean ratio using HE staining. With the aid of an optical microscope coupled to the image acquisition system (LEICA), digital images were captured. According to the modified study carried out, the following measurements were taken: (A) outer area of the tunic adventitia, (B) outer area of the tunica media, and (C) inner area of the tunica media.

The difference between the outer area of the tunica adventitia and the outer area of the tunica media (A-B) results in the total area of the tunica adventitia. The difference between the outer and inner areas of the tunica media (B-C) results in the total area of the tunica media. From these results, the adventitial/mean ratio was calculated to analyze whether there was a change in the proportion of collagen fiber deposition around the artery.

In addition, we included another form of analysis in our study. The image of the aorta was divided into 4 (four) quadrants of equal size and dimensions. In each quadrant, we selected 2 (two) different regions to measure this ratio, thus making 8 microphotographs in each aortic ring.

2.6.3. Area and Volume of Cardiomyocytes

To assess the areas and volumes of cardiomyocytes, 5 (five) photomicrographs of each heart were taken. To obtain the morphometric data from the images, longitudinal sections of muscle bundles of the heart were performed. Samples were stained in HE.

Subsequently, for volume analysis, with the aid of a mouse and ImageJ software, the smallest and largest diameters of 5 cardiomyocytes per image were measured to assess cell activity, and the values were applied in the formula v = a2. b/1.91, where a = smallest diameter, b = largest diameter, and 1.91 is a constant. To assess the area, using ImageJ software and with the aid of a mouse, the outer edge of the cardiomyocytes was outlined, thus making up the entire area.

2.7. Biochemical Analysis of Urine

Bioclin (Química Básica LTDA., Belo Horizonte, MG, Brazil) commercial kits were used for the dosages of biochemical markers. The measurement of plasma and urinary creatinine was performed using the modified Jaffé method and, for urea, a kinetic UV urea kit. The tests were carried out with BS 120 automated equipment using spectrophotometry.

2.8. Vascular Reactivity Study

Following evaluation of the functional vascular endothelium, two contractions were induced using depolarizing Tyrode's solution (K^+ 60 mmol·L^{-1}). After washing out the responses to high K^+, cumulatively increasing concentrations of PHE (0.1 nmol·L^{-1}–10 mmol·L^{-1}) were added to the aortic rings with and without endothelium to evaluate the contractile responses. To evaluate the relaxation responses, acetylcholine (ACh, 0.1 nmol·L^{-1}–10 μmol·L^{-1}) in rings with endothelium or sodium nitroprusside (SNP, 0.1 pmol·L^{-1}–1 μmol·L^{-1}) in rings without endothelium were added to the bath. Concentration-dependent contractile responses to PHE were recorded as a percentage of the maximum contraction obtained following tissue stimulation with high K^+. Relaxation responses to cumulative concentrations of ACh and SNP were calculated as a percentage of inhibition of the PHE-induced maximal contraction. In vascular reactivity experiments, we use the parameters of efficacy (MR—the maximum response, expressed in %) and potency (pEC50 = the negative logarithm to base 10 of the EC50 of an agonist, which is the molar concentration of an agonist that produces 50% of the maximal possible effect of that agonist). Relaxation responses to cumulative concentrations of ACh and SNP were calculated as a percentage of inhibition of the PHE-induced maximal contraction. In vascular smooth muscle relaxation tests, the relaxing effect (R) of the substances was calculated, for each concentration, as a function of the maximum contraction provided by the agonist, according to the expression $R = ((T_A - T_S)/T_A) \times 100$, where T_A and T_S are, respectively, the tensions resulting from the action of the agonist (PHE) and a given substance (ACh or SNP). The graphs were then created based on the average values of the magnitude of the vasodilator or vasoconstrictor effect, calculated for each concentration of the substance (after logarithmic transformation). Such data were used to construct concentration–effect curves using nonlinear regression analysis. To address this, the model that uses a sigmoid function of the type was taken as a basis, $y = a + (b - a)/(1 + 10^{((\log CE50 - x) \cdot s)})$, where y corresponds to the response measure (relaxing effect), x to the decimal logarithm of the concentration, a to the minimum response, and b to the maximum response.

2.9. Statistical Analysis

Values are expressed as mean ± SEM. Statistical analysis was performed using one-way analysis of variance ANOVA followed by Tukey's multiple comparison post hoc test. We used the Shapiro–Wilk test to verify the normality of data. The statistical analyses and graphs were performed and constructed using GraphPad Prism version 8.0 (GraphPad Software Corporation San Diego, CA, USA). The differences between groups were considered significant at $p < 0.05$.

3. Results

3.1. Effect of α-Lipoic Acid Treatment on Water Intake and Urine Levels in 2K1C Rats

The daily water intake (Figure 1A) did not change among the experimental groups, which presented the following mean values after 14 days of treatment with LA: Sham (31.00 ± 2.89 mL/24 h), Sham-LA (28.84 ± 3.25 mL/24 h), 2K1C (31.73 ± 5.50 mL/24 h), and 2K1C-LA (34.88 ± 9.48 mL/24 h) (Figure 1B).

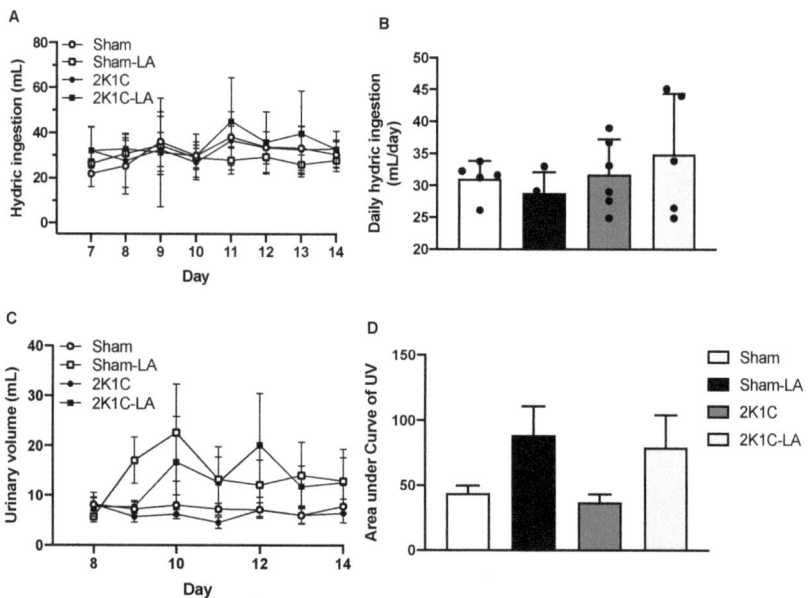

Figure 1. Effects of LA treatment on water intake and urinary levels. (**A**) Daily water intake, (**B**) mean water intake, (**C**) temporal progression of urinary volume, and (**D**) urine volume. Data are expressed as mean ± SEM. Comparisons between groups by one-way ANOVA associated with Tukey's post-test.

The daily urinary volume was analyzed (Figure 1C) counting from the 1st day that the animals remained in the metabolic cage, that is, the beginning of the 2nd week of treatment. No significant changes in the urinary volume were observed between groups (Figure 1D).

3.2. Effect of α-Lipoic Acid Treatment on Food Intake in 2K1C Rats

Daily food intake (Figure 2A) was also measured from the 1st day the animals remained in the metabolic cage, in parallel with the start of the 2nd week of treatment. No differences were observed between the analyzed groups, although the 2K1C group presented an ingestion peak on the 2nd (35.67 ± 12.72 g) and 3rd days (38.50 ± 22.76 g), before normalizing the next day.

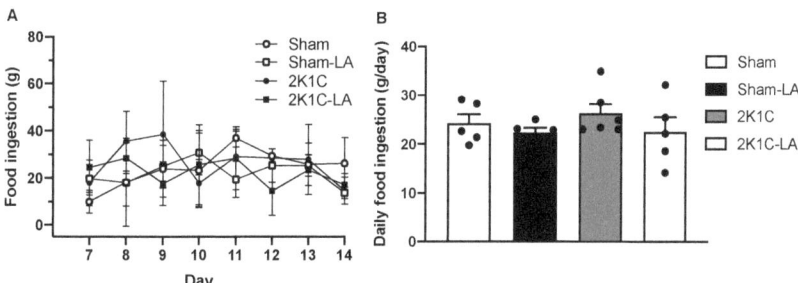

Figure 2. Effects of LA treatment on food intake. (**A**) Temporal progression of food intake and (**B**) average daily food intake. Data are expressed as mean ± SEM. Comparisons between groups by one-way ANOVA associated with Tukey's post-test.

The average daily food intake presented the following values: Sham (24.23 ± 4.21 g/24 h), Sham-LA (22.16 ± 2.46 g/24 h), 2K1C (26.27 ± 4.71 g/24 h), and 2K1C-LA (22.45 ± 6.84 g/24 h). There was a tendency of the treated animals to ingest less food during the study; however, these values did not show a significant difference (Figure 2B).

3.3. Blood Pressure Measurement and Relationship between Kidney/Body Weight and Renal Index

The successful induction of 2K1C surgery was confirmed by the blood pressure analysis using the non-invasive method of tail plethysmography. We can note that there was a greater mean arterial pressure (MAP) through the six weeks of observation in the 2K1C compared to the Sham group as observed in the area under curves of MAP (631.1 ± 19.0 vs. 536.6 ± 16.4, respectively, $p > 0.05$). The hypertensive animals treated with LA did not reverse the increase in MAP (647.2 ± 20.4 vs. 631.1 ± 19.0, respectively) (Figure 3A,B).

In order to confirm whether the reduction in blood flow in the renal artery of the clipped kidney could be a parameter to assess RVH, the weights of the kidneys in relation to the body weight of the animals and the LKw/RKw ratio were calculated (Figure 3).

After weighing the kidneys, it was found that the Sham (1.00 ± 0.05) and Sham-LA groups (0.99 ± 0.03) presented kidney weights in the proportion of 1:1, demonstrating that there was no significant change between them. Between the 2K1C (0.78 ± 0.16) and 2K1C-LA (0.73 ± 0.17) groups, there was also no significant change.

However, there was a reduction in the renal index in both the 2K1C and 2K1C-LA groups compared to the Sham group, demonstrating that hypertensive animals in the 2K1C hypertension model presented a reduction in this index in relation to normotensive animals.

Each kidney in its referred group was analyzed separately. We found that the right kidneys of the 2K1C (0.0042 ± 0.0006 g) and 2K1C-LA (0.0046 ± 0.0009 g) groups, when compared to the Sham (0.0038 ± 0.0004 g) and Sham-LA groups (0.0037 ± 0.0004 g), showed a tendency to exhibit compensatory hypertrophy, but this was not significant (Figure 4A).

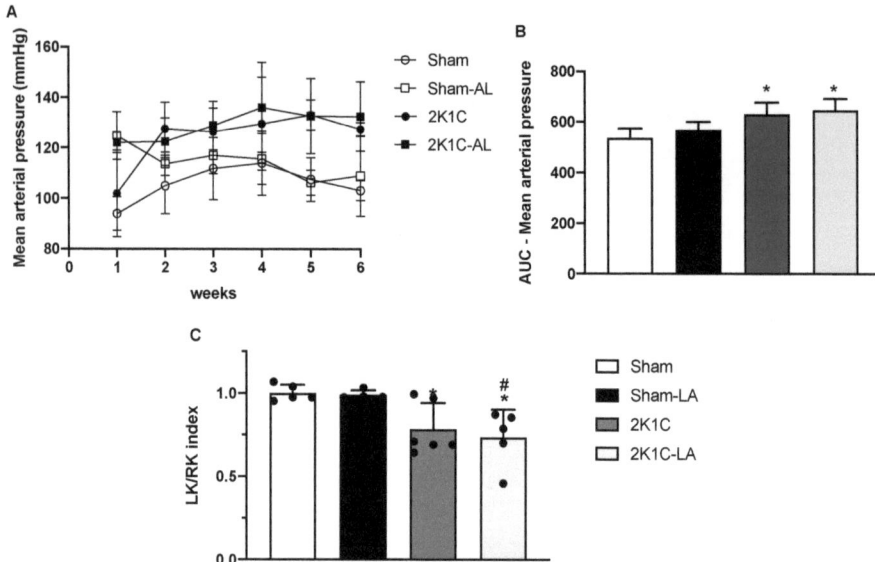

Figure 3. Effects of LA treatment on mean arterial pressure during the weeks (**A**) and the area under the curve graph of the MAP (**B**). Graph (**C**) represents the left kidney (LK)/right kidney (RK) ratio. Data expressed as mean ± SEM. Comparisons between groups by one-way ANOVA associated with Tukey's post-test. * Denotes a significant difference concerning the Sham group ($p < 0.05$). # Denotes a significant difference in relation to the Sham-LA group ($p < 0.05$).

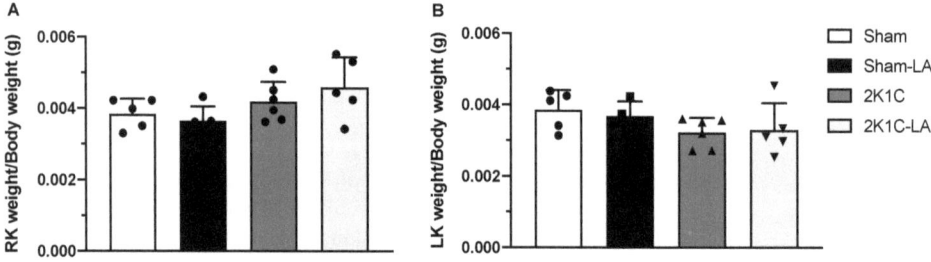

Figure 4. Effects of LA treatment on relationship between kidney weight and body weight. (**A**) Right kidney/body weight ratio and (**B**) left kidney/body weight ratio. Data expressed as mean ± SEM. Comparisons between groups by one-way ANOVA associated with Tukey's post-test.

The left kidneys of the 2K1C (0.0032 ± 0.0004 g) and 2K1C-LA (0.0033 ± 0.0008 g) groups compared to the Sham (0.0039 ± 0.0005 g) and Sham-LA (0.0037 ± 0.0004) groups showed a tendency to exhibit hypotrophy, which is explained by the RVH model used in this study, which promotes a reduction in renal blood flow without causing ischemia. However, neither these results nor those of the right kidneys were significant (Figure 4B).

3.4. Effect of α-Lipoic Acid Treatment on Weight and Cardiac Morphology in 2K1C Rats

No significant changes were observed in the relationship between heart and body weight from the animals in the 2K1C (0.0035 ± 0.0007 mg/g) and Sham (0.0029 ± 0.0001 mg/g) groups. However, animals in the 2K1C-LA group (0.0039 ± 0.0006 mg/g) compared to the Sham (0.0029 ± 0.0001 mg/g) and Sham-LA groups showed a significant increase ($p < 0.05$) in the relationship between heart/body weight, which indicates the presence of a hypertrophic process in the organ after renovascular surgery and the absence of reversal with LA treatment (Figure 5).

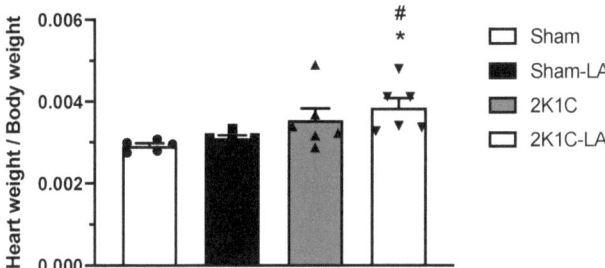

Figure 5. Effects of LA treatment on relationship between heart weight and body weight. Data expressed as mean ± standard deviation. * Denotes a significant difference with the Sham group; # vs. Sham-LA group ($p < 0.05$). Data expressed as mean ± SEM. Comparisons between groups by one-way ANOVA associated with Tukey's post-test.

3.5. Treatment with α-Lipoic Acid Promotes Changes in Vascular Reactivity in 2K1C Rats

In relation to vascular reactivity, no significant difference was observed in PHE curve related to maximum response (MR) (Figure 6A and Table 1). However, there was an increase in the potency (pD2) of PHE curve in the hypertensive animals treated with LA compared to the 2K1C group (7.52 ± 0.09 vs. 6.97 ± 0.15, respectively, $p > 0.05$, n = 6), and also compared to the Sham group (7.52 ± 0.09 vs. 6.84 ± 0.10, respectively, $p > 0.05$, n = 6) (Figure 6A and Table 1).

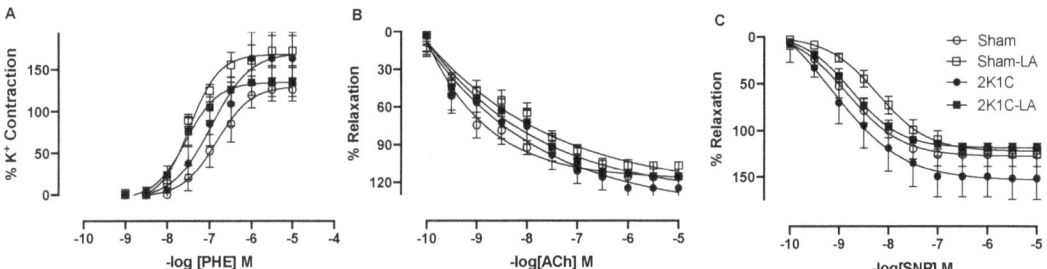

Figure 6. Effects of alpha lipoic acid treatment on endothelium-dependent and -independent relaxation and contractile response to phenylephrine in 2K1C rats. Concentration–response curves for increasing concentrations of (**A**) acetylcholine (ACh), (**B**) phenylephrine (PHE), and (**C**) sodium nitroprusside (SNP). Values are expressed as mean ± SEM. Comparisons between groups by one-way ANOVA associated with Tukey's post-test.

Table 1. Table showing the MR expressed in % as well as the pD2 values from different treatments in normotensive and hypertensive rats using PHE, Ach, and SNP.

(A) MR (%)	Sham	Sham-LA	2K1C	2K1C-LA
PHE	126.2 ± 13.9	172.3 ± 18.0	163.2 ± 11.6	134.7 ± 17.2
ACh	116.0 ± 2.1	106.8 ± 2.5	124.9 ± 8.9	115.4 ± 11.6
SNP	126.0 ± 12.3	121.0 ± 3.9	151.5 ± 22.3	118.1 ± 2.5
(B) pD2	Sham	Sham-LA	2K1C	2K1C-LA
PHE	6.84 ± 0.10	7.40 ± 0.07	6.97 ± 0.15	7.52 ± 0.09 *,&
ACh	9.05 ± 0.09	8.58 ± 0.09	8.64 ± 0.14	8.64 ± 0.13
SNP	8.70 ± 0.10	8.19 ± 0.06	8.74 ± 0.17	8.65 ± 0.04

* Denotes a significant difference in relation to the Sham group. & Denotes a significant difference in relation to the 2K1C group.

In vitro pharmacological tests were performed on isolated aortic rings with intact endothelium (ACh). Evaluation of pD2 values among the respective groups: Sham (9.05 ± 0.09), Sham-LA (8.58 ± 0.09), 2K1C (8.64 ± 0.14), and 2K1C-LA (8.64 ± 0.13). When evaluating MR (116.0 ± 2.1%, 106.8 ± 2.5%, 124.9 ± 8.9% and 115.4 ± 11.6%, respectively, n = 6), the results demonstrate that the LA treatment was not able to significantly improve vasorelaxation for the ACh in hypertensive animals (Figure 6B and Table 1).

In the aortic rings without endothelium, in the vasorelaxation curve for the SNP, no significant difference was observed when we evaluated pD2 (8.70 ± 0.10), Sham-LA (8.19 ± 0.06), 2K1C (8.74 ± 0.17), and 2K1C-LA groups (8.65 ± 0.04) or when evaluating MR in the respective groups (126.0 ± 12.3%, 121.0 ± 3.9%, 151.5 ± 22.3%, 118.1 ± 2.5%, n = 6) (Figure 6C and Table 1).

3.6. Effect of Alpha Lipoic Acid Treatment on Urinary Biochemistry in 2K1C Rats

There was no significant difference in creatinine levels between groups: Sham (4.29 ± 1.98 mg/Dl), Sham-LA (2.77 ± 1.30 mg/Dl), 2K1C (4.45 ± 1.79 mg/Dl), and 2K1C-LA (1.69 ± 1.46 mg/Dl) (Figure 7A). There were also no significant differences in urea levels: Sham (695.80 ± 61.24 mg/Dl), Sham-LA (742.30 ± 35.33 mg/Dl), 2K1C (714.40 ± 33.64 mg/Dl), and 2K1C-LA (740.60 ± 59.22 mg/Dl) (Figure 7B). This finding suggests that treatment with LA did not promote effects on renal function.

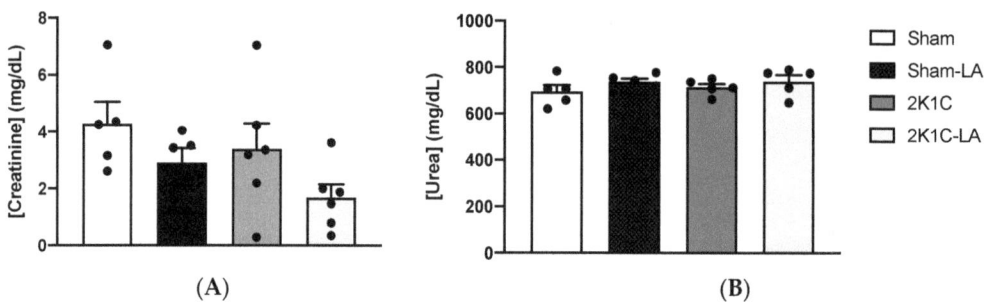

Figure 7. Analysis of renal function according to urinary creatinine and urea clearance. Creatinin concentration (**A**) and urea concentration (**B**). Comparisons between groups by one-way ANOVA associated with Tukey's post-test (2K1C-LA vs. 2K1C group, p = 0.3116; 2K1C-LA vs. Sham group, p = 0.0771).

3.7. Effect of Alpha α-Acid Treatment on Aortic and Cardiac Morphology in 2K1C Rats

To evaluate the effects of LA treatment on cardiac and aortic morphology, photomicrographs were captured and analyses were performed using ImageJ software. Although we did not find significant changes in any of the studied groups, the 2K1C group (1,008,276.00 ± 402,505.00 µm) and the 2K1C-LA group (1,220,547.00 ± 256,010.00 µm) did not present changes as many changes in the aortic lumen compared to the Sham group (1,100,162.00 ± 124,539.00 µm) (Figure 8A).

The evaluation of the proportion of collagen fiber deposition around the animals' aortas was measured by the adventitia/mean ratio through HE staining. The 2K1C group (1.39 ± 0.48 µm) and the 2K1C-LA group (1.28 ± 0.31 µm) did not show alterations in collagen deposition compared to the Sham group (1.09 µm ± 0.13) (Figures 8B and 9).

We did not find significant differences in cardiomyocyte areas between the groups: Sham (49.01 ± 4.68 µm), Sham-LA (50.90 ± 3.74 µm), 2R1C (51.75 ± 5.18 µm), and 2K1C-LA (50.90 ± 3.74 µm) (Figure 8C). Likewise, for cardiomyocyte volume results, Sham (0.50 ± 0.12 µm), Sham-LA (0.39 ± 0.12 µm), 2K1C (0.58 µm ± 0.10), and 2K1C-LA (0.48 ± 0.04 µm) (Figure 8D). While no significant differences were identified, the increased area and volume values in the hypertensive groups infer the presence of cardiac hypertrophy in these animals, unlike normotensive animals, demonstrating that the treatment was not able to reduce the area and volume of cardiomyocytes (Figure 10).

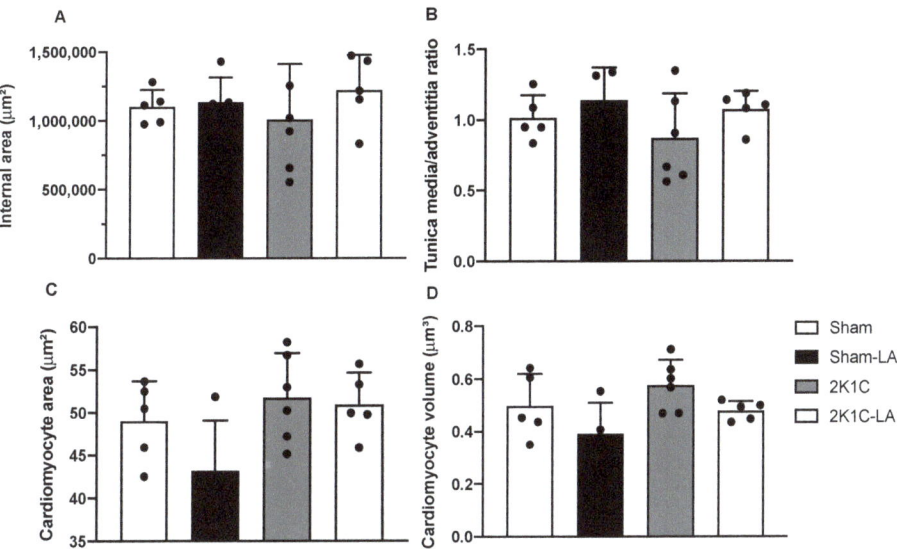

Figure 8. Histomorphometric analysis of the heart and aorta. (A) Aortic lumen area, (B) tunica adventitia/media ratio, (C) cardiomyocyte area, (D) cardiomyocyte volume. Comparisons between groups by one-way ANOVA associated with Tukey's post-test.

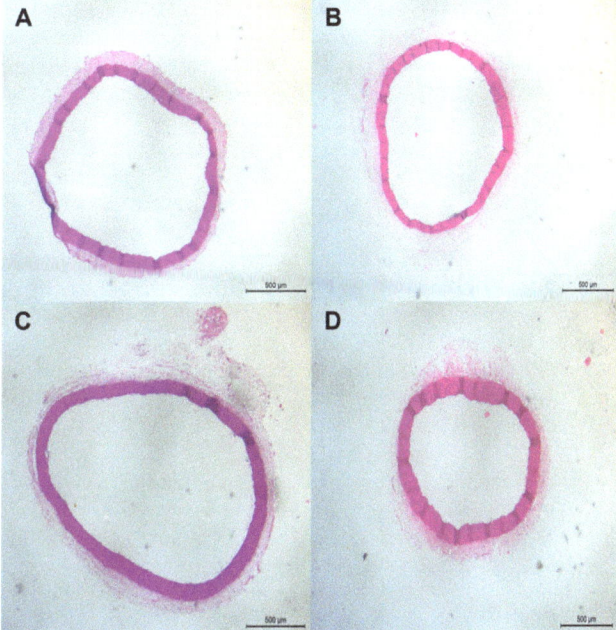

Figure 9. Histological analysis of abdominal aorta ring determined by HE. (A) Sham, (B) Sham-LA, (C) 2K1C, and (D) 2K1C-LA. Scale bar = 500 µm, 100×.

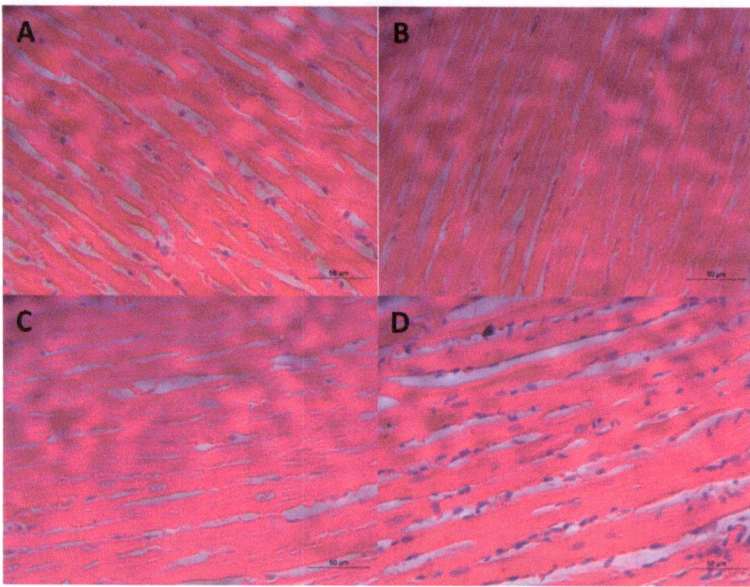

Figure 10. Histological analysis of heart determined by HE. (**A**) Sham, (**B**) Sham-LA, (**C**) 2K1C, and (**D**) 2K1C-LA. Scale bar = 50 μm, 400×.

4. Discussion

Our results demonstrated that oral treatment with LA induced an increase in the potency of PHE curve in the hypertensive animals treated with LA compared to the 2K1C group and also compared to Sham group. However, it was observed that the treatment did not promote a significant change in the concentration–response curve for ACh and SNP, indicating that LA did not restore the vasorelaxation altered by 2K1C hypertension.

The findings in the present research corroborate the study by Queiroz et al. (2012) [10], which showed that 2K1C animals treated with LA at the same dose presented a reduction in blood pressure compared to untreated 2K1C. This decrease in BP occurred due to the improvement in baroreflex sensitivity in 2K1C animals treated with the antioxidant [10].

An important finding of this study was that treatment with LA increased the contractile response in animals submitted to Sham surgery. This fact, as far as we could establish, has not yet been described in the literature. Contrary to the results for vasorelaxation obtained in this study, another study carried out with a different experimental model in diabetic rats using the streptozotocin injection method and fed with a high-fat diet, the treatment with LA was able to improve vascular reactivity in aortic rings, increasing ACh vasorelaxation, and improving vascular function through pathways that increase hydrogen sulfide, a gaseous transmitter with a beneficial effect on the vascular system, in addition to decreasing vascular smooth muscle cell autophagy through regulation of the AMPK/mTOR pathway [21].

Concerning the investigation of the renal effects of the treatment, after analyzing the ratio between the weights of the kidneys, it was verified that the 2K1C animals had a lower LKw/RKw ratio than the Sham animals. This result corroborates another study in which this relationship was evaluated in rats after 6–8 weeks, similar to our renal clipping protocol. The authors demonstrated that animals with an intermediate ratio, between the range of 0.5 to 0.8, were 100% hypertensive, with blood pressure greater than 150 mmHg, whereas for animals with a ratio lower than 0.4 and greater than 0.9, less than 50% were hypertensive [22]. Therefore, the LKw/RKw ratio found in 2K1C rats in the present study suggested changes in renal morphology after the development of RVH, as shown in other studies [23–25].

Regarding the qualitative and quantitative assessment of the excretion capacity of the kidney, the glomerular filtration rate is the most commonly used measure for such purposes, being more faithfully obtained during clinical practice from the creatinine clearance, and is performed in urine collected in precisely 24 h [25]. This is one of the most widely used methods in preclinical studies to assess the function and presence of renal dysfunction. It is performed by estimating the glomerular filtration rate, considering plasma or serum levels of urea and creatinine [26,27]. However, in the current study, we used 24 h urine for the individual evaluations. Studies have demonstrated that LA decreased the renal tubular injury scores and urinary damage markers and increased glomerular filtration [20]. The augment in glomerular filtration leads to an increase in urinary volume, which can be an explanation of the UV data presented in our study.

Chronic creatinine levels in animals treated with LA did not show a significant difference compared to untreated animals. Studies evaluating the renoprotective and cardioprotective effects of treatment with LA observed a decrease in serum creatinine levels [28]. In addition, another biomarker of kidney injury was evaluated, urea level, and showed that our treatment also did not affect the reduction in urea between the groups, demonstrating that in these animals there was no commitment of renal function. Contrary to what was observed in the study by Amat et al. (2014) [29], 2K1C hypertension promoted renal dysfunction in animals, evidenced by increased serum levels of creatinine and urea. Although we did not measure the serum values of these markers, we used the values as a basis since urea and creatinine values reported through 24 h urine analyses are scarce in the literature.

Classically, the increase in serum creatinine, oliguria, albuminuria, and electrolyte abnormalities are considered indicators of kidney injury, due to tubular changes or even structural damage that can only be visualized by imaging or histology exams [19]. However, in the kidney injury model caused by cisplatin [18], the authors reported an increase in glomerular filtration, with increased creatinine clearance and consequent reduction in plasma creatinine, in addition to the attenuation of oxidative damage. Therefore, even though there is a tendency for creatinine to decrease, it is not possible to affirm, in the studied model, that there was significant renal protection with the use of lipoic acid. Likewise, no histological changes were observed that support this finding

The 2K1C model is described as a model of hypertension independent of the increase in volume and vasopressin secretion, since the remaining non-clipped kidney acts in a compensatory way [30]. Our behavioral results involving daily food and water intake and daily diuresis did not show significant differences between groups. These findings are contrary to what was found in other studies [31,32], in which an increase in water intake and diuresis was demonstrated in 2K1C animals when compared to control animals. The divergence in behavioral patterns observed among those studies and in the present results might be due to the different rat strain used in the experiments.

Heart diseases have a common characteristic, accompanied mainly by increased myocardial mass. This hypertrophy occurs through the absolute or relative thickening of the walls of the chambers due to the increase in the dimensions of the cardiomyocytes, especially pathologically in conditions of prolonged and abnormal hemodynamic stress, such as hypertension and myocardial infarction [33,34].

It is understood that cardiac hypertrophy is expected in this experimental model of hypertension. However, this increase was not seen in the 2K1C animals. In a study using the same model, renal stenosis caused cardiac hypertrophy, accompanied by increased collagen deposition and cardiomyocyte diameter [35]. Therefore, we also decided to conduct a cardiac evaluation using the dimensions of the cardiomyocytes and concluded that the area and volume of the cardiomyocytes did not present significant alterations among the studied groups.

Similarly, in a study carried out with SHR, the 30-day treatment with the enantiomer -(−)-ALA (125 µmol/kg/day) did not reverse cardiac hypertrophy and fibrosis generated by hypertension. However, in animals treated with another enantiomer, (+)-ALA (125 µmol/kg/day), the attenuation of left ventricular fibrosis was observed [36].

The aortic wall consists of three concentric layers: the tunica intima, tunica media, and tunica adventitia [37]. The biomechanical properties of vessels, large arteries, and veins largely depend on the amount and balance between extracellular matrix constituents, such as collagen and elastin, and local proteases, such as matrix metalloproteinases and leukocyte elastase. This balance can be impaired in the presence of vascular pathologies, such as hypertension [38,39].

From this, it was decided to evaluate the adventitia/mean ratio to evaluate the collagen deposition around the aortas of the Sham and 2K1C groups. We observed that the 2K1C rats tended to accumulate more collagen fibers, whereas the treatment with LA showed a possible tendency to reduce this deposition. However, the results were not significant. Similarly, a study carried out with the descending thoracic aorta of a rabbit model with aortic valve calcification demonstrated that treatment with LA protected against medial vascular calcification but did not prevent the increase in arterial wall thickness [40].

Understanding the histological alterations caused by hypertension, we sought to analyze the area/inner lumen of the aorta in our animals. Our results showed a reduction in this lumen in 2K1C animals, while in the presence of LA, this lumen returned to its normal morphology. This alteration may be explained by smooth muscle cells, which can rearrange themselves around a reduced arterial lumen, accompanied by a more significant deposition of extracellular matrix, which can lead to a decrease in biomechanical and hemodynamic function, consequently compromising tissue perfusion [39,41], which was supposedly verified in the adventitious/mean ratio.

5. Conclusions

Our results indicate that the treatment promoted little effects on the cardiovascular system, showing that LA was not able to decrease the augment of mean arterial pressure induced by renovascular hypertension. LA also increased the vasoconstriction evoked by PHE in animals with renovascular hypertension compared to the 2K1C and Sham groups. Although we observed changes in urinary volume and creatinine levels induced by LA, these responses are not significant; this was the same for the morphology alterations. In this context, further investigation is needed to examine the detailed mechanisms related to the cardiovascular and renal actions of LA.

Author Contributions: Conceptualization, T.M.d.Q. and R.B.D.; methodology, D.F.A., D.G.R., C.F.R., R.J.B.J. and R.d.S.A.; software, M.V.B.d.S. and D.G.R.; validation, R.J.B.J. and W.T.F.e.S.; formal analysis, D.V.G.N., D.F.A. and D.G.R.; investigation, D.F.A., D.G.R. and C.F.R.; resources, T.M.d.Q., R.B.D., C.F.R., W.T.F.e.S. and R.d.S.A.; data curation, R.J.B.J. and R.B.D.; writing—original draft preparation, D.V.G.N., M.V.B.d.S., L.C.M.G. and T.M.d.Q.; writing—review and editing, M.V.B.d.S., T.M.d.Q. and L.C.M.G.; supervision, T.M.d.Q. and R.B.D. All authors have read and agreed to the published version of the manuscript.

Funding: This research was funded by the Conselho Nacional de Desenvolvimento Científico e Tecnológico (CNPq), grant number [436605/2018-0], to T.M.d.Q. This study was also financed in part by the Coordenação de Aperfeiçoamento de Pessoal de Nível Superior—Brazil (CAPES)—Finance Code 001.

Institutional Review Board Statement: The animal study protocol was approved by the Institutional Animal Care and Use Committee of the Federal University of Ceará CEUA/UFC (protocol #2867020519, 6 June 2019).

Informed Consent Statement: Not applicable.

Data Availability Statement: Data are contained within the article.

Acknowledgments: We would like to thank Francisco Vagnaldo Fechine Jamacaru for his support in the experiments involving the vascular reactivity, the "Núcleo de Estudos em Microscopia e Processamento de Imagens" (NEMPI)/Federal University of Ceará for the support of some analyses performed during this study, and the Federal University of Pernambuco for the support of this work.

Conflicts of Interest: The authors declare no conflicts of interest. The funders had no role in the design of the study; in the collection, analyses, or interpretation of data; in the writing of the manuscript; or in the decision to publish the results.

References

1. James, P.A.; Oparil, S.; Carter, B.L.; Cushman, W.C.; Dennison-Himmelfarb, C.; Handler, J.; Lackland, D.T.; Lefevre, M.L.; MacKenzie, T.D.; Ogedegbe, O.; et al. 2014 Evidence-Based Guideline for the Management of High Blood Pressure in Adults. *JAMA* **2014**, *311*, 507. [CrossRef] [PubMed]
2. Herrington, W.; Lacey, B.; Sherliker, P.; Armitage, J.; Lewington, S. Epidemiology of Atherosclerosis and the Potential to Reduce the Global Burden of Atherothrombotic Disease. *Circ. Res.* **2016**, *118*, 535–546. [CrossRef]
3. Patrick, D.M.; Van Beusecum, J.P.; Kirabo, A. The role of inflammation in hypertension: Novel concepts. *Curr. Opin. Physiol.* **2021**, *19*, 92–98. [CrossRef]
4. Mannemuddhu, S.S.; Ojeda, J.C.; Yadav, A. Renovascular Hypertension. *Prim. Care Clin. Off. Pr.* **2020**, *47*, 631–644. [CrossRef]
5. Iversen, B.M.; Heyeraas, K.J.; Sekse, I.; Andersen, K.J.; Ofstad, J. Autoregulation of renal blood flow in two-kidney, one-clip hypertensive rats. *Am. J. Physiol. Physiol.* **1986**, *251*, F245–F250. [CrossRef]
6. dos Santos, V.M.; da Silva, M.V.B.; Prazeres, T.C.M.M.; Cartágenes, M.D.S.S.; Calzerra, N.T.M.; de Queiroz, T.M. Involvement of shedding induced by ADAM17 on the nitric oxide pathway in hypertension. *Front. Mol. Biosci.* **2022**, *9*, 1032177. [CrossRef]
7. Griendling, K.K.; Minieri, C.A.; Ollerenshaw, J.D.; Alexander, R.W. Angiotensin II stimulates NADH and NADPH oxidase activity in cultured vascular smooth muscle cells. *Circ. Res.* **1994**, *74*, 1141–1148. [CrossRef]
8. Santana-Garrido, Á.; Reyes-Goya, C.; Fernández-Bobadilla, C.; Blanca, A.J.; André, H.; Mate, A.; Vázquez, C.M. NADPH oxidase-induced oxidative stress in the eyes of hypertensive rats. *Mol. Vis.* **2021**, *27*, 161–178. [PubMed]
9. Costa, C.A.; Amaral, T.A.; Carvalho, L.C.; Ognibene, D.T.; da Silva, A.F.; Moss, M.B.; Valença, S.S.; de Moura, R.S.; Resende, C. Antioxidant Treatment With Tempol and Apocynin Prevents Endothelial Dysfunction and Development of Renovascular Hypertension. *Am. J. Hypertens.* **2009**, *22*, 1242–1249. [CrossRef]
10. Queiroz, T.M.; Guimarães, D.D.; Mendes-Junior, L.G.; Braga, V.A. α-Lipoic Acid Reduces Hypertension and Increases Baroreflex Sensitivity in Renovascular Hypertensive Rats. *Molecules* **2012**, *17*, 13357–13367. [CrossRef]
11. de Queiroz, T.M.; Xia, H.; Filipeanu, C.M.; Braga, V.A.; Lazartigues, E. α-Lipoic acid reduces neurogenic hypertension by blunting oxidative stress-mediated increase in ADAM17. *Am. J. Physiol. Circ. Physiol.* **2015**, *309*, H926–H934. [CrossRef] [PubMed]
12. García-Trejo, E.M.A.; Arellano-Buendía, A.S.; Argüello-García, R.; Loredo-Mendoza, M.L.; García-Arroyo, F.E.; Arellano-Mendoza, M.G.; Castillo-Hernández, M.C.; Guevara-Balcázar, G.; Tapia, E.; Sánchez-Lozada, L.G.; et al. Effects of Allicin on Hypertension and Cardiac Function in Chronic Kidney Disease. *Oxidative Med. Cell. Longev.* **2016**, *2016*, 3850402. [CrossRef] [PubMed]
13. Gao, H.-L.; Yu, X.-J.; Hu, H.-B.; Yang, Q.-W.; Liu, K.-L.; Chen, Y.-M.; Zhang, Y.; Zhang, D.-D.; Tian, H.; Zhu, G.-Q.; et al. Apigenin Improves Hypertension and Cardiac Hypertrophy Through Modulating NADPH Oxidase-Dependent ROS Generation and Cytokines in Hypothalamic Paraventricular Nucleus. *Cardiovasc. Toxicol.* **2021**, *21*, 721–736. [CrossRef]
14. Guimaraes, D.A.; dos Passos, M.A.; Rizzi, E.; Pinheiro, L.C.; Amaral, J.H.; Gerlach, R.F.; Castro, M.M.; Tanus-Santos, J.E. Nitrite exerts antioxidant effects, inhibits the mTOR pathway and reverses hypertension-induced cardiac hypertrophy. *Free. Radic. Biol. Med.* **2018**, *120*, 25–32. [CrossRef]
15. Temiz-Resitoglu, M.; Guden, D.S.; Senol, S.P.; Vezir, O.; Sucu, N.; Kibar, D.; Yılmaz, S.N.; Tunctan, B.; Malik, K.U.; Sahan-Firat, S. Pharmacological Inhibition of Mammalian Target of Rapamycin Attenuates Deoxycorticosterone Acetate Salt–Induced Hypertension and Related Pathophysiology: Regulation of Oxidative Stress, Inflammation, and Cardiovascular Hypertrophy in Male Rats. *J. Cardiovasc. Pharmacol.* **2022**, *79*, 355–367. [CrossRef] [PubMed]
16. Su, Q.; Liu, J.-J.; Cui, W.; Shi, X.-L.; Guo, J.; Li, H.-B.; Huo, C.-J.; Miao, Y.-W.; Zhang, M.; Yang, Q.; et al. Alpha lipoic acid supplementation attenuates reactive oxygen species in hypothalamic paraventricular nucleus and sympathoexcitation in high salt-induced hypertension. *Toxicol. Lett.* **2016**, *241*, 152–158. [CrossRef]
17. Xu, C.; Li, E.; Liu, S.; Huang, Z.; Qin, J.G.; Chen, L. Effects of α-lipoic acid on growth performance, body composition, antioxidant status and lipid catabolism of juvenile Chinese mitten crab Eriocheir sinensis fed different lipid percentage. *Aquaculture* **2018**, *484*, 286–292. [CrossRef]
18. Kamt, S.F.; Liu, J.; Yan, L.-J. Renal-Protective Roles of Lipoic Acid in Kidney Disease. *Nutrients* **2023**, *15*, 1732. [CrossRef]
19. Levey, A.S.; Becker, C.; Inker, L.A. Glomerular Filtration Rate and Albuminuria for Detection and Staging of Acute and Chronic Kidney Disease in Adults. *JAMA* **2015**, *313*, 837–846. [CrossRef]
20. Zhang, J.; McCullough, P.A. Lipoic Acid in the Prevention of Acute Kidney Injury. *Nephron* **2016**, *134*, 133–140. [CrossRef]
21. Qiu, X.; Liu, K.; Xiao, L.; Jin, S.; Dong, J.; Teng, X.; Guo, Q.; Chen, Y.; Wu, Y. Alpha-lipoic acid regulates the autophagy of vascular smooth muscle cells in diabetes by elevating hydrogen sulfide level. *Biochim. Biophys. Acta (BBA)—Mol. Basis Dis.* **2018**, *1864*, 3723–3738. [CrossRef]
22. Smith, S.H.; Bishop, S.P. Selection criteria for drug-treated animals in two-kidney, one clip renal hypertension. *Hypertension* **1986**, *8*, 700–705. [CrossRef]
23. Kaur, S.; Muthuraman, A. Therapeutic evaluation of rutin in two-kidney one-clip model of renovascular hypertension in rat. *Life Sci.* **2016**, *150*, 89–94. [CrossRef]

24. Pereira, P.G.; Rabelo, K.; da Silva, J.F.R.; Ciambarella, B.T.; Argento, J.G.C.; Nascimento, A.L.R.; Vieira, A.B.; de Carvalho, J.J. Aliskiren improves renal morphophysiology and inflammation in Wistar rats with 2K1C renovascular hypertension. *Histol. Histopathol.* **2020**, *35*, 609–621. [CrossRef]
25. Lima, C.M.; Lima, A.K.; Melo MG, D.; Dória GA, A.; Serafini, M.R. Alores de referência hematológicos e bioquímicos de ratos (Rattus novergicus linhagem Wistar) provenientes do biotério da Universidade Tiradentes. *Sci. Plena* **2014**, *10*, 1–9.
26. Zhang, Q.; Davis, K.J.; Hoffmann, D.; Vaidya, V.S.; Brown, R.P.; Goering, P.L. Urinary Biomarkers Track the Progression of Nephropathy in Hypertensive and Obese Rats. *Biomark. Med.* **2014**, *8*, 85–94. [CrossRef]
27. Hojná, S.; Kotsaridou, Z.; Vaňourková, Z.; Rauchová, H.; Behuliak, M.; Kujal, P.; Kadlecová, M.; Zicha, J.; Vaněčková, I. Empagliflozin Is Not Renoprotective in Non-Diabetic Rat Models of Chronic Kidney Disease. *Biomedicines* **2022**, *10*, 2509. [CrossRef]
28. El-Beshbishy, H.A.; Bahashwan, S.A.; Aly, H.A.; Fakher, H.A. Abrogation of cisplatin-induced nephrotoxicity in mice by alpha lipoic acid through ameliorating oxidative stress and enhancing gene expression of antioxidant enzymes. *Eur. J. Pharmacol.* **2011**, *668*, 278–284. [CrossRef]
29. Amat, N.; Amat, R.; Abdureyim, S.; Hoxur, P.; Osman, Z.; Mamut, D.; Kijjoa, A. Aqueous extract of dioscorea opposita thunb. normalizes the hypertension in 2K1C hypertensive rats. *BMC Complement. Altern. Med.* **2014**, *14*, 36. [CrossRef]
30. Rabito, S.F.; Carretero, O.A.; Scicli, A.G. Evidence against a role of vasopressin in the maintenance of high blood pressure in mineralocorticoid and renovascular hypertension. *Hypertension* **1981**, *3*, 34–38. [CrossRef]
31. Lincevicius, G.S.; Shimoura, C.G.; Nishi, E.E.; Perry, J.C.; Casarini, D.E.; Gomes, G.N.; Bergamaschi, C.T.; Campos, R.R. Aldosterone Contributes to Sympathoexcitation in Renovascular Hypertension. *Am. J. Hypertens.* **2015**, *28*, 1083–1090. [CrossRef] [PubMed]
32. Roncari, C.F.; Barbosa, R.M.; Vendramini, R.C.; De Luca, L.A., Jr.; Menani, J.V.; Colombari, E.; Colombari, D.S. Enhanced angiotensin II induced sodium appetite in renovascular hypertensive rats. *Peptides* **2018**, *101*, 82–88. [CrossRef] [PubMed]
33. Anversa, P.; Ricci, R.; Olivetti, G. Quantitative structural analysis of the myocardium during physiologic growth and induced cardiac hypertrophy: A review. *J. Am. Coll. Cardiol.* **1986**, *7*, 1140–1149. [CrossRef] [PubMed]
34. Nakamura, M.; Sadoshima, J. Mechanisms of physiological and pathological cardiac hypertrophy. *Nat. Rev. Cardiol.* **2018**, *15*, 387–407. [CrossRef] [PubMed]
35. Restini, C.B.A.; Garcia, A.F.E.; Natalin, H.M.; Carmo, M.F.A.; Nowicki, V.F.; Rizzi, E.; Ramalho, L.N.Z. Resveratrol Supplants Captopril's Protective Effect on Cardiac Remodeling in a Hypertension Model Elicited by Renal Artery Stenosis. *Yale J. Biol. Med.* **2022**, *95*, 57–69. [PubMed]
36. Martinelli, I.; Tomassoni, D.; Roy, P.; Mannelli, L.D.C.; Amenta, F.; Tayebati, S.K. Antioxidant Properties of Alpha-Lipoic (Thioctic) Acid Treatment on Renal and Heart Parenchyma in a Rat Model of Hypertension. *Antioxidants* **2021**, *10*, 1006. [CrossRef] [PubMed]
37. Silver, F.H.; Christiansen, D.L.; Buntin, C.M. Mechanical properties of the aorta: A review. *Crit. Rev. Biomed. Eng.* **1989**, *17*, 323–358. [PubMed]
38. Jacob, M.; Badier-Commander, C.; Fontaine, V.; Benazzoug, Y.; Feldman, L.; Michel, J. Extracellular matrix remodeling in the vascular wall. *Pathol. Biol.* **2001**, *49*, 326–332. [CrossRef] [PubMed]
39. Wang, X.; Khalil, R.A. Matrix Metalloproteinases, Vascular Remodeling, and Vascular Disease. *Adv. Pharmacol.* **2018**, *81*, 241–330. [CrossRef]
40. Bassi, E.; Liberman, M.; Martinatti, M.; Bortolotto, L.; Laurindo, F. Lipoic acid, but not tempol, preserves vascular compliance and decreases medial calcification in a model of elastocalcinosis. *Braz. J. Med. Biol. Res.* **2014**, *47*, 119–127. [CrossRef]
41. Pereira, S.C.; Parente, J.M.; Belo, V.A.; Mendes, A.S.; Gonzaga, N.A.; Vale, G.T.D.; Ceron, C.S.; Tanus-Santos, J.E.; Tirapelli, C.R.; Castro, M.M. Quercetin decreases the activity of matrix metalloproteinase-2 and ameliorates vascular remodeling in renovascular hypertension. *Atherosclerosis* **2018**, *270*, 146–153. [CrossRef] [PubMed]

Disclaimer/Publisher's Note: The statements, opinions and data contained in all publications are solely those of the individual author(s) and contributor(s) and not of MDPI and/or the editor(s). MDPI and/or the editor(s) disclaim responsibility for any injury to people or property resulting from any ideas, methods, instructions or products referred to in the content.

Article

Dabsylated Bradykinin Is Cleaved by Snake Venom Proteases from *Echis ocellatus*

Julius Abiola [1,2], Anna Maria Berg [1], Olapeju Aiyelaagbe [2], Akindele Adeyi [3] and Simone König [1,*]

[1] IZKF Core Unit Proteomics, Interdisciplinary Center for Clinical Research, University of Münster, Röntgenstr. 21, 48149 Münster, Germany; abiolajulius005@gmail.com (J.A.)
[2] Organic Unit, Department of Chemistry, University of Ibadan, Ibadan 200005, Nigeria
[3] Animal Physiology Unit, Department of Zoology, University of Ibadan, Ibadan 200005, Nigeria
* Correspondence: koenigs@uni-muenster.de; Tel.: +49-251-8357164

Abstract: The vasoactive peptide bradykinin (BK) is an important member of the renin–angiotensin system. Its discovery is tightly interwoven with snake venom research, because it was first detected in plasma following the addition of viper venom. While the fact that venoms liberate BK from a serum globulin fraction is well described, its destruction by the venom has largely gone unnoticed. Here, BK was found to be cleaved by snake venom metalloproteinases in the venom of *Echis ocellatus*, one of the deadliest snakes, which degraded its dabsylated form (DBK) in a few minutes after Pro7 (RPPGFSP↓FR). This is a common cleavage site for several mammalian proteases such as ACE, but is not typical for matrix metalloproteinases. Residual protease activity < 5% after addition of EDTA indicated that DBK is also cleaved by serine proteases to a minor extent. Mass spectrometry-based protein analysis provided spectral proof for several peptides of zinc metalloproteinase-disintegrin-like Eoc1, disintegrin EO4A, and three serine proteases in the venom.

Keywords: mass spectrometry; peptide fragmentation; envenomation; vipers

1. Introduction

Bradykinin (BK, sequence RPPGFSPFR) has been known as a vasoactive peptide for more than 80 years and the blood pressure-lowering property of the BK system is well documented [1]. Its discovery is tightly interwoven with snake venom research, because it was first detected by Rocha e Silva in plasma following the addition of venom of the pit viper *Bothrops jararaca* [2,3]. Moreover, BK-potentiating peptides were discovered by Ferreira in *B. jararaca* venom in 1965; they enhance BK action in vivo by inhibiting angiotensin-converting enzyme (ACE) [4,5]. In fact, ACE inhibitors, the drugs used for the treatment of hypertension and congestive heart failure, were developed from the venom of this species [6,7]. Interestingly, while much research has contributed to the fact that venoms liberate BK from a serum globulin fraction, the destruction of this substance by the venom has largely gone unnoticed [8]. It was only briefly mentioned in a 1955 study of venoms from 15 different viperids that 13 of these venoms destroyed BK [8].

BK is associated with multiple roles in human pathophysiology besides blood pressure homeostasis including inflammation [9–11]. In earlier work, we investigated the neuropeptide in the context of pain [12] and COVID-19 [13]. For these studies, we developed a reporter assay which used dabsylated BK (DBK) as a substrate to test serum protease activity [14]. BK is a substrate of ACE, an enzyme which has mostly been investigated with regard to hypertensive disorder, but which has also been of interest in the recent COVID-19 pandemic, because its counter-regulator ACE2 is the SARS-CoV-2 entrance port [15]. In serum, BK is additionally cleaved by carboxypeptidase N (CPN), a pleiotropic regulator of inflammation [16]. With our neuropeptide reporter assay (NRA), we studied the formation of the cleavage products of both enzymes, ACE and CPN, namely DBK fragments 1–5 and 1–8, respectively, using thin-layer chromatography (TLC) [14].

Here, we were in need of a functional assay to test the anti-venomous activity of plant extracts and checked the possible use of the NRA. To our surprise, DBK was quickly degraded within minutes by the venom of *Echis ocellatus*. We present these data, including the identification of the cleavage product.

Snake venoms are complex mixtures of primarily peptides and proteins that are harmful to the human body, and especially the neuromuscular and circulatory systems [17,18]. Snake venom composition varies with a number of factors such as age, diet, geographic location, and seasonal changes [18,19], and is still far from being fully elucidated (for introduction and overview, see [18,20]). We analyzed the venom of *E. ocellatus* available to us for proteases, which might act on DBK. To that end, we performed proteomic data-independent (DIA) mass spectrometry (MS)-based experiments. However, omics studies depend on well-curated reference sequence databases, and for many snakes, these databases are still small and incomplete. The available genomes differ notably in assembly and annotation qualities; the most complete published snake genomes to date are those of the elapid *N. naja* and the viper *Crotalus tigris* [21]. For *E. ocellatus*, a toxin transcriptome was constructed in 2006 [22], but in a subsequent comparison with proteomics data, significant differences were observed [23]. Peptides derived from 26% of the venom proteins could not be matched to the transcriptome and 67% of the toxin clusters reported in the transcriptome did not match to peptides detected in the proteome. Thus, venom proteomic analyses try to circumvent the problem by using the database for the suborder Serpentes or the available sequences from related snakes [24–27]. The analysis of proteomics data versus non-specific databases is not optimal [28]. Since related proteins in different snakes have similar, but not identical sequences, this approach can only generate hints at the protein ID, but not a complete and correct sequence. Along with the method-inherent limitations of using peptide fragment ion mass spectra for sequence assignment in proteomics [29,30], any proposed protein sequence needs thus to be carefully validated by orthogonal methods. Target tandem MS (MS/MS) is a method for peptide sequence confirmation. We thus conducted both proteomic analysis of *E. ocellatus* venom proteins and target MS/MS of selected peptides for validation.

Nigeria records an average of 43,000 cases of snakebite annually, with about 1900 people being killed and approximately the same number losing a limb following a snakebite [31,32]. Thereby, the carpet viper *Echis ocellatus* is responsible for about 90% of bites and 60% of snakebite deaths [33,34]. Globally, approximately 2.7 million people are envenomated annually [25] and the World Health Organization has added snakebite to the list of neglected tropical diseases [35]. Many of the victims do not have access to health care facilities and ethno-medical means of treatment are still commonly used. Research on the anti-venomous activity of local plants is thus very important and we are also involved in a bilateral project to that effect. During the course of the experiments, we discovered that our well-proven DBK-based NRA is a useful tool to test venom activity, and we therefore investigated the cleavage of DBK by the venom of *E. ocellatus* and identified its product. Furthermore, we briefly examined the abundant snake proteases possibly involved and validated a number of peptide sequences using tandem MS.

2. Materials and Methods

2.1. Preparation of Snake Venom Samples

This study involved 13 male snakes. All handling protocols were followed as stipulated by the University of Ibadan Animal Care and Use Research Ethics Committee (NHREC/UIACUREC/05/12/2022A) and in agreement with ARRIVE guideline 2.0. The adult snakes used in the study were captured from the wild in Kaltungo (9°48'51" N 11°18'32" E), which is located in the northeastern part of Nigeria's Gombe State. Gombe is a guinea savannah area with many shrubs and a few tall trees. The snakes primarily feed on small rodents and reptiles, which are abundant in the region. The average length of *E. ocellatus*, measured from head to tail, was 53 cm. The snakes were housed at the serpentarium of the Department of Zoology at the University of Ibadan, where they were acclimatized and fed for two weeks before manual milking. Venoms were pooled, frozen at $-80\,°C$, lyophilized, and stored at $-20\,°C$ until further use.

2.2. NRA and TLC

The NRA was performed as described with slight changes [14]. Venom (3 µL, 0.1 mg/mL) was added to 350 pmol of dried DBK and incubated at 37 °C for 5 min. The reaction was halted by adding 18 µL of ice-cold acetone and freezing the sample at $-20\,°C$ for 2 h. Afterwards, the solution was centrifuged at $18,000\times g$ and 4 °C for 1 h and the supernatant was transferred to a new sample tube. The pellet was washed with 20 µL of ice-cold acetone. The resulting solution was combined, dried in a speedvac (Savant SPD 111 V, Thermo Scientific, Dreieich, Germany), and resuspended in methanol (MeOH, 1.5 µL) for TLC. TLC sheets were trimmed to a size of 10×10 cm. The sample was spotted onto the sheet. The sample tube was rinsed with 1 µL MeOH and the solution was added to the same spot. The mobile phase was a mixture of $CHCl_3/MeOH/H_2O/CH_3COOH$ (11:4:0.6:0.09 $v/v/v/v$). The sheets were scanned using a conventional flatbed scanner (Canon IJ Scan Utility, Krefeld, Germany) and analyzed with JustTLC (Sweday, Sodra Sandby, Sweden). To convert the scanned image to grayscale, the Photoshop plug-in Silver Efex Pro (Google, Mountain View, CA, USA) with neutral settings and a blue filter was used.

2.3. Protein Preparation

For analysis, 5.3 mg of *E. ocellatus* venom was lysed in 500 µL BCA-compatible lysis buffer (4 M urea, 50 mM tris base, 4% SDS) and 10 mM TCEP-HCl (tris-(2-chloroethyl) phosphate) and then vortexed, ultra-sonicated, and centrifuged for 15 min at $30,000\times g$ and 4 °C. The protein concentration was determined with three replicates using the Pierce BCA Protein Assay Kit–Reducing Agent Compatible (Thermo Fisher Scientific, Darmstadt, Germany) according to the manufacturer instructions. Absorbance readings were taken at 562 nm. The protein concentration was determined by comparing the sample with a BSA standard curve measured against the same buffer beforehand.

Venom proteins were prepared by filter-aided sample preparation (FASP) as described [36]. For FASP, to 50 µg of each protein, sample buffer (8 M urea, 100 mM tris base, pH 8.5) was added to give 200 µL. After vortexing the solution for 15 min, the sample was transferred to a filter unit with a 10 kDa cut-off (VWR, Darmstadt, Germany) and centrifuged for 15 min at $12,500\times g$ at room temperature (RT). The filter unit was washed with 100 µL of urea buffer by centrifugation at the same conditions. Proteins were reduced and alkylated using dithiothreitol (DTT) and iodoacetamide (IAA) in urea buffer. First, 100 µL of 50 mM DTT was added to the filter unit and incubated for 45 min at RT with gentle shaking (500 rpm). Subsequently, the unit was centrifuged (15 min, $12,500\times g$, RT) and rinsed with 100 µL urea buffer, again by centrifugation. After that, 100 µL of 50 mM IAA solution was added to the sample and incubated for 30 min at RT in the dark with gentle shaking, followed by centrifugation under the same conditions. The alkylation reaction was quenched by adding 100 µL 50 mM DTT solution onto the filter unit and incubation for 15 min at RT in the dark with gentle shaking. Finally, the filter unit was washed four times by centrifugation with 300 µL of 50 mM NH_4HCO_3 containing 10% can, and the permeate was discarded. The sample was subjected to tryptic digestion using an enzyme solution with a concentration of 0.01 µg/µL. Trypsin solution (200 µL) was added to the filter unit, which was then sealed with laboratory film to limit evaporation. The sample was incubated overnight at 37 °C and 800 rpm in a thermostatic shaker. Peptides were collected by centrifugation (15 min, $12,500\times g$, RT). The filter unit was rinsed three times with 40 µL of 0.1% formic acid (FA) containing 5% ACN and the eluate was added to the peptide solution, dried, and stored at $-20\,°C$ until further use.

2.4. Protein Analysis

For proteomic analysis, proteins were MS analyzed as described [37]. Briefly, total venom digests were dissolved in 100 µL 0.1% FA containing 5% ACN (500 ng/µL), and 3 µL was analyzed by reversed-phase LC coupled to high-resolution MS with Synapt G2 Si/M-Class nanoUPLC (Waters Corp., Manchester, UK) using C18 µPAC columns (trapping and 50 cm analytical; PharmaFluidics, Ghent, Belgium) with a 90 min gradient (solvent system 100% water versus 100% ACN, both containing 0.1% FA, 0.3 µL/min flow rate) with three technical replicates. Peptides extracted from 1D-PAGE bands of *E. ocellatus* were analyzed with the same instrumentation but using a 30 min gradient. Data were analyzed with Progenesis for Proteomics (Nonlinear Diagnostics/Waters Corp., Manchester, UK) using the Uniprot entries for Echis, Viperidae, and Colubroidea (accessed 18 October 2023). Carbamidomethylation was set as fixed modification and methionine oxidation was a variable modification; one missed cleavage site was allowed. The peptide output from Progenesis analysis was screened by peptide score; values of 8 and better were used for further considerations. The expected fragment ions were calculated by the MassLynx spectrometer software V. 4.1. The fragment ion tables for the spectra shown here are available in the Supplement for clarification.

3. Results and Discussion

3.1. DKB Cleavage by Snake Venom

The available knowledge on protease substrates to date suggests that, indeed, BK should be cleaved by venom enzymes, although it has not been experimentally demonstrated, so far. A search in the MEROPS database of proteolytic enzymes [38] for the BK sequence resulted in 66 potentially BK-cleaving enzymes including ACE and CPN (Supplementary Excel file, MEROPS table). The MEROPS output was not comprehensive (e.g., BK cleavage by ACE to BK1-5, the fragment we observe in the NRA, was not mentioned), but it did illustrate that BK could be cleaved on all positions except BK3 (no enzyme found), and BK2 and BK6 (only one enzyme found) by several enzymes.

A high-throughput screening for protease activity of snake venoms [39] presented 2160 activity profiles based on 360 tested peptide substrates for five viperids. The authors used a commercial assay (JPT Enzyme Substrate Set). The substrates sharing at least two consecutive amino acids with BK are given in the Supplementary Excel file (JPT table). The substrate of most resemblance to BK was *SPFR*SSRI derived from human kininogen-1 (KNG1; incidentally, the precursor of BK within the kinin–kallikrein system [40]) sharing the SPFR sequence unit. Six substrates of the JPT kit had three successive amino acids in common with BK (RPP, PPG) and 20 shared two residues (RP, PG, FS). Many substrates for both snake venom metalloproteinases and serine proteinases were found in this study, including inflammation mediators, coagulation factors, and collagen-integrin proteins. Of the five viperids investigated (*E. carinatus, Bothrops asper, Daboia russelii, Bitis arietans, Bitis gabonica*), *E. carinatus* had the highest abundance (~60%) of metalloproteinases and the highest serine proteinase activity [39].

When incubating the venom of *E. ocellatus* with DBK and detecting the cleavage products by TLC following the proven protocol of the NRA [14], we noted fast degradation of the starting material within 5 min to less than 10% of its original amount (Figure 1); longer periods of digestion did not change this result significantly. That is why we settled at this incubation time for our experiments. A product was formed as a result of DBK cleavage that was neither DBK1-5 (typically difficult to visualize without contrast enhancement) nor DBK1-8, both known to us from earlier work [14]. By extraction of the material from the TLC spot for the product and its analysis by MS, we demonstrated that DBK1-7 had been generated by the enzymatic activity in the venom. The peptide fragment ion spectrum is shown in Figure 2; it delivers convincing proof for the presence of this cleavage product. Evidence for DBK1-5, DBK1-8, or other fragments of DBK cleavage could not be found. BK7-8 is the cleavage site for ACE [14], but also for matrix metalloproteinase (MMP)-8, neprylisin, prolyl oligopeptidase, and a number of other enzymes, as indicated by MEROPS search (Supplementary Excel file, table MEROPS).

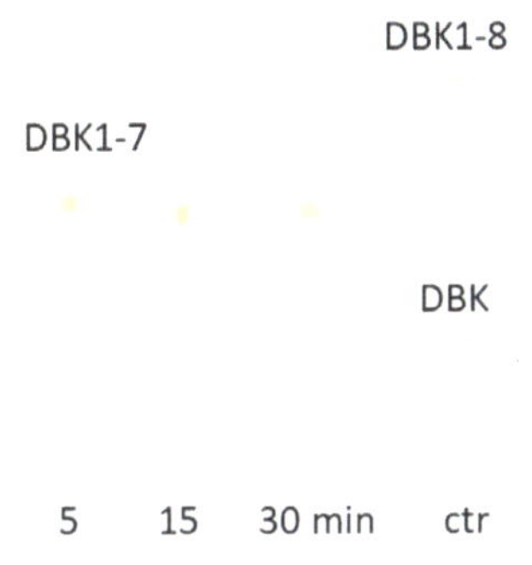

Figure 1. Scan of a TLC plate showing the results of DBK digestion by snake venom at different time points. The reaction is already complete after a few minutes. The serum control visualizes the location of DBK and its fragment DBK1-8 (fragment DBK1-5 is only visible with contrast enhancement and thus not seen here [14]). The product of venom digestion was shown by MS analysis to be fragment DBK1-7 (Figure 2).

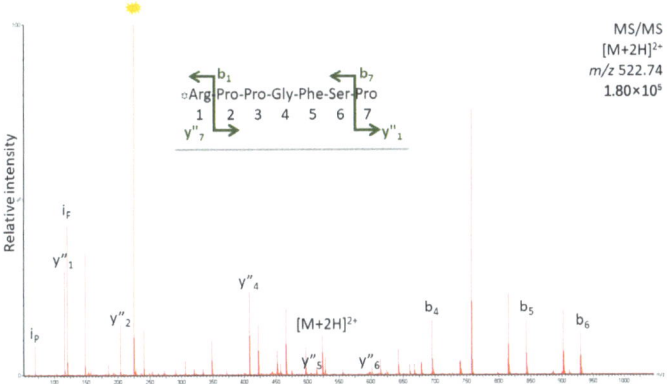

Figure 2. Peptide fragment ion spectrum for DBK1-7 detected after DBK digest by *E. ocellatus* venom. For analysis, the TLC spot was scraped off the plate, and the peptide was extracted and subjected to target MS/MS. Ions were labeled according to the b- and y-ion series for amino acid residue losses from either end of the peptide (for theoretical fragment ion masses and original spectrum, see Supplementary Figure S1). The star indicates an intense ion derived from the dabsyl label [14].

The addition of EDTA to the incubation mixture abolished DBK cleavage almost completely, indicating that most of the activity originated from metalloproteinases. Residual DBK1-7 formation of ~5% also hinted at small contributions from other enzymes.

3.2. MS-Based Protein Assignment

In an effort to identify some of the contributing proteases, we tryptically digested *E. ocellatus* venom and analysed the peptide products using reversed-phase liquid nanochromatography (LC) coupled to high-resolution MS using the Uniprot databases for Echis, Viperidae, and Colubroidea. Many hits were generated for similar proteins from related species, confusing the output considerably, so we decided to use target MS/MS of some of the DIA-detected peptides to validate the best matches (for DIA data, see Supplementary Excel file).

3.2.1. Snake Venom Metalloproteinases (SVMPs)

Experience has taught us that DIA hits assigned with scores >8 by our software tool represent reliable matches that agree with manual spectrum interpretation. Indeed, we could confirm the presence of three such peptides by manual fragmentation, as shown in Figure 3 (Supplementary Excel file, table SVMP DIA). The first, LTPGSQ-CADGECCDQCK, was a match to zinc metalloproteinase-disintegrin-like protein H3 from *Vipera ammodytes ammodytes* (R4NNL0), which exhibited similarity to the *E. ocellatus* zinc metalloproteinase-disintegrin-like Eoc1 (Q2UXR0; for alignment, see Supplementary Figure S5). However, the validated peptide differed in the last amino acid, and only one additional DIA match for a peptide with a score > 8 agreed with the Eoc1 sequence.

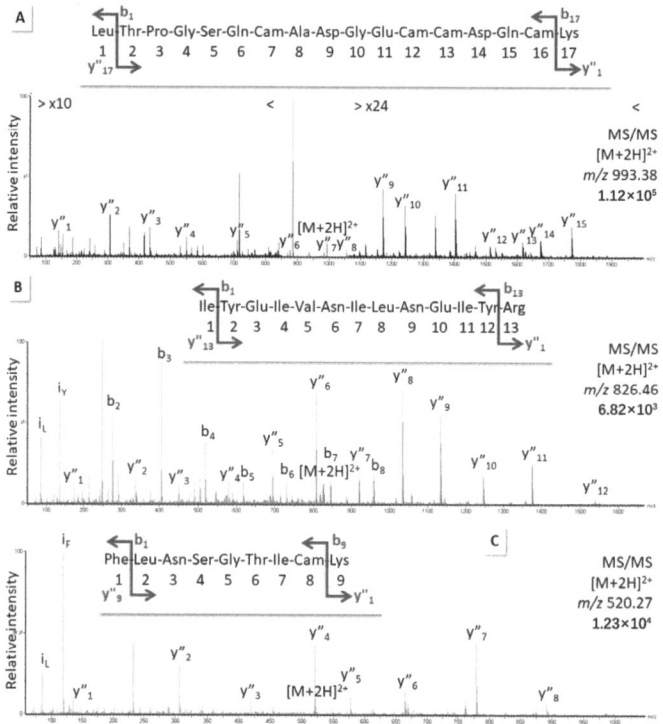

Figure 3. Fragment ion spectra for peptides measured in *E. ocellatus* venom digest using target MS/MS on the doubly charged precursor. Matches from: (**A**) zinc metalloproteinase-disintegrin-like protein H3, *Vipera ammodytes ammodytes*, R4NNL0, note zoom ranges; (**B**) metalloproteinase (fragment), *E. coloratus*, E9JG63; (**C**) disintegrin EO4A, *E. ocellatus*, Q3BER3. Ions were labeled according to the b- and the y-ion series for amino acid residue losses from either end of the peptide (for theoretical fragment ion masses and original spectra, see Supplementary Figures S2–S4).

The second manually fragmented peptide, IYEIVNILNEIYR, suggested the presence of a so-called metalloproteinase fragment from *E. coloratus* (E9JG63), also with resemblance to Eoc1 (for alignment, see Supplementary Figure S6), again, with a one-amino-acid difference to the peptide validated in the venom digests. It appears that Eoc1 is present in the venom, but that either its sequence does not match completely known information (as observed for other proteins before [23]) or it is present in at least two slightly different forms. Eoc1 is known to hydrolyze azocasein, oxidized insulin B-chain, and α/β-chain of fibrinogen, but does not cleave fibrin (see information for entry Q2UXR0).

The third sequenced peptide, FLNSGTICK, originates from disintegrin EO4A of *E. ocellatus* (Q3BER3). Disintegrins are small proteins known from viper venoms. They are potent inhibitors of both platelet aggregation and integrin-dependent cell adhesion. Some SVMPs contain a disintegrin domain [41,42].

The Uniprot database provides 53 entries for SVMPs in *E. ocellatus* (accessed 16 January 2024). Several of the entries seem to represent the same protein with only small differences in the sequences. We have not comprehensively located all possible SVMPs in our venom digest, because any scientifically sound protein identification would include proper isolation of individual proteins, their purification, and the experimental description of their biochemical properties, which is beyond the scope of the present work.

3.2.2. Snake Venom Serine Proteases (SVSPs)

In the same manner, we validated peptides from SVSPs by tandem MS in an effort to account for the residual protease activity after SVMP inhibition by EDTA (for DIA data, see Supplementary Excel file, table SVSP DIA). Figure 4 shows spectra for two peptides detected in *E. ocellatus* venom digest derived from the sequence of serine protease fragment D5KRX9 of *E. ocellatus* (A/B) and spectra for a third peptide assigned to serine protease fragment D5KRY1 of *E. ocellatus* (C). Spectral data for three more peptides are presented in Supplementary Figures S10–S12; they underline the presence of these proteases, which are similar in sequence and share some peptides (for alignment, see Supplementary Figure S13).

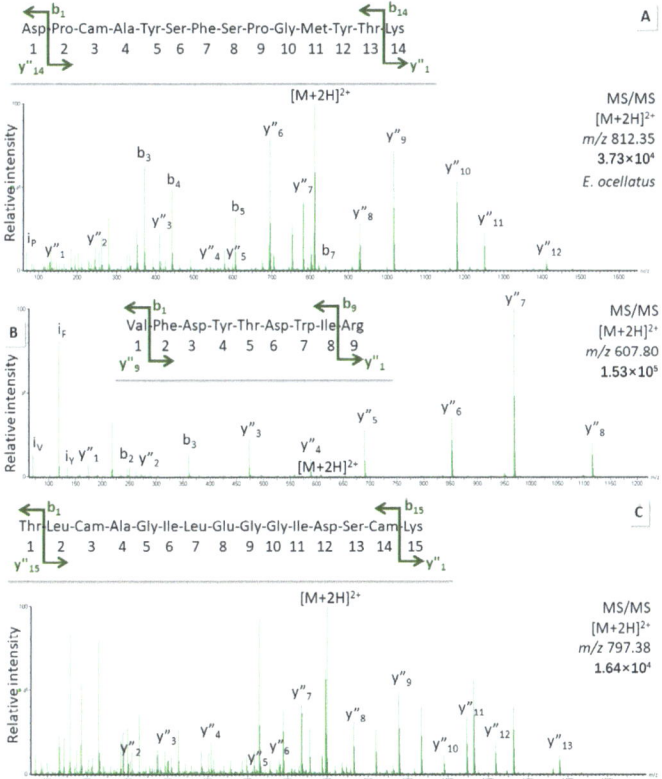

Figure 4. Fragment ion spectra for peptides measured in *E. ocellatus* venom digest using target MS/MS on the doubly charged precursor. Matches from: (**A**,**B**) serine protease (fragment, *E. ocellatus*, D5KRX9); (**C**) serine protease (fragment, *E. ocellatus*, D5KRY1). Ions were labeled according to the b- and the y-ion series for amino acid residue losses from either end of the peptide (for theoretical fragment ion masses and original spectra, see Supplementary Figures S7–S9).

Another peptide, which had been assigned to serine proteases A and B of *E. coloratus* (A0A0A1WDS7, A0A0A1WCI0), was also confirmed by target MS (Supplementary Figure S14). A known sequence from *E. ocellatus* coming closest was that of the serine protease B5U6Y3 (for alignment, see Supplementary Figure S15).

3.3. Substrate Recognition by Metalloproteinases with Respect to BK

A high-throughput study of MMP substrate recognition used substrate phage display for protease profiling [43]. From over 1300 substrates tested, only ~100 were cleaved efficiently by all of the MMPs, and almost all of these contained the canonical P-X-X-↓L motif. Transmembrane MMPs (MMP-14, -15, -16, -24) and GPI-anchored MMPs (MMP-17, -25) frequently exhibited P1' L, whereas gelatinases did not. In gelatinases (MMP-2, -9), the P3 position displayed the highest frequency residue, which was predominantly Pro [43]. We compared the substrate sequences given in this publication to the BK sequence (Supplementary Excel file, table PNAS2014). We demanded an overlap of at least two consecutive amino acid residues at the cleavage site with the substrates chosen in the study. With the knowledge that DBK was cleaved between Pro7 and Phe8 by SVMPs, we found only one substrate, which was cleaved at that position. VRPR*PF* was degraded by MMP-2, -9, and -14 very efficiently, to some degree by MMP-15, -16, -24, and -25, and not at all by MMP-17. A number of substrates in the study were cleaved after residues RP, but fragment DBK1-2 was not observed in our experiments. The presence of Leu after the cleavage site was not necessary in our case, which agreed with other authors who rather proposed the need for a large hydrophobic residue at the P1' position [44], and this was Phe in DBK. However, they also demanded Pro in the P3 position, which was not available in DBK when cleaved at Pro7. Beside PXX↓X_{Hy}, these authors described substrate motifs L/IXX↓X_{Hy}, X_{Hy}SX↓L, and HXX↓X_{Hy}, which were selective of MMP-2 over MMP-9.

It thus appears that BK is not a clear fit for any known substrate category for MMPs. The comparison of its sequence to the available substrate information indicates cleavage by gelatinases, which are common in snake venoms (for an overview of SVMPs, see [45]).

4. Conclusions

We discovered that BK can be cleaved by proteases in *E. ocellatus* venom, which degraded its labeled form DBK after Pro7 in a few minutes. This is a common cleavage site for several mammalian proteases such as ACE and neprylisin, but is not typical for MMPs according to the comparison with the results from large substrate profiling studies [43,44]. Most but not all of the protease activity was inhibited by EDTA, indicating that DBK was also cleaved by SVMPs; less than 5% of the available material was unaffected by inhibition. It remains to be seen in future studies if dedicated inhibitors of SVMPs abolish all activity, or if, indeed, SVSPs also act on BK.

The fact that venom proteases may target the vasoactive neuropeptide BK is of great interest when studying the response of the human body to snakebite. The knowledge of which proteases are involved assists in basic research of envenomation, and its influence on blood pressure and the RAS in general [46]. Moreover, with our DBK-based assay [14], we have a very good tool at our hands to test medicinal plants for active compounds. In fact, its use is a step forward from the artificial peptide substrates typically used in this line of work.

We supplemented these data by MS-based protein analyses for SVMPs and SVSPs, which we performed in the total venom digest. As a result of the insufficient availability of a species-specific database, we used collections of proteins sequences from higher orders, which provided very complex results (for limits of the method, see [28]). Therefore, we validated some of the peptides suggested by DIA with target MS/MS and screened the protein matches for known sequences in *E. ocellatus*. We provide spectral evidence for the presence of a homologous form of *E. ocellatus* zinc metalloproteinase-disintegrin-like Eoc1 in our venom. Furthermore, disintegrin EO4A of *E. ocellatus* was detected. For the SVSPs, protein fragments D5KRX9 and D5KRY1 of *E. ocellatus* were validated, as well as a homolog of serine protease B5U6Y3. The comprehensive protein identification of all available proteoforms was not the goal of this study. Classical protein isolation, purification, and biochemical investigation will be required to properly define the protein content of the venom. As long as the number of individual protein species in the venoms is not known, and, in addition, no reliable genome sequencing data are available, any attempts at protein identification or activity studies will remain superficial. Moreover, we tested pooled venom from a single geographical location only, and thus have no information on regional differences in venom composition and activity.

Snake venom is a mixture of many substances including proteins and peptides, each of which contribute in different ways to the pathophysiological results of snakebite. Our experiments provide knowledge regarding snake venom proteases and support basic research, but they, of course, cannot provide an immediate cure for snakebite.

Supplementary Materials: The following supporting information can be downloaded at: https://www.mdpi.com/article/10.3390/biomedicines12051027/s1, Figure S1. Spectrum for DBK1-7 measured after DBK digestion by venom of *E. ocellatus*; Figure S2. Fragment ion spectra and theoretical peptide fragment ions calculated using Masslynx software (Waters Corp.) for peptide measured in *E. ocellatus* venom digest using target MS/MS of the doubly-charged precursor—Match from zinc metalloproteinase-disintegrin-like protein H3; Figure S3. Fragment ion spectra and theoretical peptide fragment ions calculated using Masslynx software (Waters Corp.) for peptide measured in *E. ocellatus* venom digest using target MS/MS of the doubly-charged precursor—Match from metalloproteinase (Fragment); Figure S4. Fragment ion spectra and theoretical peptide fragment ions calculated using Masslynx software (Waters Corp.) for peptide measured in E. ocellatus venom digest using target MS/MS o the doubly-charged precursor. Match from disintegrin EO4A; Figure S5. Clustal sequence alignment of R4NL0 and Q2UXR0; Figure S6. Clustal sequence alignment of E9JG63 and Q2UXR0; Figure S7. Fragment ion spectra and theoretical peptide fragment ions calculated using Masslynx software (Waters Corp.) for peptide measured in *E. ocellatus* venom digest using target MS/MS o the doubly-charged precursor—Match from serine protease; Figure S8. Fragment ion spectra and theoretical peptide fragment ions calculated using Masslynx software (Waters Corp.) for peptide measured in *E. ocellatus* venom digest using target MS/MS o the doubly-charged precursor—Match from serine protease; Figure S9. Fragment ion spectra and theoretical peptide fragment ions calculated using Masslynx software (Waters Corp.) for peptide measured in *E. ocellatus* (top trace) venom digest using target MS/MS o the doubly-charged precursor; Figure S10. Fragment ion spectra and theoretical peptide fragment ions calculated using Masslynx software (Waters Corp.) for peptide measured in E. ocellatus venom digest using target MS/MS o the doubly-charged precursor. Match from serine protease; Figure S11. Fragment ion spectra and theoretical peptide fragment ions calculated using Masslynx software (Waters Corp.) for peptide measured in E. ocellatus venom digests using target MS/MS o the doubly-charged precursor. Match from serine protease; Figure S12. Fragment ion spectra and theoretical peptide fragment ions calculated using Masslynx software (Waters Corp.) for peptide measured in *E. ocellatus* venom digest using target MS/MS o the doubly-charged precursor. Match from serine protease; Figure S13. Clustal sequence alignment of D5KRX9 and D5KRY1; Figure S14. Fragment ion spectra and theoretical peptide fragment ions calculated using Masslynx software (Waters Corp.) for peptide measured in *E. ocellatus* venom digest using target MS/MS o the doubly-charged precursor. Match from serine proteases A and B of *E coloratus*; Figure S15. Clustal sequence alignment of B5U6Y3 and A0A0A1WDS7.

Author Contributions: Conceptualization, J.A., O.A., A.A. and S.K.; methodology, J.A., A.M.B. and S.K.; validation, S.K.; formal analysis, S.K.; resources, J.A. and S.K.; writing—original draft preparation, S.K. and J.A., writing—review and editing, S.K.; supervision, S.K., O.A. and A.A.; project administration, J.A.; funding acquisition, J.A. All authors have read and agreed to the published version of the manuscript.

Funding: This research was funded by the German Academic Exchange Service with a fellowship to J.A. (ref. 91862297).

Institutional Review Board Statement: The animal study protocol was approved by the University of Ibadan Animal Care and Use Research Ethics Committee (NHREC/UIACUREC/05/12/2022A) and in agreement with ARRIVE guideline 2.0.

Informed Consent Statement: Not applicable.

Data Availability Statement: All data are given in the results and Supplementary Materials.

Conflicts of Interest: The authors declare no conflicts of interest.

References

1. Sharma, J.N. Hypertension and the bradykinin system. *Curr. Hypertens. Rep.* **2009**, *11*, 178–181. [CrossRef] [PubMed]
2. Hawgood, B.J. Maurício Rocha e Silva MD: Snake venom, bradykinin and the rise of autopharmacology. *Toxicon* **1997**, *35*, 1569–1580. [CrossRef] [PubMed]
3. Rocha e Silva, M.; Beraldo, W.T.; Rosenfeld, G. Bradykinin, a hypotensive and smooth muscle stimulating factor released from plasma globulin by snake venoms and by trypsin. *Am. J. Physiol.-Leg. Content* **1949**, *156*, 261–273. [CrossRef] [PubMed]
4. Sciani, J.M.; Pimenta, D.C. The modular nature of bradykinin-potentiating peptides isolated from snake venoms. *J. Venom. Anim. Toxins Incl. Trop. Dis.* **2017**, *23*, 45. [CrossRef] [PubMed]
5. Ferreira, S.H. A Bradykinin-Potentiating Factor (BPF) Present in the Venom of *Bothrops Jararca*. *Br. J. Pharmacol. Chemother.* **1965**, *24*, 163–169. [CrossRef]
6. Waheed, H.; Moin, S.F.; Choudhary, M.I. Snake Venom: From Deadly Toxins to Life-saving Therapeutics. *Curr. Med. Chem.* **2017**, *24*, 1874–1891. [CrossRef] [PubMed]
7. Péterfi, O.; Boda, F.; Szabó, Z.; Ferencz, E.; Bába, L. Hypotensive Snake Venom Components—A Mini-Review. *Molecules* **2019**, *24*, 2778. [CrossRef]
8. Deutsch, H.F.; Diniz, C.R. Some proteolytic activities of snake venoms. *J. Biol. Chem.* **1955**, *216*, 17–26. [CrossRef]
9. Rex, D.A.B.; Vaid, N.; Deepak, K.; Dagamajalu, S.; Prasad, T.S.K. A comprehensive review on current understanding of bradykinin in COVID-19 and inflammatory diseases. *Mol. Biol. Rep.* **2022**, *49*, 9915–9927. [CrossRef]
10. Kaplan, A.P.; Joseph, K.; Silverberg, M. Pathways for bradykinin formation and inflammatory disease. *J. Allergy Clin. Immunol.* **2002**, *109*, 195–209. [CrossRef]
11. Jayasinghe, M.; Caldera, D.; Prathiraja, O.; Jena, R.; Coffie-Pierre, J.A.; Agyei, J.; Silva, M.S.; Kayani, A.M.A.; Siddiqui, O.S. A Comprehensive Review of Bradykinin-Induced Angioedema Versus Histamine-Induced Angioedema in the Emergency Department. *Cureus* **2022**, *14*, e32075. [CrossRef] [PubMed]
12. König, S.; Bayer, M.; Dimova, V.; Herrnberger, M.; Escolano-Lozano, F.; Bednarik, J.; Vlckova, E.; Rittner, H.; Schlereth, T.; Birklein, F. The serum protease network-one key to understand complex regional pain syndrome pathophysiology. *PAIN* **2019**, *160*, 1402–1409. [CrossRef] [PubMed]
13. König, S.; Vollenberg, R.; Tepasse, P.-R. The Renin-Angiotensin System in COVID-19: Can Long COVID Be Predicted? *Life* **2023**, *13*, 1462. [CrossRef] [PubMed]
14. Bayer, M.; König, S. A vote for robustness: Monitoring serum enzyme activity by thin-layer chromatography of dabsylated bradykinin products. *J. Pharm. Biomed. Anal.* **2017**, *143*, 199–203. [CrossRef] [PubMed]
15. Ghafouri-Fard, S.; Noroozi, R.; Omrani, M.D.; Branicki, W.; Pośpiech, E.; Sayad, A.; Pyrc, K.; Łabaj, P.P.; Vafaee, R.; Taheri, M.; et al. Angiotensin converting enzyme: A review on expression profile and its association with human disorders with special focus on SARS-CoV-2 infection. *Vasc. Pharmacol.* **2020**, *130*, 106680. [CrossRef]
16. Matthews, K.W.; Mueller-Ortiz, S.L.; Wetsel, R.A. Carboxypeptidase N: A pleiotropic regulator of inflammation. *Mol. Immunol.* **2004**, *40*, 785–793. [CrossRef] [PubMed]
17. Kini, R.M. Anticoagulant proteins from snake venoms: Structure, function and mechanism. *Biochem. J.* **2006**, *397*, 377–387. [CrossRef] [PubMed]
18. Mackessy, S.P. *Handbook of Venoms and Toxins of Reptiles*; CRC Press: Boca Raton, FL, USA, 2010; ISBN 9780429186394.
19. Bhatia, S.; Vasudevan, K. Comparative proteomics of geographically distinct saw-scaled viper (*Echis carinatus*) venoms from India. *Toxicon X* **2020**, *7*, 100048. [CrossRef]
20. Tasoulis, T.; Isbister, G.K. A current perspective on snake venom composition and constituent protein families. *Arch. Toxicol.* **2023**, *97*, 133–153. [CrossRef] [PubMed]
21. Rao, W.; Kalogeropoulos, K.; Allentoft, M.E.; Gopalakrishnan, S.; Zhao, W.; Workman, C.T.; Knudsen, C.; Jiménez-Mena, B.; Seneci, L.; Mousavi-Derazmahalleh, M.; et al. The rise of genomics in snake venom research: Recent advances and future perspectives. *GigaScience* **2022**, *11*, giac024. [CrossRef]
22. Wagstaff, S.C.; Harrison, R.A. Venom gland EST analysis of the saw-scaled viper, Echis ocellatus, reveals novel alpha9beta1 integrin-binding motifs in venom metalloproteinases and a new group of putative toxins, renin-like aspartic proteases. *Gene* **2006**, *377*, 21–32. [CrossRef] [PubMed]
23. Wagstaff, S.C.; Sanz, L.; Juárez, P.; Harrison, R.A.; Calvete, J.J. Combined snake venomics and venom gland transcriptomic analysis of the ocellated carpet viper, *Echis ocellatus*. *J. Proteom.* **2009**, *71*, 609–623. [CrossRef]
24. Nguyen, G.T.T.; O'Brien, C.; Wouters, Y.; Seneci, L.; Gallissà-Calzado, A.; Campos-Pinto, I.; Ahmadi, S.; Laustsen, A.H.; Ljungars, A. High-throughput proteomics and in vitro functional characterization of the 26 medically most important elapids and vipers from sub-Saharan Africa. *GigaScience* **2022**, *11*, giac121. [CrossRef] [PubMed]
25. Dingwoke, E.J.; Adamude, F.A.; Mohamed, G.; Klein, A.; Salihu, A.; Abubakar, M.S.; Sallau, A.B. Venom proteomic analysis of medically important Nigerian viper *Echis ocellatus* and *Bitis arietans* snake species. *Biochem. Biophys. Rep.* **2021**, *28*, 101164. [CrossRef] [PubMed]

26. Patra, A.; Kalita, B.; Chanda, A.; Mukherjee, A.K. Proteomics and antivenomics of *Echis carinatus carinatus* venom: Correlation with pharmacological properties and pathophysiology of envenomation. *Sci. Rep.* **2017**, *7*, 17119. [CrossRef]
27. Patra, A.; Mukherjee, A.K. Proteomic Analysis of Sri Lanka *Echis carinatus* Venom: Immunological Cross-Reactivity and Enzyme Neutralization Potency of Indian Polyantivenom. *J. Proteome Res.* **2020**, *19*, 3022–3032. [CrossRef]
28. König, S.; Obermann, W.M.J.; Eble, J.A. The Current State-of-the-Art Identification of Unknown Proteins Using Mass Spectrometry Exemplified on De Novo Sequencing of a Venom Protease from Bothrops moojeni. *Molecules* **2022**, *27*, 4976. [CrossRef]
29. König, S. Spectral quality overrides software score-A brief tutorial on the analysis of peptide fragmentation data for mass spectrometry laymen. *J. Mass Spectrom.* **2021**, *56*, e4616. [CrossRef]
30. Coorssen, J.R.; Yergey, A.L. Proteomics Is Analytical Chemistry: Fitness-for-Purpose in the Application of Top-Down and Bottom-Up Analyses. *Proteomes* **2015**, *3*, 440–453. [CrossRef]
31. Gutiérrez, J.M.; Maduwage, K.; Iliyasu, G.; Habib, A. Snakebite envenoming in different national contexts: Costa Rica, Sri Lanka, and Nigeria. *Toxicon X* **2021**, *9–10*, 100066. [CrossRef]
32. Habib, A.G. Venomous Snakes and Snake Envenomation in Nigeria. In *Clinical Toxinology*; Gopalakrishnakone, P., Faiz, S., Gnanathasan, C.A., Habib, A.G., Fernando, R., Yang, C.-C., Vogel, C.-W., Tambourgi, D.V., Seifert, S.A., Eds.; Springer: Dordrecht, The Netherlands, 2020; pp. 1–21. ISBN 978-94-007-6288-6.
33. Royal, D.O. Nigeria Records 20,000 Cases of Snake Bites, 2,000 Deaths Annually. *Vanguard News [Online]*. 20 September 2021. Available online: https://www.vanguardngr.com/2021/09/nigeria-records-20000-cases-of-snake-bites-2000-deaths-annually/ (accessed on 18 December 2023).
34. Habib, A.G.; Gebi, U.I.; Onyemelukwe, G.C. Snake bite in Nigeria. *Afr. J. Med. Med. Sci.* **2001**, *30*, 171–178. [PubMed]
35. Chippaux, J.-P. Estimate of the burden of snakebites in sub-Saharan Africa: A meta-analytic approach. *Toxicon: Off. J. Int. Soc. Toxinology* **2011**, *57*, 586–599. [CrossRef] [PubMed]
36. Wiśniewski, J.R.; Zougman, A.; Nagaraj, N.; Mann, M. Universal sample preparation method for proteome analysis. *Nat. Methods* **2009**, *6*, 359–362. [CrossRef] [PubMed]
37. Distler, U.; Kuharev, J.; Navarro, P.; Tenzer, S. Label-free quantification in ion mobility-enhanced data-independent acquisition proteomics. *Nat. Protoc.* **2016**, *11*, 795–812. [CrossRef] [PubMed]
38. Rawlings, N.D.; Barrett, A.J.; Thomas, P.D.; Huang, X.; Bateman, A.; Finn, R.D. The MEROPS database of proteolytic enzymes, their substrates and inhibitors in 2017 and a comparison with peptidases in the PANTHER database. *Nucleic Acids Res* **2018**, *46*, D624–D632. [CrossRef] [PubMed]
39. Kalogeropoulos, K.; Treschow, A.F.; Keller, U.a.d.; Escalante, T.; Rucavado, A.; Gutiérrez, J.M.; Laustsen, A.H.; Workman, C.T. Protease Activity Profiling of Snake Venoms Using High-Throughput Peptide Screening. *Toxins* **2019**, *11*, 170. [CrossRef] [PubMed]
40. Motta, G.; Tersariol, I.L.S.; Calo, G.; Gobeil, F.; Regoli, D. Kallikrein–Kinin System. In *eLS*; John Wiley & Sons, Ltd.: Hoboken, NJ, USA, 2018; pp. 1–9.
41. Vasconcelos, A.A.; Estrada, J.C.; David, V.; Wermelinger, L.S.; Almeida, F.C.L.; Zingali, R.B. Structure-Function Relationship of the Disintegrin Family: Sequence Signature and Integrin Interaction. *Front. Mol. Biosci.* **2021**, *8*, 783301. [CrossRef] [PubMed]
42. Almeida, G.d.O.; Oliveira, I.S.d.; Arantes, E.C.; Sampaio, S.V. Snake venom disintegrins update: Insights about new findings. *J. Venom. Anim. Toxins Incl. Trop. Dis.* **2023**, *29*, e20230039. [CrossRef] [PubMed]
43. Ratnikov, B.I.; Cieplak, P.; Gramatikoff, K.; Pierce, J.; Eroshkin, A.; Igarashi, Y.; Kazanov, M.; Sun, Q.; Godzik, A.; Osterman, A.; et al. Basis for substrate recognition and distinction by matrix metalloproteinases. *Proc. Natl. Acad. Sci. USA* **2014**, *111*, E4148–E4155. [CrossRef]
44. Chen, E.I.; Kridel, S.J.; Howard, E.W.; Li, W.; Godzik, A.; Smith, J.W. A unique substrate recognition profile for matrix metalloproteinase-2. *J. Biol. Chem.* **2002**, *277*, 4485–4491. [CrossRef]
45. Olaoba, O.T.; Karina Dos Santos, P.; Selistre-de-Araujo, H.S.; Ferreira de Souza, D.H. Snake Venom Metalloproteinases (SVMPs): A structure-function update. *Toxicon: X* **2020**, *7*, 100052. [CrossRef] [PubMed]
46. Hallberg, M.; Nyberg, F. Neuropeptide conversion to bioactive fragments An important pathway in neuromodulation. *Curr. Protein Pept. Sci.* **2003**, *4*, 31–44. [CrossRef] [PubMed]

Disclaimer/Publisher's Note: The statements, opinions and data contained in all publications are solely those of the individual author(s) and contributor(s) and not of MDPI and/or the editor(s). MDPI and/or the editor(s) disclaim responsibility for any injury to people or property resulting from any ideas, methods, instructions or products referred to in the content.

Article

Classical and Alternative Pathways of the Renin–Angiotensin–Aldosterone System in Regulating Blood Pressure in Hypertension and Obese Adolescents

Adrian Martyniak [1], Dorota Drożdż [2] and Przemysław J. Tomasik [1,*]

[1] Department of Clinical Biochemistry, Institute of Pediatrics, Jagiellonian University Medical College, 30-663 Krakow, Poland; adrian.martyniak@uj.edu.pl

[2] Department of Pediatric Nephrology and Hypertension, Institute of Pediatrics, Jagiellonian University Medical College, 30-663 Krakow, Poland; dorota.drozdz@uj.edu.pl

* Correspondence: p.tomasik@uj.edu.pl

Abstract: Primary hypertension (PH) is the leading form of arterial hypertension (AH) in adolescents. Hypertension is most common in obese patients, where 20 to 40% of the population has elevated blood pressure. One of the most effective mechanisms for regulating blood pressure is the renin–angiotensin–aldosterone system (RAAS). The new approach to the RAAS talks about two opposing pathways between which a state of equilibrium develops. One of them is a classical pathway, which is responsible for increasing blood pressure and is represented mainly by the angiotensin II (Ang II) peptide and, to a lesser extent, by angiotensin IV (Ang IV). The alternative pathway is responsible for the decrease in blood pressure and is mainly represented by angiotensin 1–7 (Ang 1–7) and angiotensin 1–9 (Ang 1–9). Our research study aimed to assess changes in angiotensin II, angiotensin IV, angiotensin 1–7, and angiotensin 1–9 concentrations in the plasma of adolescents with hypertension, with hypertension and obesity, and obesity patients. The Ang IV concentration was lower in hypertension + obesity versus control and obesity versus control, respectively $p = 0.01$ and $p = 0.028$. The Ang 1–9 concentration was lower in the obesity group compared to the control group ($p = 0.036$). There were no differences in Ang II and Ang 1–7 peptide concentrations in the hypertension, hypertension and obesity, obesity, and control groups. However, differences were observed in the secondary peptides, Ang IV and Ang 1–9. In both cases, the differences were related to obesity.

Keywords: angiotensin II; angiotensin 1–7; angiotensin IV; angiotensin 1–9; obesity; arterial hypertension

1. Introduction

Cardiovascular disease (CVD) is the leading cause of death in the world. The WHO report (2021) estimated 17.9 million deaths each year [1]. Only in the United States (USA), one in every five people dies of CVD, and heart disease costs the United States approximately USD 239.9 billion per year [2]. One of the main causes of CVD is hypertension. Arterial hypertension (AH) is a very serious disease that leads to many complications, including death, kidney failure, and myocardial infarction. Hypertension is a serious medical problem in adolescents and even in children. In the US, hypertension is estimated to be associated with 0.3 to 4.5% of the pediatric population [3]. Primary hypertension (PH) is the leading form of arterial hypertension in adolescents. Hypertension is more common in obese patients, where 20 to 40% of the population has elevated blood pressure [4]. In addition to obesity, genetic and environmental factors play an important role [4].

AH symptoms are nonspecific and often difficult to observe. Patients complain of headaches, fatigue, nosebleeds, sleep disturbances, and nervousness. Although the diagnostic criteria for hypertension in adults are well known and widespread, there is no such consensus among adolescents.

An important factor in hypertension is obesity. Although obesity is the underlying disease, the associated hypertension is still considered primary [4]. The effects of obesity on blood pressure are multiple and include hormonal, neurological, and anatomical changes. The most important are insulin resistance, increased sympathetic nervous system activity, increased renin release, increased circulating blood volume, and increased circulating blood resistance [5].

One of the most effective mechanisms to regulate blood pressure is the renin–angiotensin–aldosterone system (RAAS) [6]. The reaction cascade starts from the enzymatic fragmentation of angiotensinogen. Angiotensinogen is a peptide hormone, produced in the liver, that is a precursor of all angiotensin peptides [7]. Angiotensin peptides, despite a small molecular weight and similar structure, act in a different role. The RAAS causes the retention of water and sodium, the release of aldosterone, the contraction of blood vessels, increased heart rate, and consequently hypertension. On the other hand, the RAAS can decrease blood pressure through several reverse mechanisms, such as the vasodilation of blood vessels or the release of nitric oxide [8]. Therefore, the new approach to the RAAS talks about two opposing pathways between which a state of equilibrium develops. One of them is a classical pathway that is responsible for increasing blood pressure. The main effector of the classical pathway is angiotensin II (Ang II) and, to a lesser extent, angiotensin IV (Ang IV). The second pathway is called an alternative. The alternative pathway is responsible for decreasing blood pressure and is mainly represented by angiotensin 1–7 (Ang 1–7) and, to a lesser extent, angiotensin 1–9 (Ang 1–9) [9,10].

Our research study aimed to assess changes in angiotensin II, angiotensin IV, angiotensin 1–7, and angiotensin 1–9 concentrations in plasma of adolescents with hypertension, hypertension and obesity, and obese patients with normal blood pressure. Based on measured concentrations of the main peptides of classical and alternative RAAS pathways, the disbalance in these pathways in AH and obese adolescents was analyzed. A secondary aim was to identify a marker predictive of arterial hypertension.

2. Materials and Methods

We recruited adolescents suffering from AH, obesity, and combinations of these diseases among patients from the Department of Pediatrics Nephrology and Hypertension of the University Children's Hospital in Krakow, Poland. All patients were recruited to this study by the hypertensiologist, according to the consensus of the European Society of Hypertension (ESH) named Pediatric Hypertension Guidelines 2016 [11], and related patient data were collected from their medical records. Patients with the E66 (ICD-10) diagnosis in medical documentation were eligible for the obese group. Fasting blood samples (S-Monovette EDTA K3E/2.6 mL, Sarstedt AG & Co.KG, Numbrecht, Germany) were taken in the morning, immediately cooled, and then centrifuged. Adolescents in the control group were recruited from families and friends of the study researchers. These adolescents had no pathological clinical signs or complaints or any pharmacological treatment. In the control group, fasting blood samples were drawn in the same procedure as in the study group.

Blood samples were centrifuged and separated, and EDTA plasma was frozen at -80 centigrade until measurement. The maximum bank loan time was not longer than 12 months. Peptide concentrations were measured using commercially available enzyme-linked immunosorbent assay (ELISA) immunoassays: angiotensin II assay range: 12.5 ng/mL–800 ng/mL, angiotensin IV assay range: 1.56 ng/mL–100 ng/mL, angiotensin 1–7 assay range: 12.5 ng/mL–800 ng/mL, and angiotensin 1–9 assay range: 7.8 ng/mL–500 ng/mL (Qayee Bio-Technology Co., Ltd., Shanghai, China). The manufacturer declares that there is no cross-reaction and coefficient variation < 15%. The samples were slowly defrosted. The first step was the transfer of the samples from -80 centigrade to -20 centigrade for a night; then, they were thawed in ice-free water. According to the manufacturer's guidelines, all samples were diluted five times before analysis.

The assay procedures were performed according to the manufacturer's manuals using a Bio-Rad washer and plate reader (Bio-Rad, Hercules, CA, USA). According to the manufacturer, all of the tests used have high sensitivity and excellent specificity for the detection of the measured parameters without significant cross-reactivity or interference between the analytes and their analogues.

The study protocol was approved by the Jagiellonian University Bioethical Committee (approval no. 1072.61.20.67.2019), and informed consent was obtained from all the legal guardians of the patients and all the patients over 16 years of age enrolled in the study.

The statistical analysis was performed using IBM SPSS Statistics (v29, IBM Corporation, Armonk, NY, USA). The concentrations of angiotensin derivative peptides were expressed as median values and quartile ranges. Normality was checked using the Shapiro–Wilk test in each group. The Kruskal–Wallis test was performed for comparisons between the studied groups. If the Kruskal–Wallis test did not show any significant statistical differences, but the analysis of differences in individual groups showed possible differences, a U Mann–Whitney test was performed.

3. Results

We studied 28 patients with AH (15.05 ys $\pm$ 2.98; BMI 21.75 $\pm$ 3.44 kg/m^2), 17 patients with AH and obesity (13.95 ys $\pm$ 3.79; BMI 29.89 $\pm$ 4.64 kg/m^2), and 29 patients with obesity (13.50 ys $\pm$ 3.39; BMI 28.40 $\pm$ 5.59 kg/m^2) and normal blood pressure. As a control group, 52 healthy children were observed with normal BMI (12.95 ys $\pm$ 3.69; BMI 18.63 $\pm$ 3.9 kg/m^2) and normal blood pressure. All patient characteristics are summarized in Table 1.

Table 1. Patients' characteristics.

	Number of Patients	Age	BMI	Systolic Pressure (SP)	Diastolic Pressure (DP)
Hypertension	28	15.05 ys $\pm$ 2.98	21.75 $\pm$ 3.44 kg/m^2	132 $\pm$ 16 mmHg	79 $\pm$ 11 mmHg
Hypertension + obesity	17	13.95 ys $\pm$ 3.79	29.89 $\pm$ 4.64 kg/m^2	138 $\pm$ 19 mmHg	74 $\pm$ 14 mmHg
Obesity	29	13.50 ys $\pm$ 3.39	28.40 $\pm$ 5.59 kg/m^2	114 $\pm$ 10 mmHg	68 $\pm$ 9 mmHg
Control	52	12.95 ys $\pm$ 3.69	18.63 $\pm$ 3.9 kg/m^2	112 $\pm$ 11 mmHg	66 $\pm$ 10 mmHg

3.1. Angiotensin II

Plasma Ang II concentrations did not differ statistically in the analyzed groups. The median and quartiles 1 and 3, respectively, were 275.04 (245.89–344.33) ng/mL in hypertension, 285.12 (236.95–323.60) ng/mL in hypertension and obesity, and 281.11 (256.11–321.37) ng/mL in obesity. The concentration of Ang II in the control group was 308.05 (271.52–372.24) ng/mL. There were also no statistical differences between the study and the control group (Figure 1).

3.2. Angiotensin IV

Plasma Ang IV concentrations differed statistically in the analyzed groups ($p = 0.017$). The median and quartiles 1 and 3, respectively, were 34.51 (15.15–49.07) ng/mL in hypertension, 29.11 (19.07–38.98) ng/mL in hypertension and obesity, and 33.55 (24.52–40.71) ng/mL in obesity. The concentration of Ang IV in the control group was 39.71 (32.09–47.75) ng/mL. The significant statistical differences were between hypertension + obesity and control and obesity versus control, respectively $p = 0.01$ and $p = 0.028$ (Figure 2).

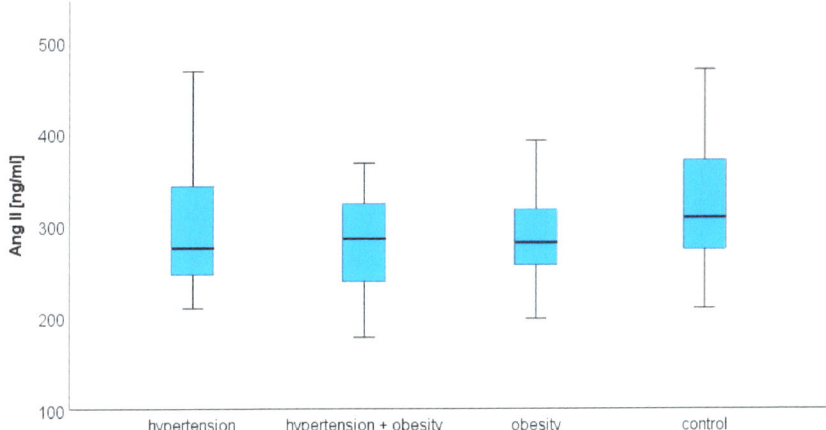

Figure 1. The median concentration of angiotensin II in the hypertension group, hypertension and obesity group, obesity group, and control group. The boxes show the median and quartile range of the measured plasma angiotensin II concentrations in the study and control group; the whiskers show the minimal and maximal measured concentration. No significant differences were observed.

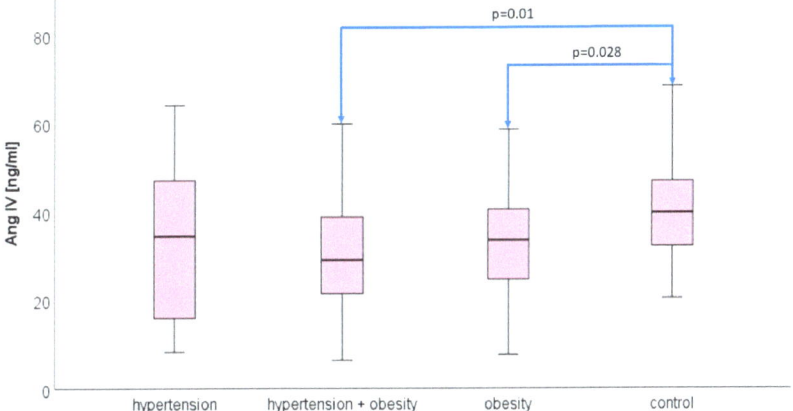

Figure 2. The median concentration of angiotensin IV in the hypertension group, hypertension and obesity group, obesity group, and control group. The boxes show the median and quartile range of the measured plasma angiotensin IV concentrations; the whiskers show the minimal and maximal measured concentration: a significant difference between adolescents with hypertension and the obesity vs. control group, $p = 0.01$; a significant difference between obese adolescents and the control group, $p = 0.028$.

3.3. Angiotensin 1–7

Plasma Ang 1–7 concentrations did not differ statistically in the analyzed groups. The median and quartiles 1 and 3, respectively, were 302.11 (255.87–367.40) ng/mL in hypertension, 268.01 (225.16–345.57) ng/mL in hypertension and obesity, and 283.42 (237.50–341.61) ng/mL in obesity. The concentration of Ang 1–7 in the control group was 268.76 (236.29–380.56) ng/mL. There were no statistical differences between the study and the control group (Figure 3).

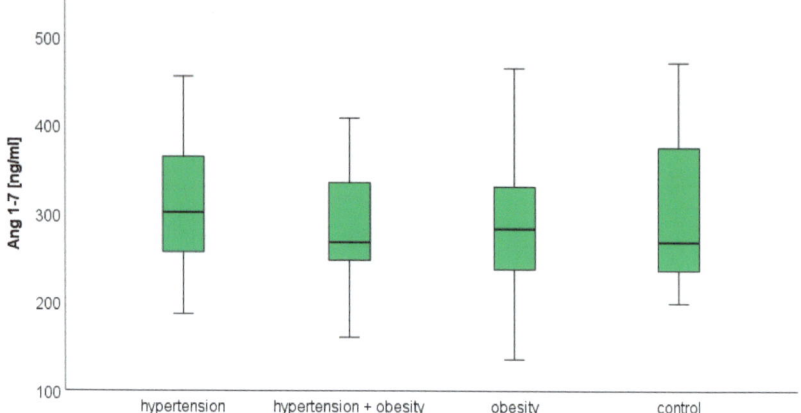

Figure 3. The median concentration of angiotensin 1–7 in the hypertension group, hypertension and obesity group, obesity group, and control group. The boxes show the median and quartile range of the measured plasma angiotensin 1–7 concentrations in the study and control group; the whiskers show the minimal and maximal measured concentration. No significant differences were observed.

3.4. Angiotensin 1–9

Plasma Ang 1–9 concentrations did not differ statistically in the analyzed groups. The median and quartiles 1 and 3, respectively, were 181.38 (159.69–205.30) ng/mL in hypertension, 169.16 (147.68–203.13) ng/mL in hypertension and obesity, and 172.07 (155.81–205.67) ng/mL in obesity. The concentration of Ang 1–9 in the control group was 193.30 (169.23–215.63) ng/mL. A statistical difference between the obesity group and the control group ($p = 0.036$) was confirmed by the U Mann–Whitney test (Figure 4).

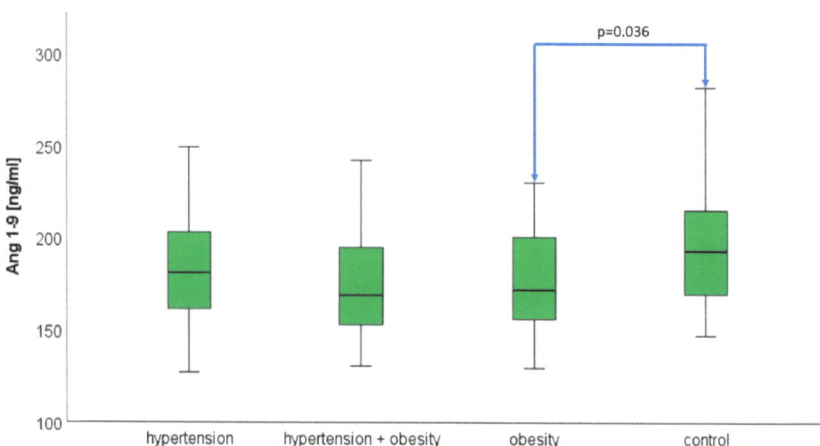

Figure 4. The median concentration of angiotensin 1–9 in the hypertension group, hypertension and obesity group, obesity group, and control group. The boxes show the median and quartile range of the measured plasma angiotensin 1–9 concentrations in the study and control group; the whiskers show the minimal and maximal measured concentration: a significant difference between obese adolescents and the control group was confirmed by the U Mann–Whitney test, $p = 0.036$.

4. Discussion

Several studies describe the concentrations of chosen angiotensin peptides. However, due to differences in the measurement methods, it is difficult to compare the subsequent peptide levels of different papers. This study is the first to comprehensively present and analyze both RAAS pathways in adolescents with hypertension.

This study is innovative for several reasons. It is the first study to comprehensively analyze the classical and alternative RAA systems in the adolescence period (10–18 years). According to Litwin et al., this is when idiopathic hypertension occurs most often. Authors analyze the RAA system as two separate pathways that should be in balance. The disbalance of these pathways may result in AH. Some angiotensin peptides such as Ang IV and Ang 1–9 have never been previously investigated as potential factors involved in the development of AH.

Angiotensin II (Asp-Arg-Val-Tyr-Ile-His-Pro-Phe) has the strongest biological activity in the RAAS classical pathway. The peptide has a strong affinity for the angiotensin type I receptor (AT1R). The AT1R is located mainly in the kidneys, vascular smooth muscle, lungs, and liver. The receptor belongs to the superfamily of G protein-bound receptors. The stimulation of the AT1R affects the release of aldosterone from the adrenal cortex, vasoconstriction, the activation of inflammatory processes, fibrosis, and myocardial hypertrophy [12,13]. In 2004, Silva et al. compared several angiotensin derivatives in children and adolescents with hypertension and normotension (3.1–16.7 ys). Similar to our study, the concentration of Ang II was almost identical in both groups (21.4 ± 8.7 vs. 22.2 ± 10.3 pg/mL [RIA]) [14]. Also, Al-Daghri et al., in 2010, compared Ang II concentrations in lean and obese children and adolescents (5–12 years). In this study, the researchers did not show a difference between the concentration of Ang II in lean and obese patients (0.70 ± 0.32 vs. 0.51 ± 0.13 boys and 0.65 ± 0.33 vs. 0.95 ± 1.0 girls) or between boys and girls (0.65 ± 0.3 vs. 0.73 ± 0.6) [15]. The results obtained in our study are compatible with previous studies. These results seem to suggest the existence of a local renin–angiotensin system. For many years, the RAAS has been described as a systemic system. This local system may influence blood pressure, while peripheral concentrations of angiotensin derivatives are at normal levels. The local renin–angiotensin system (RAS) is present in many tissues and organs such as muscle, the heart, the nervous system, bone, the kidneys, and the brain [9].

Angiotensin IV (Val-Tyr-Ile-His-Pro-Phe) has low biological activity. Ang IV has an affinity for the type IV angiotensin receptor (ATR4). The ATR4 receptor is widely distributed and is found in many tissues such as the brain, adrenal glands, kidneys, lungs, and heart. The ATR4 receptor is a transmembrane enzyme, insulin-regulated membrane aminopeptidase (IRAP). In the kidneys, Ang IV increases blood flow and decreases the transport of sodium ions to the proximal tubules in the kidney. However, the effect of arterial blood pressure on Ang IV is minimal. Ang IV may also have an influence on mental disorders. By regulating blood flow in the structure of the brain, Ang IV participates in neural plasticity, learning, and memory processes [16,17].

There are no studies that describe the concentration of Ang IV in the plasma of children and adolescents. Many studies described Ang IV as a very important brain blood regulatory peptide [18,19]. Animal studies, particularly rats, confirm the hypertensive effect of Ang IV [20]. However, due to the structure of the molecule and the activity of endopeptidases, the peripheral activity of Ang IV may be limited. However, the confirmation of this effect requires additional research, in particular on the local RAS. Ang IV and the local RAS play an important role in obesity and insulin resistance. In 2011, Wang et al. conducted studies in rats, where they showed that Ang IV causes an increase in glucose tolerance and insulin signaling [21]. The local RAS in adipose tissue has a significant impact on the development of obesity. The presence of Ang IV increases glucose uptake [22]. In our study, the concentration of Ang IV was significantly lower in the obesity and hypertension + obesity groups. This direction is in line with the expectations.

Angiotensin 1–7 (Asp-Arg-Val-Tyr-Ile-His-Pro) is formed directly from Ang II and, to a lesser extent, from different angiotensin derivatives. Ang 1–7 is the selective endogenous ligand for Mas receptors (MasRs). The MasR belongs to the G protein-coupled receptor family. The MasR is located in the endothelium of blood vessels, macrophages, and neurones. Ang 1–7 decreases blood pressure due to the vasodilation of blood vessels, the synthesis of anti-inflammatory prostaglandins, and the release of nitric oxide. Ang 1–7 also has a positive effect on the myocardium, limiting hypertrophy and cardiomyocyte proliferation. In the kidneys, Ang 1–7 regulates sodium ion transport and increases glucose resorption [23,24].

There are a few studies that describe the concentration of Ang 1–7 in children and adolescents. Silva et al., in the same study as Ang II, determined Ang 1–7 concentrations. Patients with essential hypertension have significantly higher Ang 1–7 concentrations than in the control group (78.8 ± 22.8 vs. 16.2 ± 7.9 [pg/mL] $p < 0.05$) [14]. Our results are in opposition to the work of Silva et al. because in our study, there are no differences between the hypertension and control groups, similar results to those received by Kohara et al. In their studies on hypertensive rats, they obtained a 3.7 times higher concentration of Ang 1–7 than in healthy rats ($p < 0.05$) [25]. Cambell et al. obtained different results. In their study, plasma concentrations of Ang 1–7 were similar in hypertensive and normotensive rats [26]. Large discrepancies in the test results may be the result of the in vivo and ex vivo degradation of the peptide. The authors of the studies used methods to prevent enzymatic degradation, such as cooling the sample and/or using inhibitors. The tissue RAS and related proteolytic enzymes are also of importance. They identify at least a few enzymes involved in Ang 1–7 degradation, such as angiotensin-converting enzyme 2 (ACE2), decarboxylase, aminopeptidase, or angiotensin-converting enzyme (ACE) [27]. From a clinical point of view, these differences are difficult to describe. Perhaps the increased concentration of the peptide is due to an attempt to achieve balance in the RAAS. Shifting the balance towards the alternative axis and lowering Ang II to Ang 1–7 could contribute to lower blood pressure.

Angiotensin 1–9 (Asp-Arg-Val-Tyr-Ile-His-Pro-Phe-His) has been found in many tissues and organs, such as the heart, kidneys, and testes. The highest concentrations have been observed in the endothelium of the coronary vessels. Ang 1–9 demonstrates a very similar effect to Ang 1–7. Ang 1–9 has a stronger effect on the myocardium than on the blood vessel. Angiotensin 1–9 acts through the angiotensin receptor type II (AT2R) [28].

No studies have been found that describe the concentration of Ang 1–9 in hypertensive children and adolescents. There are several studies in rats that confirm the hypotensive effect of Ang 1–9 [29–31]. Similarly to Ang IV, Ang 1–9 may play an important role in the local RAS. Due to the relatively high degradation of the peptide to Ang 1–7, it plays a secondary role. However, the reduced concentration in the group of obese patients suggests the influence of adipose tissue on the peripheral concentration of Ang 1–9 Perhaps ACE2-rich adipose tissue degrades large amounts of Ang 1–9 to Ang 1–7, shifting the balance of the RAAS toward cardioprotective and hypotensive effects. However, this requires further research.

5. Conclusions

The renin–angiotensin–aldosterone system is an extremely complex mechanism. The RAAS consists of many peptides and enzymes. In the system and among the measured peptides, Ang II and Ang 1–7 have the strongest biological activity. In this study, there are no differences in the concentrations of those peptides in the hypertension, hypertension and obesity, obesity, and control groups. However, differences are observed in the secondary peptides Ang IV and Ang 1–9. These peptides could be a predictor of AH in obese patients, but this hypothesis should be confirmed by prospective studies. This may mean that even minor disturbances in the RAAS of homeostasis lead to hypertension. On the other hand, the classical pathway of the RAAS system appears to be resistant to changes in the organism, and the alternative axis may be responsible for the development of hypertension. The disbalance found in this study is shown in Figure 5.

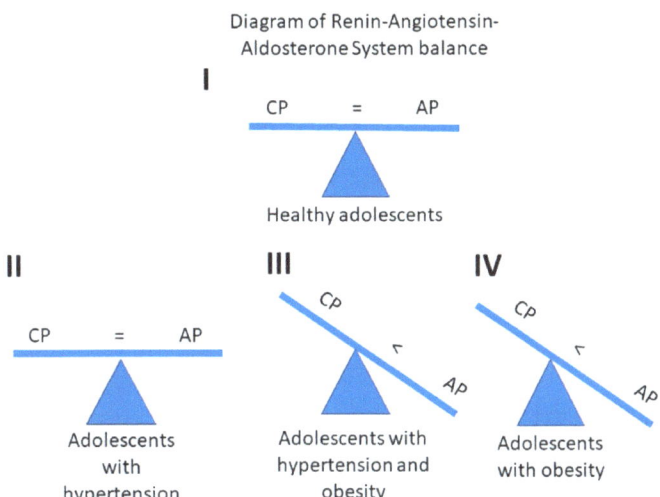

Figure 5. This diagram represents a theoretical balance between the classical pathway (CP) and alternative pathway (AP) of the renin–angiotensin–aldosterone system; (**I**) shows a fully balanced RAAS in healthy adolescents; (**II**) shows a theoretical balance in adolescents with hypertension, where the activity of the CP equals the activity of the AP; (**III**) shows a disbalance of the RAAS in obese adolescents with hypertension. The AP is more active than the CP; (**IV**) shows a disbalance of the RAAS in obese adolescents. The AP is more active than the CP.

A detailed understanding of the pathogenesis of hypertension and the impact of obesity requires further research. Only a small fraction of the RAAS was analyzed in our study. Among angiotensin derivatives, there are still many promising peptides that can be used in the diagnosis, monitoring, or even treatment of hypertension in children and adolescents. A very promising direction of development is research on Ang 1–7 and the MasR. Although Ang 1–7 is the only endogenous ligand for the MasR, its analogues are actively sought. The stimulation of the Mas receptor can bring many benefits to people with hypertension. In addition to vasodilation and NO synthesis, the MasR can induce the synthesis of anti-inflammatory prostaglandins. This is important in inflammatory processes and myocardial fibrosis. Therefore, the ACE2/Ang 1–7/MasR axis is a very promising development direction. Another conclusion of the study is the need to better understand local RAAs. Local RAAs perform very important functions in regulating blood flow and inflammatory processes. Unfortunately, its role in systemic action and clinical use is currently unknown. In the case of obese people, excessively developed adipose tissue is assumed to assume the role of the endocrine organ, which can affect the economy of the whole organism.

As shown in Figure 5, the balance of RAAS shifts to the alternative pathway in obese patients with hypertension and obese patients with normal blood pressure. Perhaps this is the result of compensatory mechanisms operating in the RAAS and/or that are associated with obesity. Maintained balance in patients with hypertension may indicate that compensation possibilities have been exhausted and other mechanisms are involved in the regulation of blood pressure.

6. Limitations of the Study

The main limitation of this study was the high biological and analytical variability of the analyzed compounds. Because the tested peptides were unstable in plasma, laboratory control over the pre-analytical phase and material storage were very important.

Author Contributions: Conceptualization, A.M.; methodology, A.M.; software, A.M.; validation, D.D. and P.J.T.; formal analysis, P.J.T.; investigation, A.M.; resources, P.J.T.; data curation, A.M.; writing—original draft preparation, A.M.; writing—review and editing, P.J.T.; visualization, A.M.; supervision, P.J.T. and D.D.; project administration, A.M.; funding acquisition, P.J.T. All authors have read and agreed to the published version of the manuscript.

Funding: This research was funded by Jagiellonian University Medical College, grant number N41/DBS/000470.

Institutional Review Board Statement: The study was conducted in accordance with the Declaration of Helsinki and approved by the Institutional Review Board (or Ethics Committee) of the Jagiellonian University Bioethical Committee (approval no. 1072.61.20.67.2019, 28 March 2019).

Informed Consent Statement: Informed consent was obtained from all subjects involved in the study.

Data Availability Statement: Data are contained within the article.

Conflicts of Interest: The authors declare no conflicts of interest.

References

1. Cardiovascular Diseases. Available online: https://www.who.int/health-topics/cardiovascular-diseases (accessed on 1 August 2023).
2. Heart Disease Facts | Cdc.Gov. Available online: https://www.cdc.gov/heartdisease/facts.htm (accessed on 1 August 2023).
3. Rao, G. Diagnosis, Epidemiology, and Management of Hypertension in Children. *Pediatrics* **2016**, *138*, e20153616. [CrossRef]
4. Litwin, M.; Kułaga, Z. Obesity, Metabolic Syndrome, and Primary Hypertension. *Pediatr. Nephrol.* **2021**, *36*, 825–837. [CrossRef]
5. Landsberg, L.; Aronne, L.J.; Beilin, L.J.; Burke, V.; Igel, L.I.; Lloyd-Jones, D.; Sowers, J. Obesity-Related Hypertension: Pathogenesis, Cardiovascular Risk, and Treatment. *J. Clin. Hypertens.* **2013**, *15*, 14–33. [CrossRef]
6. Patel, S.; Rauf, A.; Khan, H.; Abu-Izneid, T. Renin-Angiotensin-Aldosterone (RAAS): The Ubiquitous System for Homeostasis and Pathologies. *Biomed. Pharmacother.* **2017**, *94*, 317–325. [CrossRef]
7. Cruz-López, E.O.; Ye, D.; Wu, C.; Lu, H.S.; Uijl, E.; Mirabito Colafella, K.M.; Danser, A.H.J. Angiotensinogen Suppression: A New Tool to Treat Cardiovascular and Renal Disease. *Hypertension* **2022**, *79*, 2115–2126. [CrossRef] [PubMed]
8. Poznyak, A.V.; Bharadwaj, D.; Prasad, G.; Grechko, A.V.; Sazonova, M.A.; Orekhov, A.N. Renin-Angiotensin System in Pathogenesis of Atherosclerosis and Treatment of CVD. *Int. J. Mol. Sci.* **2021**, *22*, 6702. [CrossRef]
9. Vargas Vargas, R.A.; Varela Millán, J.M.; Fajardo Bonilla, E. Renin-Angiotensin System: Basic and Clinical Aspects—A General Perspective. *Endocrinol. Diabetes Nutr.* **2022**, *69*, 52–62. [CrossRef] [PubMed]
10. Martyniak, A.; Tomasik, P.J. A New Perspective on the Renin-Angiotensin System. *Diagnostics* **2022**, *13*, 16. [CrossRef] [PubMed]
11. Lurbe, E.; Agabiti-Rosei, E.; Cruickshank, J.K.; Dominiczak, A.; Erdine, S.; Hirth, A.; Invitti, C.; Litwin, M.; Mancia, G.; Pall, D.; et al. 2016 European Society of Hypertension Guidelines for the Management of High Blood Pressure in Children and Adolescents. *J. Hypertens.* **2016**, *34*, 1887–1920. [CrossRef]
12. Thomas, W.G.; Mendelsohn, F.A.O. Angiotensin Receptors: Form and Function and Distribution. *Int. J. Biochem. Cell Biol.* **2003**, *35*, 774–779. [CrossRef] [PubMed]
13. Benigni, A.; Cassis, P.; Remuzzi, G. Angiotensin II Revisited: New Roles in Inflammation, Immunology and Aging. *EMBO Mol. Med.* **2010**, *2*, 247–257. [CrossRef]
14. e Silva, A.C.S.; Diniz, J.S.S.; Regueira Filho, A.; Santos, R.A.S. The Renin Angiotensin System in Childhood Hypertension: Selective Increase of Angiotensin-(1–7) in Essential Hypertension. *J. Pediatr.* **2004**, *145*, 93–98. [CrossRef]
15. Al-Daghri, N.M.; Al-Attas, O.S.; Alokail, M.S.; Alkharfy, K.M.; Draz, H.M. Relationship between Resistin and APAI-1 Levels with Insulin Resistance in Saudi Children. *Pediatr. Int.* **2010**, *52*, 551–556. [CrossRef]
16. Chai, S.Y.; Fernando, R.; Peck, G.; Ye, S.-Y.; Mendelsohn, F.A.O.; Jenkins, T.A.; Albiston, A.L. The Angiotensin IV/AT4 Receptor. *Cell. Mol. Life Sci.* **2004**, *61*, 2728–2737. [CrossRef] [PubMed]
17. Wright, J.W.; Harding, J.W. The Brain Angiotensin System and Extracellular Matrix Molecules in Neural Plasticity, Learning, and Memory. *Prog. Neurobiol.* **2004**, *72*, 263–293. [CrossRef]
18. Molina-Van den Bosch, M.; Jacobs-Cachá, C.; Vergara, A.; Serón, D.; Soler, M.J. The renin-angiotensin system and the brain. *Hipertens. Riesgo Vasc.* **2021**, *38*, 125–132. [CrossRef] [PubMed]
19. Jackson, L.; Eldahshan, W.; Fagan, S.C.; Ergul, A. Within the Brain: The Renin Angiotensin System. *Int. J. Mol. Sci.* **2018**, *19*, 876. [CrossRef]
20. Yang, R.; Smolders, I.; De Bundel, D.; Fouyn, R.; Halberg, M.; Demaegdt, H.; Vanderheyden, P.; Dupont, A.G. Brain and Peripheral Angiotensin II Type 1 Receptors Mediate Renal Vasoconstrictor and Blood Pressure Responses to Angiotensin IV in the Rat. *J. Hypertens.* **2008**, *26*, 998–1007. [CrossRef]
21. Wong, Y.-C.; Sim, M.-K.; Lee, K.-O. Des-Aspartate-Angiotensin-I and Angiotensin IV Improve Glucose Tolerance and Insulin Signalling in Diet-Induced Hyperglycaemic Mice. *Biochem. Pharmacol.* **2011**, *82*, 1198–1208. [CrossRef] [PubMed]

22. Slamkova, M.; Zorad, S.; Krskova, K. Alternative Renin-Angiotensin System Pathways in Adipose Tissue and Their Role in the Pathogenesis of Obesity. *Endocr. Regul.* **2016**, *50*, 229–240. [CrossRef]
23. Olkowicz, M.; Chlopicki, S.; Smolenski, R.T. Perspectives for Angiotensin Profiling with Liquid Chromatography/Mass Spectrometry to Evaluate ACE/ACE2 Balance in Endothelial Dysfunction and Vascular Pathologies. *Pharmacol. Rep.* **2015**, *67*, 778–785. [CrossRef] [PubMed]
24. Bader, M.; Alenina, N.; Andrade-Navarro, M.A.; Santos, R.A. MAS and Its Related G Protein-Coupled Receptors, Mrgprs. *Pharmacol. Rev.* **2014**, *66*, 1080–1105. [CrossRef] [PubMed]
25. Kohara, K.; Bridget^Brosnihan, K.; Ferrario, C.M. Angiotensin(1–7) in the Spontaneously Hypertensive Rat. *Peptides* **1993**, *14*, 883–891. [CrossRef] [PubMed]
26. Campbell, D.J.; Duncan, A.-M.; Kladis, A.; Harrap, S.B. Angiotensin Peptides in Spontaneously Hypertensive and Normotensive Donryu Rats. *Hypertension* **1995**, *25*, 928–934. [CrossRef]
27. Santos, R.A.S.; Sampaio, W.O.; Alzamora, A.C.; Motta-Santos, D.; Alenina, N.; Bader, M.; Campagnole-Santos, M.J. The ACE2/Angiotensin-(1–7)/MAS Axis of the Renin-Angiotensin System: Focus on Angiotensin-(1–7). *Physiol. Rev.* **2018**, *98*, 505–553. [CrossRef]
28. Moraes, P.L.; Kangussu, L.M.; da Silva, L.G.; Castro, C.H.; Santos, R.A.S.; Ferreira, A.J. Cardiovascular Effects of Small Peptides of the Renin Angiotensin System. *Physiol. Rep.* **2017**, *5*, e13505. [CrossRef]
29. Paz Ocaranza, M.; Riquelme, J.A.; García, L.; Jalil, J.E.; Chiong, M.; Santos, R.A.S.; Lavandero, S. Counter-Regulatory Renin-Angiotensin System in Cardiovascular Disease. *Nat. Rev. Cardiol.* **2020**, *17*, 116–129. [CrossRef]
30. Ocaranza, M.P.; Moya, J.; Barrientos, V.; Alzamora, R.; Hevia, D.; Morales, C.; Pinto, M.; Escudero, N.; García, L.; Novoa, U.; et al. Angiotensin-(1–9) Reverses Experimental Hypertension and Cardiovascular Damage by Inhibition of the Angiotensin Converting Enzyme/Ang II Axis. *J. Hypertens.* **2014**, *32*, 771–783. [CrossRef]
31. Ali, Q.; Wu, Y.; Hussain, T. Chronic AT2 Receptor Activation Increases Renal ACE2 Activity, Attenuates AT1 Receptor Function and Blood Pressure in Obese Zucker Rats. *Kidney Int.* **2013**, *84*, 931–939. [CrossRef]

Disclaimer/Publisher's Note: The statements, opinions and data contained in all publications are solely those of the individual author(s) and contributor(s) and not of MDPI and/or the editor(s). MDPI and/or the editor(s) disclaim responsibility for any injury to people or property resulting from any ideas, methods, instructions or products referred to in the content.

Article

Enalapril Is Superior to Lisinopril in Improving Endothelial Function without a Difference in Blood–Pressure–Lowering Effects in Newly Diagnosed Hypertensives

Attila Nagy [1], Réka Májer [2,3], Judit Boczán [2], Sándor Sipka, Jr. [4], Attila Szabó [5], Enikő Edit Enyedi [5], Ottó Tatai [5], Miklós Fagyas [4,5], Zoltán Papp [5], László Csiba [2,3,†] and Attila Tóth [5,*,†]

[1] Department of Health Informatics, Institute of Health Sciences, Faculty of Health Sciences, University of Debrecen, 4032 Debrecen, Hungary; attilanagy@med.unideb.hu
[2] Department of Neurology, Faculty of Medicine, University of Debrecen, 4032 Debrecen, Hungary; majer.reka@med.unideb.hu (R.M.); boczan@med.unideb.hu (J.B.); csiba@med.unideb.hu (L.C.)
[3] MTA–DE Cerebrovascular and Neurodegenerative Research Group, 4032 Debrecen, Hungary
[4] Division of Cardiology, Department of Cardiology, Faculty of Medicine, University of Debrecen, 4032 Debrecen, Hungary; sandrosz@gmail.com (S.S.J.); fagyasmiklos@med.unideb.hu (M.F.)
[5] Division of Clinical Physiology, Department of Cardiology, Faculty of Medicine, University of Debrecen, 4032 Debrecen, Hungary; szabo.attila@med.unideb.hu (A.S.); enyedi.eniko@med.unideb.hu (E.E.E.); tataiotto2003@gmail.com (O.T.); pappz@med.unideb.hu (Z.P.)
* Correspondence: atitoth@med.unideb.hu; Tel.: +36-52-255-978
† These authors contributed equally to the study.

Abstract: Angiotensin–converting enzyme (ACE) inhibitors are the primarily chosen drugs to treat various cardiovascular diseases, such as hypertension. Although the most recent guidelines do not differentiate among the various ACE inhibitory drugs, there are substantial pharmacological differences. Goal: Here, we tested if lipophilicity affects the efficacy of ACE inhibitory drugs when used as the first therapy in newly identified hypertensives in a prospective study. Methods: We tested the differences in the cardiovascular efficacy of the hydrophilic lisinopril (8.3 ± 3.0 mg/day) and the lipophilic enalapril (5.5 ± 2.3 mg/day) ($n = 59$ patients). The cardiovascular parameters were determined using sonography (flow-mediated dilation (FMD) in the brachial artery, intima-media thickness of the carotid artery), 24 h ambulatory blood pressure monitoring (peripheral arterial blood pressure), and arteriography (aortic blood pressure, augmentation index, and pulse wave velocity) before and after the initiation of ACE inhibitor therapy. Results: Both enalapril and lisinopril decreased blood pressure. However, lisinopril failed to improve arterial endothelial function (lack of effects on FMD) when compared to enalapril. Enalapril-mediated improved arterial endothelial function (FMD) positively correlated with its blood–pressure–lowering effect. In contrast, there was no correlation between the decrease in systolic blood pressure and FMD in the case of lisinopril treatment. Conclusion: The blood–pressure–lowering effects of ACE inhibitor drugs are independent of their lipophilicity. In contrast, the effects of ACE inhibition on arterial endothelial function are associated with lipophilicity: the hydrophilic lisinopril was unable to improve, while the lipophilic enalapril significantly improved endothelial function. Moreover, the effects on blood pressure and endothelial function did not correlate in lisinopril-treated patients, suggesting divergent mechanisms in the regulation of blood pressure and endothelial function upon ACE inhibitory treatment.

Keywords: Angiotensin–converting enzyme (ACE); ACE inhibitor; lisinopril; enalapril; clinical study; endothelial function; hypertension; carotis IMT; FMD

1. Introduction

Angiotensin–converting enzyme inhibitors (ACEi) are the primarily chosen drugs to treat hypertension [1,2] and heart failure [3,4]. There are ten approved ACEi drugs, which are generally considered to have similar medical efficacies [5]. In particular, two widely used drugs, lisinopril and enalapril, had similarly positive effects on heart failure mortality [6] or hospitalization [7]. A uniform class effect was also noted in patients with myocardial infarction [8], congestive heart failure [9], and hypertension [10].

It is important to consider that circulating ACE seems to be completely inhibited by endogenous inhibitors [11,12] such as serum albumin [13], suggesting that only tissue-bound enzymes can be modulated by ACE inhibitory drugs [13]. This later was supported by the observation that physiological serum albumin completely inhibited human serum ACE, while only partial inhibition was observed in isolated human blood vessels [13]. These findings pinpoint vascular endothelium as the site of action for ACEi drugs. Therefore, the apparent class effect for ACEi drugs is rather surprising in light of marked differences in their pharmacokinetics [5], lipophilicities [14], and elimination pathways [14,15]. Here, we tested two drugs, originally developed by Merck [16]. Enalapril and its lysine analog have a similar efficacy in a wide range of clinical studies [17–19], albeit lisinopril was co-developed by Merck and Zeneca in the clinical phase. Both drugs are approved for the treatment of hypertension and congestive heart failure.

Here we performed a prospective clinical study to investigate the effects of two ACEi drugs with different lipophilicities. Lisinopril represented hydrophilic, low-protein binding ACEi [20], which was contrasted by enalapril, a lipophilic, relatively high-protein binding drug of the same class [14,21]. ACEi drugs were initiated at the first diagnosis of hypertension and the biochemical efficacy of the treatment was confirmed. The effects of lisinopril and enalapril were tested on blood pressure (24 h ambulatory blood pressure monitoring), arterial stiffness (arteriography), and endothelial function (flow-mediated dilation, brachial artery) before initiation and after the administration (at least 30 days) of the ACEi drugs.

2. Materials and Methods

2.1. Patients

Our research took place at the Department of Neurology, Clinical Center of the University of Debrecen. Based on the study protocol, we included patients presenting at their primary care provider with newly diagnosed primary hypertension (ICD code I10H0)—GPs and occupational health physicians aided the patient enrollment process. Asymptomatic and untreated patients whose hypertension was confirmed by ABPM and who had not yet received antihypertensive treatment were included. All subjects were asymptomatic, predominantly middle-aged (active age), as identified by screening. Following the ABPM, a CT scan was also performed to detect asymptomatic abnormalities (e.g., silent brain infarction).

The exclusion criteria included extreme obesity (body mass index—BMI greater than 35 kb/m^2), previous stroke, TIA, a poor general condition, a life expectancy of fewer than 5 years, and co-morbidities that may significantly affect the hypertensive patients: diabetes, severe heart disease, psychiatric disorders (including alcohol dependence), dementia, Parkinson's disease, neuromuscular disorders, autonomic nervous system syndromes, inflammatory diseases, stroke, and TIA. A "silent" infarction or other organic abnormalities detected on cranial CT also resulted in exclusion. Pregnant or post-partum candidates were also not recruited for the investigation.

2.2. Methods

Medical history, demographic variables, the results of a physical investigation, and laboratory tests (serum electrolytes, renal function, blood glucose level, HbA1C, lipid profile, complete blood count, CRP, fibrinogen level, and urine analysis) were recorded at the initiation of the treatment.

These assessments were followed by 24 h ambulatory blood pressure monitoring (ABPM). Blood pressure was measured every 15 min during the daytime (from 6:00 to 22:00) and every 30 min at night (from 22:00 to 6:00). Based on the data, we determined the daytime and nighttime mean systolic and diastolic pressures and systolic and diastolic hyperbaric index. For the ABMP measurements, Cardiospy ABPM equipment from Labtech Ltd. (Debrecen, Hungary, Model: EC-ABP) was used.

A flow-mediated dilatation (FMD) measurement of the brachial artery was performed using a HP Sonos 5500 ultrasound device with a 10 MHz linear test transducer (National Utrasound, Tampa, FL, USA). A B-mode longitudinal section was obtained from the brachial artery above the antecubital fossa. A forearm cuff was inflated to 10–40 mmHg above the patient's systolic pressure for 5 min. Upon the cuff release, the induced hyperemia promoted an increase in the shear stress-mediated NO release and subsequent vasodilatation. The FMD was expressed as the percentage increase in the resting diameter of the artery after the cuff release with the baseline arterial diameter as a reference.

Arterial stiffness measurements were performed using TensioClinic arteriography (TensioMed Ltd., Budapest, Hungary) [22]. This technique is based on the fact that the contraction of the heart initiates pulse waves in the aorta. The first wave becomes reflected from the aortic wall at the bifurcation, therefore a second reflected wave appears as a late systolic peak. The cuff detects both waves. The morphology of this second reflected wave depends on the stiffness of the large artery, the reflection time at 35 mmHg supra systolic pressure of the brachial artery, and the peripheral resistance-dependent amplitude. Arterial stiffness was assessed by determining the Augmentation Index (AIx) and the Pulse Wave Velocity (PWV). The AIx was calculated from the amplitudes of the first and second waves and represents the pressure difference between the late systolic peak and the early systolic peak divided by the pulse pressure. The PWV is the ratio of the jugular fossa-symphysis distance (which is anatomically identical to the distance between the aortic trunk and the bifurcation) to the reflection time at 35 mmHg supra systolic pressure on the brachial artery. Brachial artery FMD is a technique for estimating the endothelial function in large arteries [23].

The data were evaluated using pairwise and correlation analyses. Assuming a normal distribution was generally avoided in the statistical analysis. One reason for this is that some sets of data did not show a normal distribution. The second is that we wanted to avoid false positive correlations caused by outliers in the datasets with relatively small observation numbers. The Kruskal–Wallis test was used for parameters from two populations (such as those treated with enalapril or lisinopril), while the Wilcoxon test was performed when the parameters before and after (paired) were evaluated. Spearman's correlation analysis was performed when correlations were addressed.

3. Results

Individuals with primary hypertension were recruited immediately after their diagnosis. Two groups were formed according to the initiated medical therapy. One group was treated with lisinopril ($n = 31$), while the other was treated with enalapril ($n = 28$). The general clinical parameters are summarized in Table 1. Unfortunately, many patients did not volunteer for the multiple-day follow-up study.

The effects of the ACEi medications were tested by measuring the flow-mediated dilation (FMD) in the brachial artery after prolonged cuff use. The treatment of patients with enalapril improved the endothelial-mediated dilation (an increase of FMD from 6.7 ± 0.6 to $8.8 \pm 0.8\%$, mean $\pm$ SEM, $n = 17$, Figure 1A), while the treatment with lisinopril was without effects (FMD before 7.5 ± 0.7 vs. after $7.7 \pm 0.6\%$ lisinopril treatment, mean $\pm$ SEM, $n = 16$, Figure 1A). The intima-media thickness of the carotid artery was unaffected by both enalapril (intima-media thickness before 0.55 ± 0.02 vs. after 0.57 ± 0.02 mm, mean $\pm$ SEM, $n = 24$, Figure 1B) and lisinopril (intima-media thickness before 0.60 ± 0.02 vs. after 0.59 ± 0.02 mm, mean $\pm$ SEM, $n = 30$, Figure 1B). Vascular stiffness was also tested with functional measurements. No significant effects were noted on the augmentation index (enalapril before: 20.1 ± 6.8, after: 23.9 ± 5.9, mean $\pm$ SEM, $n = 28$; lisinopril before: 8.2 ± 5.5, after: 12.8 ± 5.8, mean $\pm$ SEM, $n = 31$; Figure 1C) or pulse wave velocity determined using arteriography (enalapril before: 9.4 ± 0.3, after: 9.1 ± 0.3, mean $\pm$ SEM, $n = 28$; lisinopril before: 10.3 ± 0.4, after: 10.4 ± 0.5, mean $\pm$ SEM, $n = 31$; Figure 1C).

Table 1. Medical characteristics of study populations.

Medical Drug	Enalapril		Lisinopril	
Visit	First Visit	Follow Up	First Visit	Follow Up
Involved patients (n)	43	28	44	30
Age (years)	45.7 ± 11.1	N/A	45.1 ± 11.3	N/A
BMI (kg/m^2)	28.9 ± 5.6	N/A	28.3 ± 3.8	N/A
Dose (mg/day)	0	5.5 ± 2.3	0	8.3 ± 3.0
Na$^+$ (mM)	140.6 ± 2.0	140.0 ± 2.5	140.3 ± 2.1	139.7 ± 2.5
K$^+$ (mM)	4.3 ± 0.3	4.3 ± 0.3	4.2 ± 0.3	4.4 ± 0.4
Glucose (mM)	5.5 ± 1.0	5.2 ± 0.6	5.2 ± 1.3	5.2 ± 1.3
HbA1C (%)	5.3 ± 0.5	5.2 ± 0.3	5.4 ± 0.8	5.4 ± 0.8
Urea (mmol/L)	5.2 ± 1.4	5.4 ± 1.4	4.8 ± 1.2	5.0 ± 1.4
Creatinine (mg/dL)	75 ± 15	74 ± 15	78 ± 17	79 ± 16
Trigliceride (mmol/L)	1.5 ± 0.9	1.6 ± 1.5	2.1 ± 3.0	1.5 ± 0.8
Cholesterol (mmol/L)	5.1 ± 0.9	5.1 ± 0.9	5.2 ± 1.1	5.3 ± 1.1
GOT (IU/L)	25.2 ± 11.8	26.4 ± 9.6	22.8 ± 7.5	23.1 ± 7.2
Systolic blood pressure (active period, mmHg, SD)	147 ± 11	138 ± 11	148 ± 11	139 ± 8
Diastolic blood pressure (active period, mmHg, SD)	90 ± 7	83 ± 7	90 ± 9	84 ± 6
Systolic blood pressure (night period, mmHg, SD)	128 ± 13	125 ± 14	132 ± 23	118 ± 27
Diastolic blood pressure (night period, mmHg, SD)	76 ± 8	75 ± 8	80 ± 10	70 ± 17
FMD (%, SD)	7.4 ± 3.1	8.8 ± 3.4	8.7 ± 4.2	7.8 ± 2.4
IMT (mm, SD)	0.6 ± 0.1	0.6 ± 0.1	0.6 ± 0.1	0.6 ± 0.1

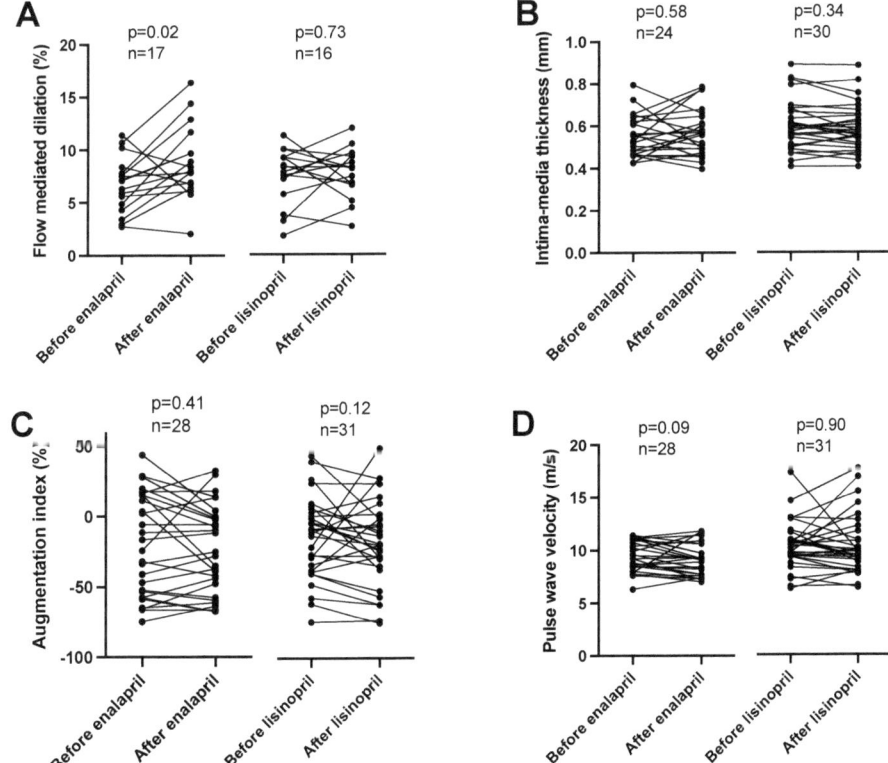

Figure 1. Different effects of enalapril and lisinopril on arterial endothelial function in hypertensive patients. Patients were tested for hypertension in the outpatient facility of the Department of Neurology.

Newly identified hypertensive patients were enrolled in the study. Vascular parameters, such as flow-mediated dilation (panel (**A**)), intima-media thickness (panel (**B**)), augmentation index (panel (**C**)), and pulse wave velocity (panel (**D**)) were determined. Each symbol represents an individual patient. Lines connect values before and after enalapril or lisinopril treatment. Significant differences between values determined before and after the initiation of ACEi medication are labeled by the *p* values. Statistical differences were calculated by the Wilcoxon test.

Aortic pulse pressure was similarly affected by both ACE inhibitors (enalapril before: 62 ± 3, after: 52 ± 2 mmHg, mean $\pm$ SEM, $n = 28$; lisinopril before: 61 ± 2, after: 54 ± 3 mmHg, mean $\pm$ SEM, $n = 30$; Figure 2A), being lowered by both lisinopril and enalapril (Figure 2B). Peripheral pulse pressure values were also similar among the groups (enalapril before: 58 ± 2, after: 54 ± 2 mmHg, mean $\pm$ SEM, $n = 21$; lisinopril before: 56 ± 1, after: 54 ± 1 mmHg, mean $\pm$ SEM, $n = 23$; Figure 2C). However, decreases at the level of the individual patients were only significant in the enalapril-treated patients (Figure 2D).

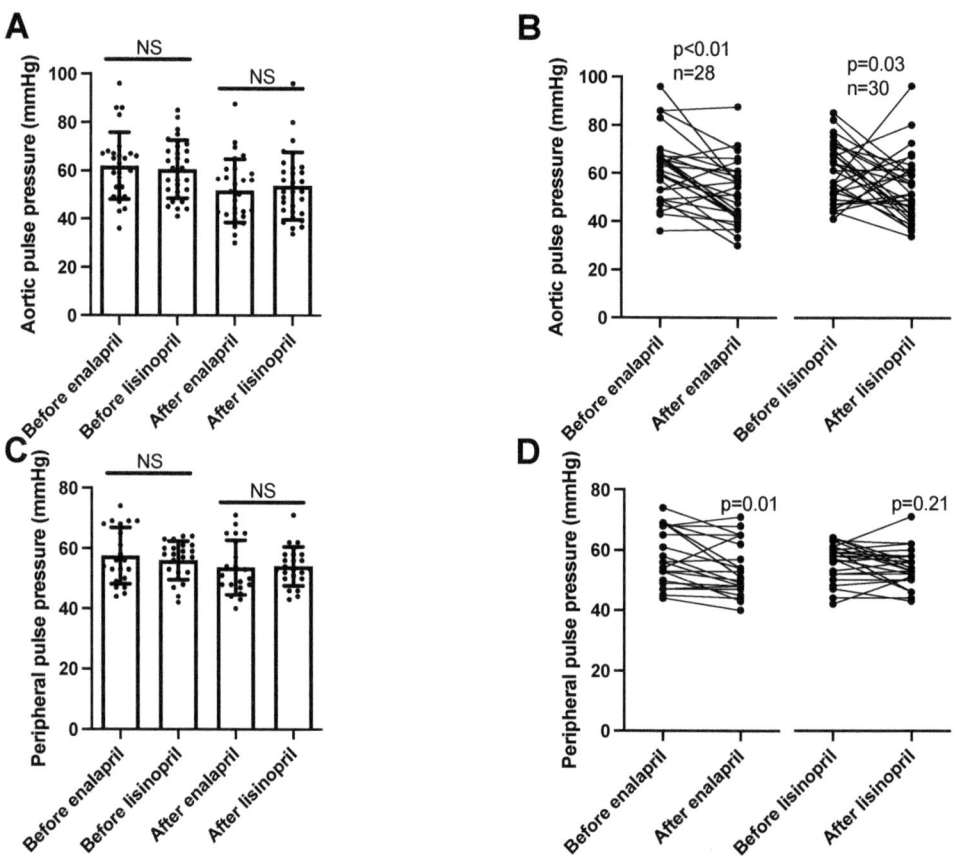

Figure 2. Similar effects of enalapril and lisinopril on pulse pressure in hypertensive patients. Newly identified hypertensive patients were enrolled in the study at the outpatient facility of the Department of Neurology. Systolic and diastolic blood pressure values were determined using 24 h ambulatory blood pressure monitoring (peripheral blood pressure) and arteriography (aortic blood pressure). The difference between the systolic and diastolic blood pressure values was calculated to yield pulse pressure. Each symbol represents an individual patient. Bars represent the mean and SD for aortic (panel (**A**)) and

peripheral (panel (**C**)) pulse pressure values. No significant differences were found among the groups (indicated by NS) by Kruskal–Wallis tests. Lines connect values before and after enalapril or lisinopril treatment in the case of aortic (panel (**B**)) and peripheral (panel (**D**)) pulse pressure values. Significant differences between values determined before and after the initiation of ACEi medication are labeled by the *p* values. Statistical differences were calculated by the Wilcoxon test.

Aortic systolic blood pressure (determined by the arteriography) levels were similar among the groups (enalapril before: 148 ± 3, after: 138 ± 2 mmHg, mean $\pm$ SEM, $n = 21$; lisinopril before: 149 ± 2, after: 138 ± 1 mmHg, mean $\pm$ SEM, $n = 26$; Figure 3A) and were similarly reduced by both ACEi drugs (Figure 3B). Peripheral systolic blood pressure values were measured using 24 h ambulatory blood pressure monitoring (ABPM). There were no significant differences among the patient groups (enalapril before: 146 ± 4, after: 136 ± 3 mmHg, mean $\pm$ SEM, $n = 28$; lisinopril before: 153 ± 4, after: 134 ± 4 mmHg, (panel (**A**))mean $\pm$ SEM, $n = 28$; Figure 3C), while both ACEi treatments uniformly resulted in significant reductions in systolic blood pressure values (Figure 3D).

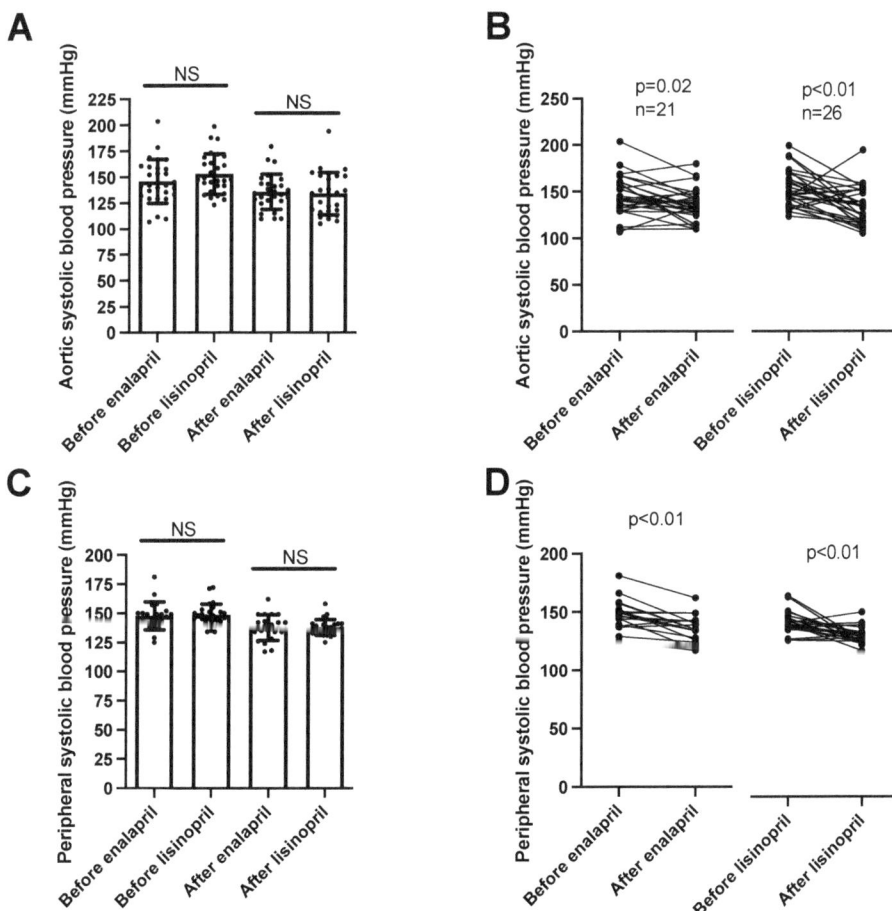

Figure 3. Same anti-hypertensive effects of enalapril and lisinopril on systolic blood pressure in hypertensive patients. Newly identified hypertensive patients were enrolled in the study at the outpatient

facility of the Department of Neurology. Systolic blood pressure values were determined using 24 h ambulatory blood pressure monitoring (peripheral blood pressure) and arteriography (aortic blood pressure). Each symbol represents an individual patient. Bars represent the mean and SD for aortic (panel (**A**)) and peripheral (panel (**C**)) systolic blood pressure values. No significant differences were found among the groups (indicated by NS) by Kruskal–Wallis tests. Lines connect values before and after enalapril or lisinopril treatment in the case of aortic (panel (**B**)) and peripheral (panel (**D**)) systolic blood pressure values. Significant differences between values determined before and after the initiation of ACEi medication are labeled by the *p* values. Statistical differences were calculated by the Wilcoxon test.

Diastolic pressure values mimicked the systolic ones. There were no differences among the patient groups in aortic (enalapril before: 90 ± 2, after: 85 ± 1 mmHg, mean ± SEM, $n = 27$; lisinopril before: 94 ± 2, after: 84 ± 2 mmHg, mean ± SEM, $n = 27$; Figure 4A) or peripheral diastolic blood pressure values (enalapril before: 90 ± 2, after: 83 ± 1 mmHg, mean ± SEM, $n = 20$; lisinopril before: 89 ± 1, after: 84 ± 1 mmHg, mean ± SEM, $n = 20$; Figure 4C). Both ACE inhibitory drugs reduced the aortic (Figure 4B) and peripheral (Figure 4D) blood pressure values.

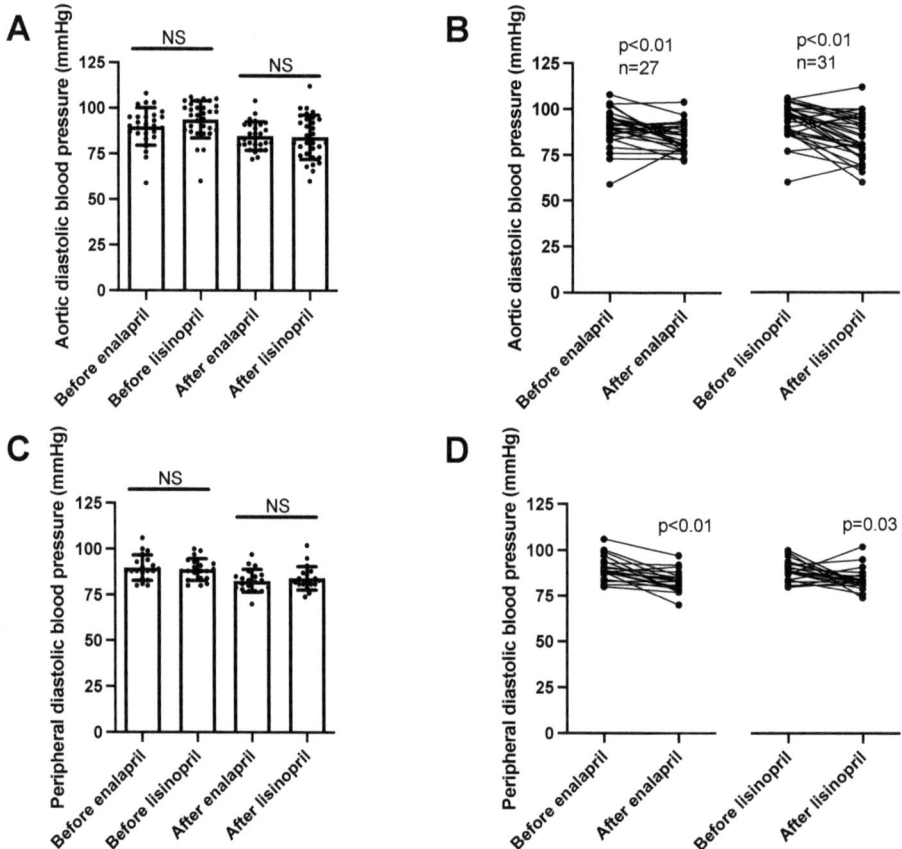

Figure 4. Same anti-hypertensive effects of enalapril and lisinopril on systolic blood pressure in hypertensive patients. Newly identified hypertensive patients were enrolled in the study at the outpatient

facility of the Department of Neurology. Diastolic blood pressure values were determined using 24 h ambulatory blood pressure monitoring (peripheral blood pressure) and arteriography (aortic blood pressure). Each symbol represents an individual patient. Bars represent the mean and SD for aortic (panel (**A**)) and peripheral (panel (**C**)) diastolic blood pressure values. No significant differences were found among the groups (indicated by NS) by Kruskal–Wallis tests. Lines connect values before and after enalapril or lisinopril treatment in the case of aortic (panel (**B**)) and peripheral (panel (**D**)) diastolic blood pressure values. Significant differences between values determined before and after the initiation of ACEi medication are labeled by the *p* values. Statistical differences were calculated by the Wilcoxon test.

Finally, the reduction in systolic blood pressure values was correlated with the improvement (increase) in flow-mediated dilation values in the case of enalapril. A strong ($rho = 0.72$) and significant ($p = 0.016$) correlation was established between the decrease in systolic blood pressure and increase in flow-mediated dilation when treated with enalapril (Figure 5A). In contrast, no significant correlation ($p = 0.23$) was observed in patients treated with lisinopril (Figure 5B).

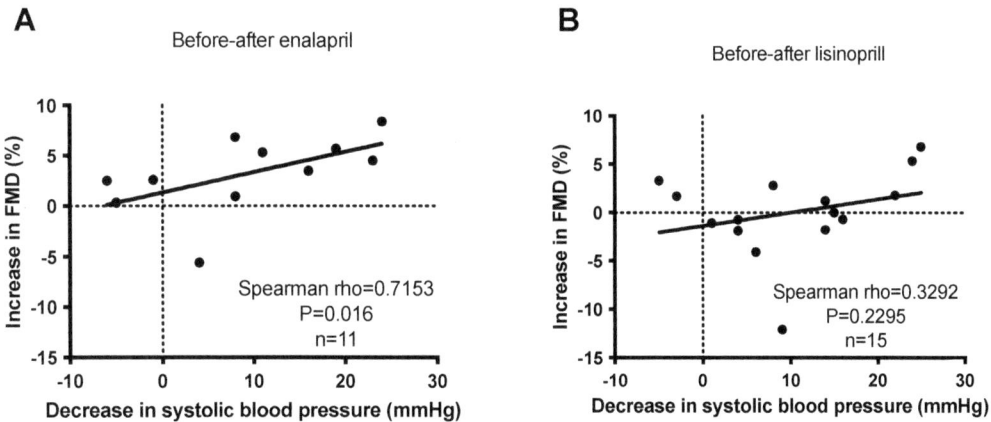

Figure 5. No correlation between anti-hypertensive effects and endothelial function improvement (FMD) upon short-term ACEi treatment in hypertensive patients. Newly identified hypertensive patients were enrolled in the study at the outpatient facility of the Department of Neurology. Increases in flow-mediated dilation (FMD) were plotted as the function of the decrease in systolic blood pressure before and after enalapril (panel (**A**)) and lisinopril (panel (**B**)) treatments. Each symbol represents an individual patient. The correlation between the parameters was tested by Spearman's test. Correlations were characterized statistically by the rho and *p* values. Correlation is considered to be significant when $p < 0.05$.

4. Discussion

ACEi drugs represent one of the most frequently prescribed medications, being the primarily chosen drugs in chronic diseases affecting large populations, like hypertension [1,2] and heart failure [3,4]. ACEi prescriptions increase from 11.4% (40–59 years) to 21.3% (60–79 years) upon aging in the population [24]. In spite of the clinical success of ACE inhibition, the mechanism of action is still not fully understood [25,26]. An important example of this is the site of action of ACEi medical drugs. Obviously, the basis of the beneficial effects of ACEi drugs is the inhibition of ACE activity. However, recently, it was shown that physiological ACE activity is regulated by endogenous inhibitors in the circulation (blood) [11], as well as in cardiac and lung tissues [27]. Serum albumin was identified as a major endogenous inhibitor in the human blood, completely inhibiting circulating ACE activity under physiological conditions [13]. This suggested that ACEi drugs cannot act on the circulating enzyme, since ACE activity is negligible under physiological serum albumin concentrations. Indeed, the serum albumin sensitivity of the vascular tissue-bound form of ACE was lower, suggesting that tissular ACE may be the target of ACEi drugs [13]. In line with these, we hypothesized that ACEi drugs with different lipophilicities and affinities for carrier proteins have different physiological effects. To investigate this hypothesis, we tested the clinical effects of the most hydrophilic ACEi drug lisinopril (with low binding to carrier proteins) and contrasted it to the lipophilic ACEi drug enalapril.

A strength of this study was that we recruited patients upon a diagnosis of hypertension. Basal clinical parameters were recorded upon the initiation of ACEi treatment and after the administration of the drugs for at least 30 days. Accordingly, we prospectively recruited our patient population, in which the effects of the ACEi drugs could be tested before and after treatment in the same individuals. This setup is particularly useful in addressing delicate processes, such as vascular responsiveness. Endothelial function was tested by measuring the flow-mediated dilation in the brachial artery (FMD) [28], while arterial stiffness was tested using arteriography [29]. The major finding of this study is that, while enalapril dramatically improved endothelial function, the hydrophilic lisinopril was without effects. It was reported that blood pressure control correlates with an improvement in endothelial function [30]. Some of our results confirmed this observation by showing a strong correlation between the blood-pressure-lowering effect of enalapril and a parallel improvement in endothelial function, as measured by FMD [31]. However, we also found that the blood-pressure-lowering effect is independent of an improvement in FMD in the case of the hydrophilic ACEi inhibitor lisinopril. These findings suggest that an improvement in endothelial function is not uniquely mediated by the blood-pressure-lowering effects of ACE inhibition. There appears to be selective targeting of the tissular ACE population linked to endothelial dysfunction by the lipophilic enalapril.

This is in accordance with findings correlating vascular ACE inhibition (measured by changes in angiotensin peptide levels) [32] or vascular ACE expression (determined by genetic polymorphism) [33] with endothelial function. Moreover, the inhibition of ACE (and type 1 angiotensin 2 receptor) is associated with better endothelial function than that in patients treated with calcium channels or beta receptor blockers [31,34]. Others have also reported an improvement in endothelial function in patients treated with enalapril [35]. They hypothesized that the over-activation of vascular ACE may contribute to the increased production of superoxide, which neutralizes the endothelial NO, antagonizing endothelium-dependent dilations [35]. ACE may also interfere with inflammation [36], affecting endothelial functions (probably by regulating reactive oxygen radical formation). Finally, it should be noted that, quinapril, an exceptionally lipophilic ACEi, had a higher efficacy than enalapril in improving endothelial function [36,37]. Taken together, the reversal of endothelial function by ACEi drugs may depend on the tissue affinity (lipophilicity).

Our results suggested that short-term ACEi treatment does not affect the vascular structure, as there was no difference in carotid artery intima-media thickness before and after the ACEi drug treatment. Similar findings were noted for arterial stiffness, as the ACEi treatment was without effects on the augmentation index and pulse wave velocity determined using arteriography. In general, it can be concluded that the initiation of ACE inhibitor therapy resulted in highly significant changes in functional parameters (such as blood pressure or endothelial function), while it did not affect structural parameters significantly at this time point. One would speculate that functional improvements would be followed by structural remodeling during treatment in the long term when all drugs are up titrated [5].

We noted a significant blood-pressure-lowering effect of both enalapril and lisinopril. Both ACEi drug treatments resulted in a robust decrease in both peripheral (24 h ambulatory blood pressure monitoring) and aortic blood pressure (measured using arteriography). Both systolic and diastolic blood pressure values were decreased by both ACEi drugs, resulting in a similar decline in both the aortic and peripheral blood pressure values. Nonetheless, there was some difference in the decrease in pulse pressure (difference in systolic and diastolic blood pressure values). While both ACEi drugs reduced aortic pulse pressure, only enalapril was able to significantly reduce peripheral pulse pressure. These apparent differences between enalapril and lisinopril may also reflect the different levels of tissular ACE inhibition, which may be attributed to the more lipophilic nature of enalapril.

Our results showing a robust anti-hypertensive effect for both enalapril and lisinopril are in accordance with previous reports, suggesting a uniform class effect for ACEi drugs [5,10,30,38]. This is important in light of the differences noted in vascular responsiveness. Both ACEi drugs had robust and similar anti-hypertensive effects. The anti-hypertensive effects of ACE inhibition were attributed to various mechanisms, including improved endothelial function, increased bradykinin levels, suppressed vascular smooth muscle contraction, and decreased volume overload, mediated by facilitated natriuresis [39]. Our results showed that the anti-hypertensive effects of ACEi drugs are not primarily mediated by improving endothelial function or arterial stiffness. This suggests that the modulation of kidney function (increased natriuresis and consequent reduction in volume overload) is the primary mechanism behind the anti-hypertensive effects of ACEi.

ACEIs can easily cause side effects such as dry cough, angioedema, rash, changes in taste, increased blood potassium, and decreased kidney function [5]. Some may note that the results of this paper do not seem to mention these phenomena. These side effects most probably did not have time to develop, since the study represents the initial examination (freshly identified untreated hypertensives) and their first visit after the prescription of low-dose ACE inhibitors (randomly, enalapril and lisinopril). It is most likely that these side effects would develop during the up-titration and long-term use of ACE inhibitory drugs in these patients.

In summary, here we showed that, while enalapril and lisinopril have similar anti-hypertensive effects, they have different effects on endothelial function. In particular, the lipophilic enalapril improved endothelial function, while the hydrophilic lisinopril was without effects. These findings suggest that the blood-pressure-lowering class effect of ACEi drugs is mediated by improvements in kidney function and decreased volume overload. On the other hand, ACEi drugs do not have a class effect on improving endothelial functions (either mediated by reduced angiotensin 2 levels or by increased bradykinin levels).

A strength of this report over previous investigations is that we verified the ACEi treatment efficacy using objective biochemical enzyme activity measurements. Another advantage is that we performed a prospective study to identify differences among ACEi drugs with different pharmacological properties. Note that the study was randomized in a way that patients were recruited consecutively (odd numbers selected for one drug even numbers for the other), although patients and doctors were not blinded. This randomization is very important, since otherwise, one may have a potential problem of confounding by indication. When a drug appears to be associated with an outcome, the outcome may, in fact, be caused by the indication for which the drug was used, or some factor associated with the indication. This can only be avoided by randomization.

Having said that, all clinical studies have limitations. Here, we investigated a patient population in a single-center study, involving a homogenous (Caucasian) population. Accordingly, our results cannot be extrapolated to all human populations. Moreover, the study consisted of patients first identified with hypertension and patients in the early stages of their anti-hypertensive treatment. The relevance of this is that the treatment regimen started with a low dose of ACE inhibitors, and there was no up-titration in the upcoming visits (since the study represents the initial visit and the first upcoming visit). Note that initial low-dose ACE inhibitors usually provide the full anti-hypertensive effects, while the up-titration of the drug is important to improve long-term efficacy and surrogate endpoints, such as in heart failure. We encountered a significant dropout rate in our study, resulting in a relatively small patient population in certain instances, thereby impacting its statistical power. Future clinical investigations could address these limitations and build on our methodological framework.

5. Conclusions

These data suggest that hydrophilicity defines the site of action of ACE inhibitors. Lipophilic inhibitors improve endothelial function, while hydrophilic ones do not. The blood pressure-lowering effect of ACE inhibitors is independent of hydrophilicity. The blood pressure lowering is independent of improvement in endothelial function.

Author Contributions: Conceptualization, J.B., M.F., Z.P., L.C. and A.T.; Methodology, R.M., A.S., E.E.E., O.T., M.F. and A.T.; Validation, A.S. and E.E.E.; Formal analysis, A.N., R.M. and A.T.; Investigation, R.M., M.F., L.C. and A.T.; Resources, M.F. and L.C.; Data curation, A.N.; Writing—original draft, S.S.J., A.S., L.C. and A.T.; Writing—review & editing, S.S.J., M.F., L.C. and A.T.; Visualization, A.N., J.B. and A.T.; Supervision, Z.P. and L.C.; Funding acquisition, Z.P., L.C. and A.T. All authors have read and agreed to the published version of the manuscript.

Funding: This study was funded by the Hungarian Academy of Sciences (POST-COVID2021-16, POST-COVID2021-33, NKM2022-30) and grants from the National Research, Development, and Innovation Office (K132623 to AT, K147243 to M.F.). Project nos. TKP2021-EGA-19 and TKP2021-EGA-20 have been implemented with the support provided by the National Research, Development, and Innovation Fund of Hungary, financed under the TKP2021-EGA funding scheme. The Thematic Excellence Programme of the Ministry for Innovation and Technology was also supported by the National Research, Development and Innovation Fund of Hungary (TKP2020-NKA-04) within the frameworks of the preclinical thematic program of the University of Debrecen. MF was supported by the János Bolyai Research Scholarship of the Hungarian Academy of Sciences (BO/00069/21/5) and by the ÚNKP-23-5-DE-482 New National Excellence Program of the Ministry for Innovation and Technology from the source of the National Research, Development and Innovation Fund.

Institutional Review Board Statement: The study was conducted in accordance with the Declaration of Helsinki, and approved by the Institutional Ethics Committee of the University of Debrecen (registration number: 4700168520, file number: KK/186/2016).

Informed Consent Statement: Informed consent was obtained from all subjects involved in the study.

Data Availability Statement: Data are contained within the article.

Conflicts of Interest: The authors declare no conflict of interest.

Abbreviations

ABPM	Ambulatory blood pressure monitoring
ACE	Angiotensin-converting enzyme
ACEi	Angiotensin-converting enzyme inhibitors
AIx	Augmentation Index
FMD	Flow-mediated dilation
PWV	Pulse Wave Velocity

References

1. Whelton, P.K.; Carey, R.M.; Aronow, W.S.; Casey, D.E.; Collins, K.J.; Dennison Himmelfarb, C.; DePalma, S.M.; Gidding, S.; Jamerson, K.A.; Jones, D.W.; et al. 2017 ACC/AHA/AAPA/ABC/ACPM/AGS/APhA/ASH/ASPC/NMA/PCNA Guideline for the Prevention, Detection, Evaluation, and Management of High Blood Pressure in Adults. *J. Am. Coll. Cardiol.* **2018**, *71*, e127–e248. [CrossRef] [PubMed]
2. Williams, B.; Mancia, G.; Spiering, W.; Agabiti Rosei, E.; Azizi, M.; Burnier, M.; Clement, D.L.; Coca, A.; de Simone, G.; Dominiczak, A.; et al. 2018 ESC/ESH Guidelines for the management of arterial hypertension. *Eur. Heart J.* **2018**, *39*, 3021–3104. [CrossRef] [PubMed]
3. McDonagh, T.A.; Metra, M.; Adamo, M.; Gardner, R.S.; Baumbach, A.; Böhm, M.; Burri, H.; Butler, J.; Čelutkienė, J.; Chioncel, O.; et al. 2021 ESC Guidelines for the diagnosis and treatment of acute and chronic heart failure. *Eur. Heart J.* **2021**, *42*, 3599–3726. [CrossRef] [PubMed]
4. Yancy, C.W.; Jessup, M.; Bozkurt, B.; Butler, J.; Casey, D.E.; Colvin, M.M.; Drazner, M.H.; Filippatos, G.S.; Fonarow, G.C.; Givertz, M.M.; et al. 2017 ACC/AHA/HFSA Focused Update of the 2013 ACCF/AHA Guideline for the Management of Heart Failure. *J. Card. Fail.* **2017**, *23*, 628–651. [CrossRef] [PubMed]
5. Brown, N.J.; Vaughan, D.E. Angiotensin-Converting Enzyme Inhibitors. *Circulation* **1998**, *97*, 1411–1420. [CrossRef] [PubMed]

6. Fröhlich, H.; Henning, F.; Täger, T.; Schellberg, D.; Grundtvig, M.; Goode, K.; Corletto, A.; Kazmi, S.; Hole, T.; Katus, H.A.; et al. Comparative effectiveness of enalapril, lisinopril, and ramipril in the treatment of patients with chronic heart failure: A propensity score-matched cohort study. *Eur. Heart J. Cardiovasc. Pharmacother.* **2018**, *4*, 82–92. [CrossRef] [PubMed]
7. Chitnis, A.S.; Aparasu, R.R.; Chen, H.; Johnson, M.L. Comparative effectiveness of different angiotensin-converting enzyme inhibitors on the risk of hospitalization in patients with heart failure. *J. Comp. Eff. Res.* **2012**, *1*, 195–206. [CrossRef] [PubMed]
8. Tu, K.; Gunraj, N.; Mamdani, M. Is Ramipril Really Better Than Other Angiotensin-Converting Enzyme Inhibitors After Acute Myocardial Infarction? *Am. J. Cardiol.* **2006**, *98*, 6–9. [CrossRef]
9. Tu, K.; Mamdani, M.; Kopp, A.; Lee, D. Comparison of angiotensin-converting enzyme inhibitors in the treatment of congestive heart failure. *Am. J. Cardiol.* **2005**, *95*, 283–286. [CrossRef]
10. Coca, A.; Sobrino, J.; Módol, J.; Soler, J.; Mínguez, A.; Plana, J.; De la Sierra, A. A multicenter, parallel comparative study of the antihypertensive efficacy of once-daily lisinopril vs enalapril with 24-h ambulatory blood pressure monitoring in essential hypertension. *J. Hum. Hypertens.* **1996**, *10*, 837–841. [CrossRef]
11. Fagyas, M.; Úri, K.; Siket, I.M.; Daragó, A.; Boczán, J.; Bányai, E.; Édes, I.; Papp, Z.; Tóth, A. New Perspectives in the Renin-Angiotensin-Aldosterone System (RAAS) I: Endogenous Angiotensin Converting Enzyme (ACE) Inhibition. *PLoS ONE* **2014**, *9*, e87843. [CrossRef] [PubMed]
12. Fagyas, M.; Úri, K.; Siket, I.M.; Daragó, A.; Boczán, J.; Bányai, E.; Édes, I.; Papp, Z.; Tóth, A. New Perspectives in the Renin-Angiotensin-Aldosterone System (RAAS) III: Endogenous Inhibition of Angiotensin Converting Enzyme (ACE) Provides Protection against Cardiovascular Diseases. *PLoS ONE* **2014**, *9*, e93719. [CrossRef] [PubMed]
13. Fagyas, M.; Úri, K.; Siket, I.M.; Fülöp, G.Á.; Csató, V.; Daragó, A.; Boczán, J.; Bányai, E.; Szentkirályi, I.E.; Maros, T.M.; et al. New Perspectives in the Renin-Angiotensin-Aldosterone System (RAAS) II: Albumin Suppresses Angiotensin Converting Enzyme (ACE) Activity in Human. *PLoS ONE* **2014**, *9*, e87844. [CrossRef] [PubMed]
14. Trbojevic-Stankovic, J.; Aleksic, M.; Odovic, J. Estimation of angiotensin-converting enzyme inhibitors protein binding degree using chromatographic hydrophobicity data. *Srp. Arh. Za Celok. Lek.* **2015**, *143*, 50–55. [CrossRef] [PubMed]
15. White, C.M. Pharmacologic, pharmacokinetic, and therapeutic differences among ACE inhibitors. *Pharmacotherapy* **1998**, *18*, 588–599. [CrossRef] [PubMed]
16. Brunner, D.; Desponds, G.; Biollaz, J.; Keller, I.; Ferber, F.; Gavras, H.; Brunner, H.; Schelling, J. Effect of a new angiotensin converting enzyme inhibitor MK 421 and its lysine analogue on the components of the renin system in healthy subjects. *Br. J. Clin. Pharmacol.* **1981**, *11*, 461–467. [CrossRef]
17. Bolzano, K.; Arriaga, J.; Bernal, R.; Bernardes, H.; Calderon, J.L.; Debruyn, J.; Dienstl, F.; Drayer, J.; Goodfriend, T.L.; Gross, W. The antihypertensive effect of lisinopril compared to atenolol in patients with mild to moderate hypertension. *J. Cardiovasc. Pharmacol.* **1987**, *9* (Suppl. S3), S43–S47. [CrossRef] [PubMed]
18. Gavras, H.; Biollaz, J.; Waeber, B.; Brunner, H.R.; Gavras, I.; Sackel, H.; Charocopos, F.; Davies, R.O. Effects of the oral angiotensin-converting enzyme inhibitor MK-421 in human hypertension. *Clin. Sci.* **1981**, *61* (Suppl. S7), 281s–283s. [CrossRef]
19. Pool, J.L.; Gennari, J.; Goldstein, R.; Kochar, M.S.; Lewin, A.J.; Maxwell, M.H.; McChesney, J.A.; Mehta, J.; Nash, D.T.; Nelson, E.B. Controlled multicenter study of the antihypertensive effects of lisinopril, hydrochlorothiazide, and lisinopril plus hydrochlorothiazide in the treatment of 394 patients with mild to moderate essential hypertension. *J. Cardiovasc. Pharmacol.* **1987**, *9* (Suppl. S3), S36–S42. [CrossRef]
20. Olvera Lopez, E.; Parmar, M.; Pendela, V.S.; Terrell, J.M. *Lisinopril*; StatPearls Publishing: Treasure Island, FL, USA, 2023.
21. Faruqi, A.; Jain, A. Enalapril. In *StatPearls [Internet]*; StatPearls Publishing: Treasure Island, FL, USA, 2022.
22. Baulmann, J.; Schillings, U.; Rickert, S.; Uen, S.; Düsing, R.; Illyes, M.; Cziraki, A.; Nickenig, G.; Mengden, T. A new oscillometric method for assessment of arterial stiffness: Comparison with tonometric and piezo-electronic methods. *J. Hypertens.* **2008**, *26*, 523–528. [CrossRef]
23. Thijssen, D.H.J.; Black, M.A.; Pyke, K.E.; Padilla, J.; Atkinson, G.; Harris, R.A.; Parker, B.; Widlansky, M.E.; Tschakovsky, M.E.; Green, D.J. Assessment of flow-mediated dilation in humans: A methodological and physiological guideline. *Am. J. Physiol. Heart Circ. Physiol.* **2011**, *300*, H2–H12. [CrossRef]
24. Hales, C.M.; Servais, J.; Martin, C.B.; Kohen, D. Prescription Drug Use Among Adults Aged 40–79 in the United States and Canada. *NCHS Data Brief* **2019**, *347*, 1–8.
25. Danilov, S.M.; Tikhomirova, V.E.; Kryukova, O.V.; Balatsky, A.V.; Bulaeva, N.I.; Golukhova, E.Z.; Bokeria, L.A.; Samokhodskaya, L.M.; Kost, O.A. Conformational fingerprint of blood and tissue ACEs: Personalized approach. *PLoS ONE* **2018**, *13*, e0209861. [CrossRef] [PubMed]
26. Nádasy, G.L.; Balla, A.; Szekeres, M. From Living in Saltwater to a Scarcity of Salt and Water, and Then an Overabundance of Salt—The Biological Roller Coaster to Which the Renin–Angiotensin System Has Had to Adapt: An Editorial. *Biomedicines* **2023**, *11*, 3004. [CrossRef] [PubMed]
27. Bánhegyi, V.; Enyedi, A.; Fülöp, G.Á.; Oláh, A.; Siket, I.M.; Váradi, C.; Bottyán, K.; Lódi, M.; Csongrádi, A.; Umar, A.J.; et al. Human Tissue Angiotensin Converting Enzyme (ACE) Activity Is Regulated by Genetic Polymorphisms, Posttranslational Modifications, Endogenous Inhibitors and Secretion in the Serum, Lungs and Heart. *Cells* **2021**, *10*, 1708. [CrossRef]
28. Weissgerber, T.L. Flow-Mediated Dilation: Can New Approaches Provide Greater Mechanistic Insight into Vascular Dysfunction in Preeclampsia and Other Diseases? *Curr. Hypertens. Rep.* **2014**, *16*, 487. [CrossRef]
29. Townsend, R.R. Arterial Stiffness: Recommendations and Standardization. *Pulse* **2016**, *4* (Suppl. S1), 3–7. [CrossRef] [PubMed]

30. Martynowicz, H.; Gać, P.; Kornafel-Flak, O.; Filipów, S.; Łaczmański, Ł.; Sobieszczańska, M.; Mazur, G.; Poręba, R. The Relationship Between the Effectiveness of Blood Pressure Control and Telomerase Reverse Transcriptase Concentration, Adipose Tissue Hormone Concentration and Endothelium Function in Hypertensives. *Heart Lung Circ.* **2020**, *29*, e200–e209. [CrossRef] [PubMed]
31. Ding, H.; Liu, S.; Zhao, K.-X.; Pu, J.; Xie, Y.-F.; Zhang, X.-W. Comparative Efficacy of Antihypertensive Agents in Flow-Mediated Vasodilation of Patients with Hypertension: Network Meta-Analysis of Randomized Controlled Trial. *Int. J. Hypertens.* **2022**, *2022*, 2432567. [CrossRef]
32. Srivastava, P.; Badhwar, S.; Chandran, D.S.; Jaryal, A.K.; Jyotsna, V.P.; Deepak, K.K. Improvement in Angiotensin 1-7 precedes and correlates with improvement in Arterial stiffness and endothelial function following Renin-Angiotensin system inhibition in type 2 diabetes with newly diagnosed hypertension. *Diabetes Metab. Syndr. Clin. Res. Rev.* **2020**, *14*, 1253–1263. [CrossRef]
33. Lv, Y.; Zhao, W.; Yu, L.; Yu, J.-G.; Zhao, L. Angiotensin-Converting Enzyme Gene D/I Polymorphism in Relation to Endothelial Function and Endothelial-Released Factors in Chinese Women. *Front. Physiol.* **2020**, *11*, 951. [CrossRef] [PubMed]
34. Shahin, Y.; Khan, J.A.; Samuel, N.; Chetter, I. Angiotensin converting enzyme inhibitors effect on endothelial dysfunction: A meta-analysis of randomised controlled trials. *Atherosclerosis* **2011**, *216*, 7–16. [CrossRef] [PubMed]
35. Javanmard, S.H.; Sonbolestan, S.A.; Heshmat-Ghahdarijani, K.; Saadatnia, M.; Sonbolestan, S.A. Enalapril improves endothelial function in patients with migraine: A randomized, double-blind, placebo-controlled trial. *J. Res. Med. Sci. Off. J. Isfahan Univ. Med. Sci.* **2011**, *16*, 26–32.
36. Kovacs, I.; Toth, J.; Tarjan, J.; Koller, A. Correlation of flow mediated dilation with inflammatory markers in patients with impaired cardiac function. Beneficial effects of inhibition of ACE. *Eur. J. Heart Fail.* **2006**, *8*, 451–459. [CrossRef] [PubMed]
37. Eržen, B.; Gradišek, P.; Poredoš, P.; Šabovič, M. Treatment of Essential Arterial Hypertension with Enalapril Does Not Result in Normalization of Endothelial Dysfunction of the Conduit Arteries. *Angiology* **2006**, *57*, 187–192. [CrossRef]
38. Diamant, M.; Vincent, H. Lisinopril versus enalapril: Evaluation of trough: Peak ratio by ambulatory blood pressure monitoring. *J. Hum. Hypertens.* **1999**, *13*, 405–412. [CrossRef]
39. Arendse, L.B.; Danser, A.H.J.; Poglitsch, M.; Touyz, R.M.; Burnett, J.C.; Llorens-Cortes, C.; Ehlers, M.R.; Sturrock, E.D. Novel Therapeutic Approaches Targeting the Renin-Angiotensin System and Associated Peptides in Hypertension and Heart Failure. *Pharmacol. Rev.* **2019**, *71*, 539–570. [CrossRef]

Disclaimer/Publisher's Note: The statements, opinions and data contained in all publications are solely those of the individual author(s) and contributor(s) and not of MDPI and/or the editor(s). MDPI and/or the editor(s) disclaim responsibility for any injury to people or property resulting from any ideas, methods, instructions or products referred to in the content.

Article

Association of Elevated Serum Aldosterone Concentrations in Pregnancy with Hypertension

Robin Shoemaker [1,*], Marko Poglitsch [2], Dolph Davis [1], Hong Huang [3], Aric Schadler [3], Neil Patel [4], Katherine Vignes [4], Aarthi Srinivasan [4], Cynthia Cockerham [4], John A. Bauer [3] and John M. O'Brien [4]

1. Department of Dietetics and Human Nutrition, University of Kentucky, Lexington, KY 40506, USA
2. Attoquant Diagnostics GmbH, 1110 Vienna, Austria
3. Department of Pediatrics, University of Kentucky, Lexington, KY 40536, USA
4. Department of Obstetrics and Gynecology, University of Kentucky, Lexington, KY 40536, USA
* Correspondence: robin.shoemaker@uky.edu

Abstract: Emerging evidence indicates a previously unrecognized, clinically relevant spectrum of abnormal aldosterone secretion associated with hypertension severity. It is not known whether excess aldosterone secretion contributes to hypertension during pregnancy. We quantified aldosterone concentrations and angiotensin peptides in serum (using liquid chromatography with tandem mass spectrometry) in a cohort of 128 pregnant women recruited from a high-risk obstetrics clinic and followed prospectively for the development of gestational hypertension, pre-eclampsia, superimposed pre-eclampsia, chronic hypertension, or remaining normotensive. The cohort was grouped by quartile of aldosterone concentration in serum measured in the first trimester, and blood pressure, angiotensin peptides, and hypertension outcomes compared across the four quartiles. Blood pressures and body mass index were greatest in the top and bottom quartiles, with the top quartile having the highest blood pressure throughout pregnancy. Further stratification of the top quartile based on increasing (13 patients) or decreasing (19 patients) renin activity over gestation revealed that the latter group was characterized by the highest prevalence of chronic hypertension, use of anti-hypertensive agents, pre-term birth, and intrauterine growth restriction. Serum aldosterone concentrations greater than 704 pmol/L, the 75th percentile defined within the cohort, were evident across all categories of hypertension in pregnancy, including normotensive. These findings suggest that aldosterone excess may underlie the development of hypertension in pregnancy in a significant subpopulation of individuals.

Keywords: hypertension; aldosteronism; renin; pregnancy; RAAS profiling; mass spectrometry; biomarkers

1. Introduction

An estimated 15.3% of pregnancies are affected by hypertensive disorders of pregnancy (HDP) [1,2], and the prevalence is rising [3]. HDP and related adverse outcomes are major contributors to maternal and fetal morbidity and mortality [3]. HDP are a group of disorders collectively characterized by elevated blood pressure during pregnancy with multiple (mostly unknown) underlying causes, including pre-existing hypertension during pregnancy (chronic hypertension), gestational hypertension, and pre-eclampsia. Rational strategies for identifying subtypes of HDP could improve the management and/or treatment of hypertension during pregnancy.

Aldosterone plays a key role in the physiological regulation of fluid homeostasis, including the expansion of blood volume to meet the demands of pregnancy, and modulates other physiological processes in the heart and blood vessels [4]. Aldosterone excess is considered pathological, contributing to vascular injury and endothelial dysfunction, as well as cardiac inflammation and hypertrophy [5]. Aldosteronism is a group of conditions characterized by abnormal production of aldosterone that can be primary (caused by adrenal adenoma, uni- or bilateral adrenal hyperplasia, or, rarer, familial hyperaldosteronism, an inherited condition), with aldosterone production inappropriately high relative to sodium status, renin activity, and potassium levels, and non-suppressible in response to volume or sodium loading [6], or secondary (attributed to conditions such as renal artery stenosis [7], or reflecting fluid or electrolyte depletion [8]), which may include activation of the renin–angiotensin system. Detection of uncontrolled aldosterone secretion in individuals is important because this condition is linked with a higher incidence of target organ damage and cardiovascular events than age- and sex-matched patients with essential hypertension with the same degree of blood pressure elevation; treatment with mineralocorticoid receptor antagonists improves outcomes for these patients [8]. Emerging studies indicate a high prevalence of aldosteronism (multiple etiologies) among adults (perhaps up to 12% of the US population), especially among patients with severe hypertension [9,10]. Given the high rates of hypertension in reproductive-age women [11], it stands to reason that unrecognized aldosteronism may contribute to hypertension during pregnancy in a subpopulation of individuals, but there are little published data about this condition in pregnancy.

In a healthy pregnancy, plasma renin activity and aldosterone concentrations increase to accommodate blood volume expansion and the development of the placental-fetal unit [12,13]. Previous studies by our lab and others have reported that gestational hypertension and/or pre-eclampsia is accompanied by attenuated renin activity and lower aldosterone concentrations compared to normotensive pregnancies [14–17], presumably reflecting a response to increased systemic vascular resistance. Few studies have investigated the condition of aldosterone excess in pregnancy; the literature comprises only a handful of retrospective studies and case reports of pregnant patients with confirmed aldosteronism. Collectively, pregnancies with confirmed primary aldosteronism are indicated to be high-risk, with a highly variable disease course, and associated with high rates of maternal and fetal complications, such as pre-eclampsia and intrauterine growth restriction [18–20]. Outside of pregnancy, primary aldosteronism is characterized by an elevated aldosterone-to-renin ratio in blood that exceeds a defined cut-off value [8]; clear criteria for detection of aldosteronism where the renin–angiotensin system is also activated are not defined. Detection of any form of aldosteronism during pregnancy is complicated by the fact that renin activity and aldosterone concentrations increase in normal pregnancy, and reference values for these biomarkers in normotensive pregnancy have not been defined. Further, accumulating evidence in non-pregnant individuals indicates that a previously unrecognized and clinically relevant spectrum of abnormal aldosterone production may exist [10]; this has not been examined at all in pregnancy.

To being to address these gaps, we used a previously described [17], liquid chromatography with tandem mass spectrometry (LC-MS/MS)-based method to quantify angiotensin peptides and aldosterone concentrations in serum in a cohort of pregnant women followed prospectively for the development of HDP (remaining normotensive, chronic hypertension, gestational hypertension, pre-eclampsia, superimposed pre-eclampsia on chronic hypertension). The aims of our study were to define associations among aldosterone concentrations in serum with blood pressure and hypertension outcomes in pregnancy, and to characterize activity of the RAAS in those with the highest levels of aldosterone. Recognition of the scope of potential abnormal aldosterone production in pregnancy could better define the prevalence of this condition, and inform the understanding of the pathogenesis and treatment of hypertension during pregnancy.

2. Materials and Methods

2.1. Study Population

This was a secondary analysis of 128 pregnant individuals enrolled in an on-going study at the University of Kentucky. All subjects gave informed consent to participate in this study, which was approved on 17 February 2019 by the University of Kentucky Institutional Review Board (#47841), and all study procedures were performed in accordance with relevant guidelines and regulations. Patients with moderate to high clinical risk factors for pre-eclampsia (presence of more than one of the following clinical risk factors: history of preterm pre-eclampsia, type 1 or 2 diabetes, renal or autoimmune disease; or more than two of the following clinical risk factors: nulliparity, obesity (body mass index > 30), family history of pre-eclampsia, Black race, lower income, age 35 or older, previous pregnancy with small birth weight or adverse outcome, or in vitro conception) [21] were recruited from high-risk obstetrics clinics in the Division of Maternal Fetal Medicine. The age range was between 18 and 45 years. Patients were excluded if they had multifetal gestation or anomalous fetus, allergy to aspirin, gastrointestinal bleeding, severe peptic ulcer or liver dysfunction, or were taking anticoagulant medications.

2.2. Data Collection

Study clinicians (who are also providers for the patients in the cohort) in the Maternal Fetal Medicine unit identified potential patients, performed consenting, and enrolled participants at the time of first trimester screening (routine pre-natal visit or ultrasound appointment). Data were collected by study clinicians as part of routine clinical care over the course of pregnancy, and patients were treated according to clinical guidelines. Demographic information was collected upon enrollment. Clinical data and maternal blood were collected from routine visits in the first (12 weeks) and third (28 weeks) trimesters of pregnancy. Patients were followed prospectively for the development of pre-eclampsia, gestational hypertension, and chronic hypertension. Clinical data and outcomes were obtained from electronic medical records by a trained clinical coordinator, and input into a REDCap database by study personnel.

Systolic blood pressure (SBP) and diastolic blood pressure (DBP) were measured following the American Heart Association Guidelines [22], and were determined as the average of two consecutive measurements per arm assessed in a seated, resting position by one observer. Maternal blood was collected during routine prenatal laboratory evaluations in the first trimester (mean of 12 gestational weeks) and in the third trimester (mean of 28 gestational weeks). Blood was drawn by trained phlebotomists from patients in a seated position and transferred into red-top serum tubes. Study personnel transferred the whole blood samples to the laboratory for processing and storage. Whole blood was allowed to rest at room temperature for 60 min, followed by centrifugation, transferred into 200 uL aliquots, and barcoded by study personnel. Samples were stored at $-80\ ^\circ$C until analysis for angiotensin peptides and aldosterone.

Pregnancy outcomes were determined during pregnancy and after delivery by trained clinical staff following guidelines published by the American College of Gynecologists [23]. Gestational hypertension was diagnosed as either SBP or DBP greater than 140 or 90 mmHg, respectively, after 20 weeks gestation in women with previously normal blood pressure, and chronic hypertension was determined by hypertension pre-dating pregnancy, or presenting before the 20th week of gestation. Pre-eclampsia was diagnosed with either SBP or DBP greater than 140 or 90 mmHg, respectively, after 20 weeks gestation, and one of the following signs indicating proteinuria, renal or kidney damage, low platelets, edema, or other neurologic features; pre-eclampsia in individuals with chronic hypertension was described as super-imposed pre-eclampsia. The following outcomes were also recorded: gestational diabetes, intrauterine growth restriction (defined as fetal weight below the 10th percentile), and pre-term birth (<37 weeks).

2.3. Quantification of Components of the RAAS: Angiotensin Peptides, Aldosterone, and Equilibrium-Based Biomarkers

Quantification of angiotensin I, angiotensin II, and aldosterone from serum samples of study patients was performed using LC-MS/MS-based methodology that has been previously described [17,24,25]. Serum samples were incubated for one hour at 37 $^\circ$C to generate a controlled ex vivo equilibration, followed by stabilization through addition of an enzyme inhibitor cocktail (i.e., equilibrium analysis). Samples, calibrators, and quality controls were spiked with stable isotope-labeled internal standards (200 pg/mL for each angiotensin metabolite, used to correct for peptide recovery), subjected to C-18-based solid-phase extraction, followed by ultra-high-pressure liquid chromatography-based separation on a reversed-phase analytical column (Acquity UPLC C18, Waters, Milford, MA, USA) operating in line with a Xevo TQ-S triple quadruple mass spectrometer (Waters, Milford, MA, USA). Multiple reaction monitoring mode was used for generation of total ion chromatograms obtained from sums of quantifier transitions for each analyte, with previously optimized conditions for sensitivity, specificity, and signal-to-noise ratio. Integrated signals were used to determine analyte concentrations based on via linear calibration (Software: MassLynx/Target/Lynx, Version 4.2, Waters, Milford, MA, USA). Calculated levels of quality controls and calibrators were used to evaluate batch performance. The lower limits of quantification for angiotensin I and angiotensin II were 5 and 4 pmol/L, respectively, and 13 pmol/L for aldosterone.

Biomarkers of plasma renin activity and the aldosterone-to-Ang II ratio (AA2-R) were calculated from equilibrium concentrations of angiotensin peptides I and II and aldosterone [17,26,27]. The plasma renin activity surrogate (PRA-S) is the calculated sum of the equilibrium concentrations of angiotensins I and I ((eqAngiotensin I) + (eqAngiotensin II), pmol/L), and the AA2-R is the ratio of the concentrations of aldosterone to angiotensin II ((Aldosterone)/(eqAngiotensin II), (pmol/L)/(pmol/L)). These biomarkers are well-published in the literature as correlating with clinical characteristics and/or prediction outcomes in studies of hypertension [28,29], aldosteronism [26,27], heart failure [24,30], and more [31].

2.4. Statistical Analysis

Statistical analyses were performed via GraphPad Prism version 9.5.1 (San Diego, CA, USA). All data were assessed for normality, and the appropriate parametric/nonparametric tests were used. A p-value less than 0.05 was considered statistically significant. Paired clinical and biochemical data from the first and third trimester were obtained for 128 pregnant women. RAAS components were analyzed within-group for change over time (first trimester to third trimester), using Wilcoxon matched pairs signed-rank tests. Between-group differences in categorical variables (demographics) were analyzed using chi-square or Fisher's exact test. Between-group differences were analyzed using one-way analysis of variance (ANOVA) followed by Dunnett's multiple comparisons test for continuous, parametric variables (blood pressure, age, body mass index, gestational age at delivery) for more than two groups or t-tests for between-group comparison of two groups. The Kruskal–Wallis test followed by Dunn's multiple comparisons test was used for between-group differences for non-parametric variables (RAAS components).

3. Results

3.1. Description of the Population

The study population included 128 pregnant women, where 52 remained normotensive throughout pregnancy, 42 were determined to have chronic hypertension, 22 developed gestational hypertension, 5 developed pre-eclampsia, and 7 with chronic hypertension developed pre-eclampsia (superimposed pre-eclampsia). In Figure 1, the first trimester serum aldosterone concentration from each patient in the cohort is ordered from lowest to highest within each outcome. There is currently no defined reference range for normal aldosterone concentrations in blood in pregnancy, so aldosterone excess in this study was defined as a first trimester aldosterone concentration above the 75th percentile of the cohort. The cohort was divided into four groups of 32 patients each, based on quartile of aldosterone concentration in serum measured in the first trimester of pregnancy, and comparisons made across quartiles. The aldosterone value for the 25th percentile, median, 75th percentile, and maximum are indicated on the y-axis, indicating the range for each quartile, Q1–Q4, respectively.

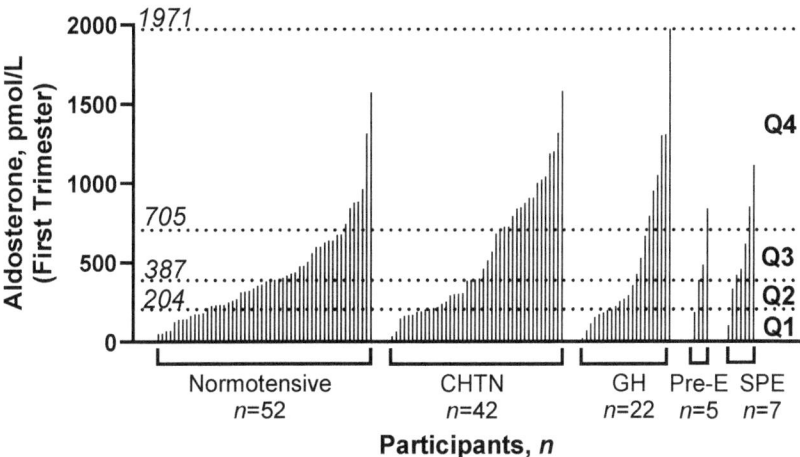

Figure 1. Distribution of serum aldosterone concentrations in the first trimester of pregnancy in the whole cohort. Serum aldosterone concentrations in the first trimester of pregnancy for each individual patient, ordered from lowest to highest, and grouped by hypertension outcome. Dotted lines indicate the value of aldosterone in pmol/L corresponding to the 25th percentile, median, 75th percentile, and maximum, for the 128 participants in the cohort. Q1: quartile one; Q2: quartile two; Q3: quartile three; Q4: quartile four; CHTN: chronic hypertension; GH: gestational hypertension; Pre-E: pre-eclampsia; SPE: superimposed pre-eclampsia.

Demographics and clinical characteristics of each quartile are displayed in Table 1. The cohort was primarily Caucasian, with Black, Hispanic, and mixed-race individuals distributed fairly equally across quartiles, with the exception of Q1, where there were no Hispanics and more Black patients compared to the other groups. The mean age of the whole cohort was 29.5 years, and did not differ widely across quartiles. The mean body mass index of the whole cohort was 32.9 kg/m^2, likely reflecting the high-risk patient population, and there was a trend of greater body mass index in Q1 and Q4. Q1 had the most type 1 diabetics, but otherwise diabetes (and gestational diabetes) was distributed fairly equally among the four quartiles.

There was a notable difference in blood pressure across the quartiles, where SBP and DBP were significantly elevated in Q4 in both the first and third trimesters (Table 1), with Q1 having the next highest mean blood pressure compared to Q2 and Q3. Q4 contained the most patients with chronic hypertension and the fewest normotensive pregnancies compared to the other quartiles, and the most patients taking labetalol or nifedipine for blood pressure control (Table 1). Previous studies of pregnancy reported an association of low activity of the RAAS with development of HDP [15,17]. In the current study, gestational hypertension was most prevalent in Q1 and Q4 (combined prevalence of 21.8%) compared to Q2 and Q3 (combined prevalence of 15.6%), but pre-eclampsia was evident across all four quartiles. Q1 had a slightly greater number of patients with previous pre-eclampsia. Preterm birth was the most prevalent in Q1 and Q4 (combined prevalence of 25%) compared to Q2 and Q3 (combined prevalence of 10.9%). IUGR was reported only in patients in Q1 and Q4 (combined prevalence of 10.9%), and not in Q2 or Q3. Gestational age at delivery and infant birth weight trended towards lower in Q1 and Q4, but there was no significant difference between groups.

Table 1. Patient demographics, clinical characteristics, and outcomes of pregnant women based on quartile of first trimester aldosterone concentration in serum.

Parameter	Q1 n = 32	Q2 n = 32	Q3 n = 32	Q4 n = 32
n (%) or mean ± SD				
Race				
Caucasian	25 (78.1)	20 (62.5)	22 (68.8)	23 (71.9)
Black or African American	7 (21.9)	4 (12.5)	4 (12.5)	4 (12.5)
Hispanic	0	5 (15.6)	4 (12.5)	3 (9.4)
Asian	0	1 (3.1)	0	1 (3.1)
Mixed race/other	0	2 (6.3)	1 (3.1)	1 (3.1)
Primiparous	9 (28.1)	10 (31.3)	9 (28.1)	7 (21.9)
Previous pre-e	7 (21.9)	2 (6.3)	5 (15.6)	2 (6.3)
Type 1 diabetes	10 (31.3)	5 (15.6)	2 (6.3)	4 (12.5)
Type 2 diabetes	3 (9.4)	6 (18.8)	9 (9.4)	3 (9.4)
Gestational diabetes	4 (12.5)	4 (12.5)	2 (6.3)	5 (15.6)
Medications				
Aspirin	25 (78.1)	21 (65.6)	18 (56.3)	27 (84.4)
Labetalol or nifedipine	4 (12.5)	1 (3.13)	3 (9.4)	10 (31.3) [@]
Anti-diabetic agents	14 (43.8)	10 (31.3)	7 (21.9)	9 (28.1)
Age, years	27 ± 4 [#]	30 ± 6	31 ± 7	30 ± 6
BMI, kg/m^2	35.0 ± 10.3	31.4 ± 7.8	31.5 ± 10.3	34.0 ± 10.4
1st trimester blood pressure				
SBP, mmHg	125 ± 13	119 ± 13	120 ± 15	129 ± 13 *
DBP, mmHg	79 ± 7	75 ± 7	77 ± 9	82 ± 4 **
3rd trimester BP				
SBP, mmHg	125 ± 12	122 ± 14	120 ± 14	132 ± 17 ^
DBP, mmHg	78 ± 6	76 ± 8	77 ± 9	83 ± 9 ~
Gestational age at delivery, weeks	36.8 ± 2.3	38.1 ± 1.7	38.1 ± 1.7	37.1 ± 2.2
Infant birth weight, grams	3123 ± 711	3394 ± 550	3366 ± 559	3122 ± 776
Normotensive	12 (37.5)	16 (50.0)	17 (53.1)	7 (21.9)
Chronic HTN	9 (28.1)	10 (31.3)	7 (21.9)	16 (50.0) [@]
Gestational HTN	8 (25.0)	5 (15.7)	3 (9.4)	6 (18.8)
Pre-eclampsia or SPE	3 (9.4)	1 (3.13)	5 (15.7)	3 (9.4)
Pre-term birth	9 (28.0)	2 (6.3)	5 (15.6)	7 (21.9)
IUGR	3 (9.4)	0	0	4 (12.5)

Abbreviations BMI: body mass index, first trimester; SBP: systolic blood pressure; DBP: diastolic blood pressure; HTN: hypertension; SPE: super-imposed pre-eclampsia; IUGR: intrauterine growth restriction. [@], $p < 0.05$ difference between groups using chi-square analysis; [#], $p < 0.05$ vs. Q3; *, $p < 0.05$, and **, $p < 0.01$ vs. Q2, Q3; ^, $p < 0.05$, and ~, $p < 0.01$ vs. Q2, Q3 analyzed using one-way analysis of variance followed by Dunnett's multiple comparisons test.

3.2. Biochemical Analysis of the Serum RAAS in Pregnancy Grouped by Quartile of First Trimester Aldosterone Concentrations in Serum

We investigated activity of the RAAS in serum samples over gestation across the four quartiles, and the median values and interquartile range for all measured components of the RAAS are displayed in Table 2. In the first trimester, the median concentrations of aldosterone in each group progressively increased across quartiles, and there was a six-fold difference between the median values of Q1 and Q4 (Figure 2B). This trend corresponded with concentrations of the upstream components, angiotensin I and angiotensin II (and correspondingly, the calculated biomarker PRA-S, Figure 2A, Table 2), where these values progressively increased across quartiles, but only a two-fold difference in the respective values between Q1 and Q4 was observed. The aldosterone-to-angiotensin II ratio was increased four-fold between Q4 and Q1 in the first trimester (Table 2). The median and interquartile range for measured components of the RAAS in the first and third trimester in the whole cohort of 128 patients are included as Supplementary Table S1.

Table 2. Concentrations of equilibrium angiotensin peptides, aldosterone, and biomarkers determined by liquid chromatography with tandem mass spectrometry in the first and third trimesters of pregnancy by quartile of aldosterone.

Parameter	Q1 $n = 32$	Q2 $n = 32$	Q3 $n = 32$	Q4 $n = 32$
Median (Interquartile Range)				
Aldosterone, pmol/L				
1st Trimester	149.8 (70.58–178.2)	279.9 (232.8–328.0)	482.1 (416.1–612.1)	928.6 (840.2–1194.0) #
3rd Trimester	437.2 (247.4–802.8) ****	562.5 (336.8–1126) ****	944.8 (642.6–1359) ****	1041.0 (585.9–1742.0) #
Angiotensin I, pmol/L				
1st Trimester	41.8 (25.3–67.6)	69.9 (49.9–92.1)	86.09 (59.1–125.9)	107.0 (89.85–161.9) #
3rd Trimester	63.6 (45.1–131.7) ***	95.8 (55.7–148.7) *	121.6 (69.9–176.1) *	108.5 (75.25–157.3) #
Angiotensin II, pmol/L				
1st Trimester	104.3 (71.33–160.1)	145.1 (110.9–189.4)	172.6 (140.3–266.9)	217.0 (169.8–350.3) #
3rd Trimester	107.6 (70.21–180.2)	158.9 (90.97–281.9)	176.0 (132.2–277.0)	172.5 (122.8–280.6) *,#
PRA-S, pmol/L				
1st Trimester	151.8 (101.7–225.7)	219.1 (148.6–280.4)	256.8 (197.8–612.1)	326.3 (250.5–494.3) #
3rd Trimester	187.3 (123.2–322.8) *	262.8 (139.5–422.9)	292.5 (221.9–466.0)	290.7 (205.6–456.3) #
AA2-R, (pmol/L)/(pmol/L)				

Table 2. Cont.

Parameter	Q1 n = 32	Q2 n = 32	Q3 n = 32	Q4 n = 32
1st Trimester	0.95 (0.71–1.62)	1.99 (1.41–2.45)	3.28 (2.10–4.29)	4.63 (2.58–6.84) #
3rd Trimester	3.90 (2.3–7.0) ****	3.50 (2.43–6.70) ****	4.81 (3.20–7.72) ****	5.25 (3.22–9.79) *,#

Abbreviations: PRA-S: surrogate biomarker for plasma renin activity, calculated from sum of the equilibrium concentrations of angiotensin I ([Ang I]) and angiotensin II ([Ang II]); AA2-R: aldosterone-to-angiotensin II ratio. *, $p < 0.05$, ***, $p < 0.001$, ****, $p < 0.0001$ for third trimester vs. first trimester within group analyzed by Wilcoxon signed-rank test. Between-group analysis using the Kruskal–Wallis test followed by Dunn's multiple comparisons test: Aldosterone: #, $p < 0.01$ vs. Q3, $p < 0.001$ vs. Q2, and $p < 0.0001$ vs. Q1 in the first trimester, and $p < 0.001$ vs. Q1 in the third trimester. Angiotensin I: #, $p < 0.01$ vs. Q2 in the first trimester, and $p < 0.05$ vs. Q1 in the third trimester. Angiotensin II: #, $p < 0.05$ vs. Q3 in the first trimester, and $p < 0.05$ vs. Q1 in the third trimester. PRA-S: #, $p < 0.05$ vs. Q3, $p < 0.01$ vs. Q2, and $p < 0.001$ vs. Q1 in the first trimester, and $p < 0.05$ vs. Q1 in the third trimester. AA2-R: #, $p < 0.001$ vs. Q2 and $p < 0.0001$ vs. Q1 in the first trimester.

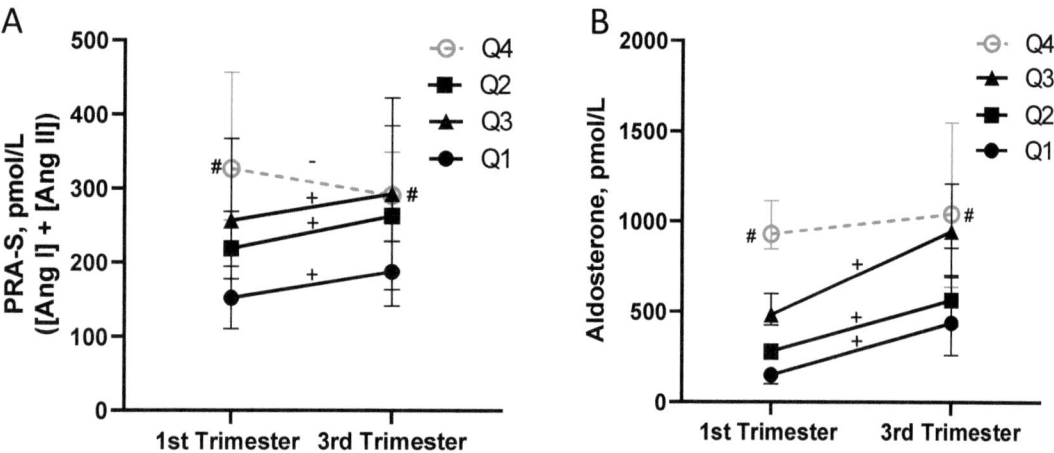

Figure 2. Activation of the RAAS in pregnancy is attenuated in the quartile (Q) with the highest concentrations of aldosterone in serum. (A) PRA-S, surrogate for plasma activity, calculated from the sum of the equilibrium concentrations of angiotensin I ([Ang I]) and angiotensin II ([Ang II]). #, $p < 0.05$ vs. Q3, $p < 0.01$ vs. Q2, and $p < 0.001$ vs. Q1 in the first trimester, and $p < 0.05$ vs. Q1 in the third trimester analyzed using the Kruskal–Wallis test followed by Dunn's multiple comparisons test; + or −, $p < 0.05$ within-group increase or decrease in angiotensin I, angiotensin II, and/or PRA-S from first to third trimester analyzed using Wilcoxon matched pairs signed-rank tests. (B) Aldosterone concentrations. #, $p < 0.01$ vs. Q3, $p < 0.001$ vs. Q2, and $p < 0.0001$ vs. Q1 in the first trimester, and $p < 0.001$ vs. Q1 in the third trimester analyzed using the Kruskal–Wallis test followed by Dunn's multiple comparisons test; +, $p < 0.0001$ positive change over gestation within group analyzed using Wilcoxon matched pairs signed-rank tests. Data are median with 95th confidence interval in 128 patients with 32 patients in each quartile.

Previous studies demonstrated increased activity of PRA-S and aldosterone concentrations over gestation in normotensive pregnancies [12,17]. In the current study, the concentrations of angiotensin I and aldosterone (and a trend for PRA-S) significantly increased from the first to the third trimester in Q1, Q2, and Q3 (Table 2, Figure 2A,B. In contrast, the concentrations of angiotensin I and aldosterone did not increase over gestation in the Q4 group (Table 2; Figure 2A,B). In fact, there was a significant decrease over gestation in the concentrations of angiotensin II in the Q4 group ($p < 0.05$), and this was reflected by a reduction in the PRA-S in the third vs. first trimester in this group only ($p = 0.120$; Table 2, Figure 2A).

3.3. Association of Declining Renin Activity over Gestation with Blood Pressure in Patients with the Highest Concentrations of Serum Aldosterone

Outside of pregnancy, aldosterone excess is often accompanied by suppressed renin activity [8], but there are no criteria to define suppressed renin activity in pregnancy. We recently demonstrated that declining PRA-S over gestation was linked to development of hypertension [17], so we performed a subanalysis of patients in Q4 based on whether PRA-S increased or decreased from the first to the third trimester (ΔPRA-S = [PRA-S, third trimester]—[PRA-S, first trimester]). Of the 32 patients in the Q4 group, there were 13 patients with increasing PRA-S, and 19 patients with decreasing PRA-S (Figure 3A). The SBP and DBP in the Q4 for these groups are depicted in Figure 3 and Table 3. In patients with high concentrations of aldosterone in the first trimester (Q4), decreasing PRA-S over gestation was associated with elevated SBP ($p = 0.08$; Figure 3B) and DBP ($p < 0.01$, Figure 3C) in the third trimester, compared to those where PRA-S increased.

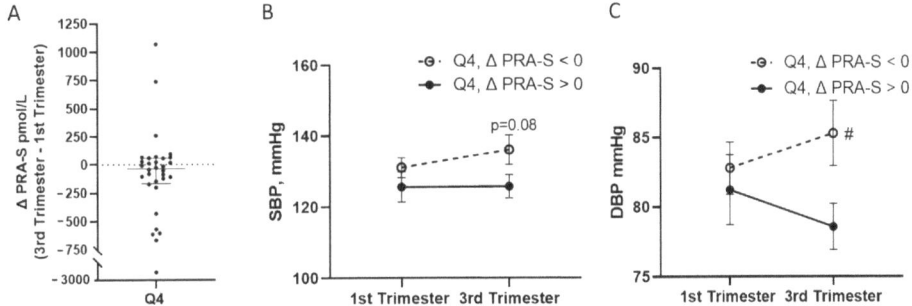

Figure 3. Systolic and diastolic blood pressure in in the quartile with the highest aldosterone concentrations in serum stratified by increasing or decreasing PRA-S from the first to third trimester. (**A**) ΔPRA-S, the change in the PRA-S (surrogate biomarker for plasma renin activity, calculated from sum of the equilibrium concentrations of angiotensin I and angiotensin II) value from the first to the third trimester: [PRA-S, third trimester]—[PRA-S, first trimester], in quartile four (Q4). Patients were stratified by increasing (13 patients) or decreasing (19 patients) ΔPRA-S. (**B**) Systolic blood pressure (SBP) and (**C**) diastolic blood pressure (DBP) in patients in Q4 with increasing vs. decreasing PRA-S. Data are median with 95th confidence interval. #, $p < 0.05$ between groups, analyzed using paired *t*-test.

For comparison, a similar analysis was performed to examine the association of declining PRA-S over gestation with third trimester blood pressure in the remaining quartiles, Q1 + Q2 + Q3 (sum of patients with first trimester aldosterone concentrations below the 75th percentile). There were 37 patients in Q1 + Q2 + Q3 where PRA-S decreased (including one patient with ΔPRA-S = 0), and 59 patients in quartiles Q1 + Q2 + Q3 where PRA-S increased over gestation. SBP and DBP for these groups are lower than those reported for Q4 with declining PRA-S (Table 3).

In addition to elevated blood pressure, the group in the highest quartile for first trimester aldosterone concentration with declining PRA-S over pregnancy had the highest prevalence of chronic hypertension, the most patients taking antihypertensive agents for blood pressure control, a high incidence of pre-term birth, and the most cases of IUGR (Table 3). There was not an apparent association among gestational hypertension and elevated aldosterone concentrations or declining renin activity. Interestingly, there were no cases of pre-eclampsia in the Q4 group with increasing PRA-S, and pre-eclampsia was most prevalent in the combined two groups were PRA-S decreased over gestation (Table 3). Within Q4, there were notably more pre-term birth cases with decreasing PRA-S, and the Incidence of pre-term birth was overall greater with declining renin (Table 3).

Table 3. Characteristics of patients based on increasing or decreasing renin activity over pregnancy, in the top aldosterone quartile compared to the sum of the remaining quartiles.

	Q4 (>75th Percentile)		Q1 + Q2 + Q3 (<75th Percentile)	
	ΔPRA-S < 0 $n = 19$	ΔPRA-S > 0 $n = 13$	ΔPRA-S < 0 $n = 37$	ΔPRA-S > 0 $n = 59$
Mean ± SD or n (%)				
BMI, kg/m^2	34.5 ± 9.2	33.3 ± 12.3	33.9 ± 11.1	31.8 ± 8.5
SBP, mmHg	136 ± 18 [#]	126 ± 12	126 ± 15	120 ± 11
DBP, mmHg	85 ± 10 [#]	79 ± 6	78 ± 7	76 ± 8
Chronic hypertension	13 (68.4) ^	5 (38.5)	14 (38)	18 (30.5)
Antihypertensive agents	7 (36.8) ^	3 (23.1)	5 (13.5)	3 (5.1)
Pre-eclampsia	3 (15.8)	0	8 (21.6)	2 (3.4)
Gestational hypertension	3 (15.8)	3 (23.1)	8 (21.6)	10 (16.9)
Pre-term birth	6 (31.6) ^	1 (7.7)	10 (27.0)	6 (16.2)
IUGR	3 (15.8) ^	1 (7.7)	1 (2.7)	2 (3.4)

Abbreviations: ΔPRA-S: [PRA-S, third trimester]—[PRA-S, first trimester] (i.e., increasing or decreasing renin activity over gestation); BMI: body mass index, first trimester; SBP: systolic blood pressure in the third trimester; DBP: diastolic blood pressure in the third trimester; IUGR: intrauterine growth restriction. [#], $p < 0.01$ vs. Q1 + Q2 + Q3 with ΔPRA-S less than zero for SBP; [#], $p < 0.01$ vs. all other groups for DBP analyzed using one-way analysis of variance followed by Dunnett's multiple comparisons test; ^, $p < 0.05$ difference between groups using Fisher's exact test.

4. Discussion

A recent study by Brown et al. across four academic medical centers reported a previously unrecognized high prevalence of primary aldosteronism present across all categories of hypertension, including normotensive individuals [10]. These data suggest aldosteronism might underlie the development of HDP in a subpopulation of pregnant individuals, but few studies have investigated pathological aldosteronism in pregnancy. We quantified concentrations of aldosterone and other upstream components of the RAAS in a cohort of 128 pregnant women with risk factors for HDP to determine whether high aldosterone concentrations in serum in the first trimester (defined within-study as values greater than the 75th percentile of the cohort) were associated with elevated blood pressure and/or hypertension outcomes in pregnancy. The main findings from this study are: (1) Aldosterone concentrations in serum greater than 704 pmol/L (the 75th percentile within cohort) were evident across all categories of hypertension in pregnancy, including normotensive; (2) systolic and diastolic blood pressures were highest in the top quartile of aldosterone concentration compared to the remaining quartiles, (3) especially when accompanied by declining renin activity, and this group was characterized by a high prevalence of chronic hypertension, use of anti-hypertensive agents, pre-term birth, and slightly higher prevalence of IUGR. The lowest quartile of aldosterone concentrations (less than 204 pmol/L; below the 25th percentile of the cohort) also had high-risk characteristics, such as elevated blood pressure, higher prevalence of type 1 diabetes, pre-term birth, and development of gestational hypertension. Taken together, these findings agree with those of previous studies that demonstrated lower renin activity and aldosterone levels in hypertensive pregnancies [14–16,32,33], and add to the literature evidence of a significant portion of pregnant women in our predominately high-risk cohort with very high concentrations of aldosterone in serum, which was associated with pre-pregnancy (chronic) hypertension and elevated blood pressure during pregnancy.

While it has been known for decades that pregnancies that develop gestational hypertension or pre-eclampsia have reduced activity of the RAAS compared to normotensive pregnancies [32,34], studies have not examined associations among high aldosterone levels and blood pressure in HDP. One reason for this is that elevated aldosterone concentrations are known to be a normal and necessary feature of healthy pregnancy: e.g., maternal aldosterone secretion was required for plasma volume expansion in pregnant ewes [35]; and aldosterone deficiency reduced placental function, litter size, and pup weight in mice [36]. The concept of aldosterone as a mediator of cardiovascular disease has emerged more recently, where attributes of modern lifestyles (physical inactivity, chronic stress, and energy overabundance) have led to an increase in the number of individuals with elevated secretion of aldosterone and/or activation of mineralocorticoid receptors with adverse effects on vascular function [4]. Females may be biologically primed for aldosterone secretion [37], which may have provided an evolutionary advantage for successful reproduction, but the recent literature indicates that females may be especially susceptible to cardiac and vascular dysfunction mediated by elevated angiotensin II [38] or aldosterone levels [39–41]. Being overweight and obesity are also associated with increased aldosterone production [42]. Increasing rates of obesity in pregnant and reproductive-age women may have resulted in the relatively recent emergence of a subpopulation of individuals with abnormally elevated secretion of during pregnancy.

A limitation to the study of aldosteronism during pregnancy is the absence of thresholds or ranges of concentrations of serum aldosterone considered to be normal vs. abnormal in pregnancy. Sanga et al. aimed to establish reference ranges for plasma aldosterone concentrations (in ng/dL) and direct renin concentration (in mU/L) in pregnancy, where values were derived from a historical study of 18 patients followed weekly throughout pregnancy and derived using immunoassay-based detection methods [43]. Mean plasma aldosterone concentrations were approximately 20 ng/dL (or converted to 555 pmol/L) in normal, healthy pregnancy at the 12th gestational week, and approximately 45 ng/dL (or converted to 1250 pmol/L) at the 28th gestational week [20]. We report here median and interquartile ranges of aldosterone concentrations in a cohort of 128 high-risk pregnant patients (a considerable sample size, comparatively [44–46]) at gestational weeks 12 (387 pmol/L, 204–705 pmol/L) and 28 (724 pmol/L, 376–1197 pmol/L) of pregnancy quantified via LC-MS/MS (the gold standard for determination of aldosterone in clinical assays; aldosterone concentrations derived from immunoassay-based methods are typically higher and more variable that when derived via mass spectrometry-based methods [47]). For reference, aldosterone concentrations outside of pregnancy measured using LC-MS/MS range between 100–300 pmol/L [47], and the median and interquartile range for aldosterone concentrations in 33 non-pregnant patients with confirmed aldosteronism were 392 pmol/L and 331 to 468 pmol/L [26].

There is not an internationally accepted, standardized methodology for quantification of renin activity. Plasma renin activity is the most sensitive and accurate method for assessing renin activity, especially in the clinically relevant low ranges, but direct renin concentration is often used because measurement is more convenient [48]. Mean direct renin concentrations at gestational weeks 12 and 28 reported by Sanga et al. from 18 patients in the literature were approximately 35 mU/L and 42 mU/L, respectively [20]. Comparatively, the median and interquartile range of direct renin concentration in 33 non-pregnant patients with confirmed aldosteronism were 3.8 mU/L and 2.9 to 7.5 mU/L, compared to the mean and interquartile range of 20.3 mU/L and 10.3 to 33 mU/L in 77 non-pregnant patients with essential hypertension [26]. While there is not a meaningful conversion rate between direct renin concentration and plasma renin activity, the latter study of non-pregnant patients also quantified PRA-S using the same methodology as the current study, and the median and interquartile range for 33 patients with confirmed aldosterone were 40 pmol/L and 18–58 pmol/L, compared to 165 pmol/L and 80–328 pmol/L in the 77 patients with essential hypertension. The median and interquartile ranges for PRA-S were reported in our study of pregnancy were comparatively higher (246 pmol/L and 153–353 pmol/L in the first trimester, and 264 pmol/L and 175–397 pmol/L in the third trimester). Results from our study confirm that PRA-S and aldosterone concentrations are elevated in pregnant compared to non-pregnant patients and add to the literature reference values for aldosterone and PRA-S concentrations, as well as concentrations of angiotensin I and angiotensin II (and other calculated biomarkers of RAAS activity) at equilibrium, measured using LC-MS/MS and reported in pmol/L, in the first and third trimesters of pregnancy in 128 patients (Supplementary Table S1).

The literature describing aldosteronism in pregnancy mostly comprises individual case reports, small retrospective studies based on medical chart review, and narrative reviews. Sanga et al. recently published the most comprehensive-to-date systematic review describing features of 83 cases available in the literature of patients with confirmed primary aldosteronism who underwent pregnancy. There were 56 cases of sporadic primary aldosteronism (uni- or bilateral hyperplasia) and 27 cases of familial hyperaldosteronism type 1 (inappropriate production of aldosterone by adrenocorticotropic hormone). In the latter, most cases were diagnosed before pregnancy, blood pressure was effectively managed, and there were fewer poor outcomes. Among the unilateral and bilateral hyperplasia cases, approximately 60% were characterized by known existence of pre-pregnancy hypertension, but less than half had a confirmed diagnosis of aldosteronism before pregnancy, and there was a complication rate of 50% in previous pregnancies. Of the 83 cases reviewed, 61% had complications in the current pregnancy, 46% had pre-term birth due to pre-eclampsia and/or fetal distress, 22% of the fetuses had intrauterine growth restriction, 7% of the neonates needed the intensive care unit, and 7% died.

Vidyasagar et al. recently described clinical characteristics and aldosterone concentrations (measured via LC-MS/MS) and direct renin concentration (via chemiluminescent immunoassay) in nine pregnant women with confirmed primary aldosteronism and 33 pregnant women with chronic hypertension via retrospective chart analysis [18]. In the nine cases of confirmed aldosteronism, 50% experienced pre-term delivery due to uncontrolled blood pressure, 37.5% had pre-eclampsia (compared to 15% in the chronic hypertension group), and there were two small for gestational age neonates, and two NICU admissions. The biochemical analysis of the RAAS in the nine patients with primary aldosteronism was performed in different weeks of gestation, ranging from 7–36 gestational weeks (mean of 28), and the mean and range of aldosterone values were 1038 pmol/L and 378–1870 pmol/L. All nine patients with confirmed aldosteronism had a direct renin concentration less than 20 mU/L, and this was significantly lower compared to the chronic hypertension patients. Aldosterone concentrations were comparatively similar, and direct renin concentrations were slightly lower in the nine patients with confirmed aldosterone compared to those reported by Sanga et al. at gestational week 28, but both biomarkers were elevated compared to non-pregnant patients with and without confirmed aldosteronism as described above. Results from our study are comparable, where the median and interquartile range of aldosterone concentrations at week 28 in the Q4 group were 1041 pmol/L and 586–1742 pmol/L, and in patients with declining renin activity over gestation, similar clinical characteristics were observed to those described in aldosteronism patients with lower direct renin concentrations.

A limitation to our study is that it did not include testing for, or confirmation of, aldosteronism as primary or secondary. However, several insights can be gleaned from our study when considered in context with the characteristics described in the preceding studies [18,20] and prevalence of aldosteronism reported by Brown et al. in the four-site study of 1015 patients receiving care for the treatment of hypertension [10]. The latter study demonstrated rates of 11–22% of renin-independent (primary) aldosteronism, which increased over the categories of normotensive, untreated stage 1 hypertension, untreated stage 2 hypertension, and treated resistant hypertension; the rates increased among a subset with suppressed renin activity, especially in the treated resistant group (51.6% prevalence of confirmed aldosteronism). In our study, the aldosterone excess group was defined as the top 25% of aldosterone concentrations, and when patients within this group were selected for declining renin activity over gestation (considering that suppressed renin activity is a feature of primary aldosteronism outside of pregnancy), there were 19 out of 128 patients, about 15%, which is consistent with current estimates of primary aldosteronism in the population. Similar to those reviewed above [20], 68% of these patients in our study had chronic hypertension; mean blood pressure was elevated despite more antihypertensive use, there were three intrauterine growth restriction cases (15% prevalence, which was greater than the other groups), and the highest prevalence (31.6%) of pre-term birth was noted.

The clinical utility of the aldosterone-to-renin ratio for detection of primary aldosteronism in pregnancy is unclear. First, the timing of determination of aldosterone and renin for detection of aldosteronism matters during pregnancy; there is an upward shift in the aldosterone and renin ranges compared to non-pregnant states [44], which rises over gestation, but the rise in aldosterone concentrations is proportionally greater than the rise in renin activity over the course of pregnancy (reflected in our study by a doubling in the aldosterone-to-angiotensin II ratio from the first to third trimester of pregnancy in the whole cohort). Second, outside of pregnancy, the elevated aldosterone-to-renin ratio in patients with primary aldosteronism results from renin-independent secretion of aldosterone leading to suppression of renin (as part of a regulatory negative feedback loop) [8]. Whether mechanisms of the normal RAAS response to pregnancy (increased renin activity and aldosterone secretion) are altered in patients with pre-existing aldosteronism (of multiple etiologies) is simply not known. We did note a small number of patients in Q4 that also had very elevated values of PRA-S, which could also be attributed to secondary aldosteronism resulting from renovascular hypertension [7], which has been reported to present as pre-eclampsia [49]. Moreover, abnormalities in the mineralocorticoid effector mechanisms (such as reduced mineralocorticoid receptor number in mononuclear leukocytes in the absence of abnormal plasma aldosterone levels) have been reported as a potential contributor to the development of pre-eclampsia [50,51]. Further, highly elevated levels of progesterone in pregnancy may promote aldosterone production independently from the renin–angiotensin system [52] or even interfere with aldosterone signaling through the mineralocorticoid receptor [53]. Results from a recent trial, treatment for mild chronic hypertension during pregnancy, underscore the important of blood pressure control in pregnancy for the prevention of adverse outcomes [54], and patients with hypertension in our study were treated with labetalol (a beta blocker) or nifedipine (a calcium channel blocker in the dihydropyridine class), where the former can decrease and the latter can increase plasma renin activity [55], potentially confounding interpretation of RAAS in pregnancy. Likewise, the ratio of uric acid to creatinine in serum is associated with the development of pre-eclampsia [56]. While our study design did not include collection of these variables, it is important to note that hyperuricemia may lead to increased renin activity in some hypertensive individuals [57].

In the current study, our analysis was based on aldosterone concentrations measured in the first trimester, prior to the peak pregnancy-mediated response. In all quartiles, plasma renin activity (estimated using a surrogate value for plasma renin activity, PRA-S, calculated from the sum of the concentrations of angiotensin I and angiotensin II) was proportional to aldosterone concentrations in the first trimester, and was highest in the top quartile. However, only in the top quartile did median PRA-S values decrease from first to third trimester, and aldosterone concentrations did not significantly rise over gestation (in contrast to the other quartiles). These data suggest that renin activity and aldosterone secretion may increase initially in response to pregnancy in patients with excess aldosterone (values are much higher compared to non-pregnant with or without confirmed aldosteronism), and declining renin activity over gestation with prolonged secretion of high levels of aldosterone may be indicative of feedback suppression on renin as part of the pathology of aldosteronism. A limitation to our study is that we did not compare PRA-S values to traditional plasma renin activity assays, although previous studies outside of pregnancy reported a strong correlation among values derived via PRA-S and plasma renin activity [26,58]. Advantages of the LC-MS/MS-based methods used in our study are that multiple effectors are quantified in a single analytic run, calculated biomarkers (e.g., for PRA-S) are derived from directly measured concentrations of effectors at high sensitivity, and the reproducibility of mass spectrometry allows for comparison across multiple studies.

Additional limitations in our study to consider include a high prevalence of overweight and obesity in our cohort, where body fat percentage is linked to risk of aldosteronism [59]. Additionally, sodium or fluid status was not controlled in our study. Differences in sodium, renin, and aldosterone relationships in pregnancy hypertension vs. normotensive are not well understood [60–62]. Moreover, other hormones with important roles in fluid adaptation, such as vasopressin, have been associated with development of pre-eclampsia [63,64].

5. Conclusions

Results from our study agree with those of previous studies where lower renin activity and aldosterone concentrations were associated with pre-eclampsia/hypertension in pregnancy, and report new evidence of an association between very high concentrations of aldosterone in serum in the first trimester of pregnancy and a phenotype of elevated blood pressure. Further stratification by increasing or decreasing renin activity over gestation revealed a subset of patients with a higher prevalence of chronic hypertension, use of antihypertensive medications, pre-term birth, and slightly higher prevalence of IUGR. These findings suggest that aldosterone excess may underlie the development of hypertension in pregnancy in a significant subpopulation of individuals.

Supplementary Materials: The following supporting information can be downloaded at: https://www.mdpi.com/article/10.3390/biomedicines11112954/s1, Table S1: Concentrations of RAAS components measured in a cohort of 128 patients in the first and third trimesters of pregnancy.

Author Contributions: Conceptualization, R.S., N.P., J.A.B. and J.M.O.; methodology, R.S., M.P. and J.M.O.; data collection, R.S., A.S. (Aarthi Srinivasan), K.V., H.H. and C.C.; formal analysis, A.S. (Aric Schadler), D.D. and R.S.; data curation, D.D. and R.S.; writing, R.S.; supervision, J.A.B., J.M.O. and R.S.; project administration, C.C.; funding acquisition, J.A.B., J.M.O. and R.S. All authors have read and agreed to the published version of the manuscript.

Funding: This research was funded by the Kentucky BIRCWH Program, National Institutes of Health, grant number 5K12DA035150 (R.S.) and the Kentucky Children's Hospital Children's Miracle Network Research Fund (J.A.B.). The APC was funded by J.A.B.

Institutional Review Board Statement: This study was conducted in accordance with the Declaration of Helsinki, and approved by the Institutional Review Board on 02/07/2019 at the University of Kentucky (approval #47841). All study procedures/methods were performed in accordance with relevant guidelines and regulations.

Informed Consent Statement: Informed consent was obtained from all subjects involved in this study.

Data Availability Statement: Data may be made available by investigators upon reasonable request to the corresponding author: robin.shoemaker@uky.edu.

Acknowledgments: We would like to acknowledge the UK Cardiovascular Research Priority Area, and the UK RAAS Analytical Laboratory for their support in measurement of RAAS components.

Conflicts of Interest: M. Poglitsch was an employee at Attoquant Diagnostics, a company developing angiotensin-based biomarkers for hypertension, at the time this study was carried out. The other authors report no conflicts.

References

1. Garovic, V.D.; White, W.M.; Vaughan, L.; Saiki, M.; Parashuram, S.; Garcia-Valencia, O.; Weissgerber, T.L.; Milic, N.; Weaver, A.; Mielke, M.M. Incidence and Long-Term Outcomes of Hypertensive Disorders of Pregnancy. *J. Am. Coll. Cardiol.* **2020**, *75*, 2323–2334. [CrossRef] [PubMed]
2. Garovic, V.D.; Dechend, R.; Easterling, T.; Karumanchi, S.A.; McMurtry Baird, S.; Magee, L.A.; Rana, S.; Vermunt, J.V.; August, P.; American Heart Association Council on Hypertension; et al. Hypertension in Pregnancy: Diagnosis, Blood Pressure Goals, and Pharmacotherapy: A Scientific Statement From the American Heart Association. *Hypertension* **2022**, *79*, e21–e41. [CrossRef] [PubMed]

3. Wen, T.; Schmidt, C.N.; Sobhani, N.C.; Guglielminotti, J.; Miller, E.C.; Sutton, D.; Lahtermaher, Y.; D'Alton, M.E.; Friedman, A.M. Trends and outcomes for deliveries with hypertensive disorders of pregnancy from 2000 to 2018: A repeated cross-sectional study. *BJOG* **2022**, *129*, 1050–1060. [CrossRef] [PubMed]
4. Biwer, L.A.; Wallingford, M.C.; Jaffe, I.Z. Vascular Mineralocorticoid Receptor: Evolutionary Mediator of Wound Healing Turned Harmful by Our Modern Lifestyle. *Am. J. Hypertens.* **2019**, *32*, 123–134. [CrossRef] [PubMed]
5. Buffolo, F.; Tetti, M.; Mulatero, P.; Monticone, S. Aldosterone as a Mediator of Cardiovascular Damage. *Hypertension* **2022**, *79*, 1899–1911. [CrossRef] [PubMed]
6. Funder, J.W. Primary aldosteronism as a public health issue. *Lancet Diabetes Endocrinol.* **2016**, *4*, 972–973. [CrossRef]
7. Robertson, J.I.; Morton, J.J.; Tillman, D.M.; Lever, A.F. The pathophysiology of renovascular hypertension. *J. Hypertens. Suppl.* **1986**, *4*, S95–S103.
8. Funder, J.W.; Carey, R.M.; Mantero, F.; Murad, M.H.; Reincke, M.; Shibata, H.; Stowasser, M.; Young, W.F., Jr. The Management of Primary Aldosteronism: Case Detection, Diagnosis, and Treatment: An Endocrine Society Clinical Practice Guideline. *J. Clin. Endocrinol. Metab.* **2016**, *101*, 1889–1916. [CrossRef]
9. Monticone, S.; Burrello, J.; Tizzani, D.; Bertello, C.; Viola, A.; Buffolo, F.; Gabetti, L.; Mengozzi, G.; Williams, T.A.; Rabbia, F.; et al. Prevalence and Clinical Manifestations of Primary Aldosteronism Encountered in Primary Care Practice. *J. Am. Coll. Cardiol.* **2017**, *69*, 1811–1820. [CrossRef]
10. Brown, J.M.; Siddiqui, M.; Calhoun, D.A.; Carey, R.M.; Hopkins, P.N.; Williams, G.H.; Vaidya, A. The Unrecognized Prevalence of Primary Aldosteronism: A Cross-sectional Study. *Ann. Intern. Med.* **2020**, *173*, 10–20. [CrossRef]
11. Tsao, C.W.; Aday, A.W.; Almarzooq, Z.I.; Anderson, C.A.M.; Arora, P.; Avery, C.L.; Baker-Smith, C.M.; Beaton, A.Z.; Boehme, A.K.; Buxton, A.E.; et al. Heart Disease and Stroke Statistics-2023 Update: A Report From the American Heart Association. *Circulation* **2023**, *147*, e93–e621. [CrossRef] [PubMed]
12. Langer, B.; Grima, M.; Coquard, C.; Bader, A.M.; Schlaeder, G.; Imbs, J.L. Plasma active renin, angiotensin I, and angiotensin II during pregnancy and in preeclampsia. *Obstet. Gynecol.* **1998**, *91*, 196–202. [CrossRef] [PubMed]
13. Verdonk, K.; Visser, W.; Van Den Meiracker, A.H.; Danser, A.H. The renin-angiotensin-aldosterone system in pre-eclampsia: The delicate balance between good and bad. *Clin. Sci.* **2014**, *126*, 537–544. [CrossRef] [PubMed]
14. Tsai, Y.L.; Wu, S.J.; Chen, Y.M.; Hsieh, B.S. Changes in renin activity, aldosterone level and electrolytes in pregnancy-induced hypertension. *J. Formos. Med. Assoc.* **1993**, *92*, 514–518.
15. Elsheikh, A.; Creatsas, G.; Mastorakos, G.; Milingos, S.; Loutradis, D.; Michalas, S. The renin-aldosterone system during normal and hypertensive pregnancy. *Arch. Gynecol. Obstet.* **2001**, *264*, 182–185. [CrossRef]
16. Malha, L.; Sison, C.P.; Helseth, G.; Sealey, J.E.; August, P. Renin-Angiotensin-Aldosterone Profiles in Pregnant Women With Chronic Hypertension. *Hypertension* **2018**, *72*, 417–424. [CrossRef]
17. Shoemaker, R.; Poglitsch, M.; Huang, H.; Vignes, K.; Srinivasan, A.; Cockerham, C.; Schadler, A.; Bauer, J.A.; O'Brien, J.M. Activation of the Renin-Angiotensin-Aldosterone System Is Attenuated in Hypertensive Compared with Normotensive Pregnancy. *Int. J. Mol. Sci.* **2023**, *24*, 12728. [CrossRef]
18. Vidyasagar, S.; Kumar, S.; Morton, A. Screening for primary aldosteronism in pregnancy. *Pregnancy Hypertens.* **2021**, *25*, 171–174. [CrossRef]
19. Morton, A. Primary aldosteronism and pregnancy. *Pregnancy Hypertens.* **2015**, *5*, 259–262. [CrossRef]
20. Sanga, V.; Rossitto, G.; Seccia, T.M.; Rossi, G.P. Management and Outcomes of Primary Aldosteronism in Pregnancy: A Systematic Review. *Hypertension* **2022**, *79*, 1912–1921. [CrossRef]
21. LeFevre, M.L.; U.S. Preventive Services Task Force. Low-dose aspirin use for the prevention of morbidity and mortality from preeclampsia: U.S. Preventive Services Task Force recommendation statement. *Ann. Intern. Med.* **2014**, *161*, 819–826. [CrossRef] [PubMed]
22. Pickering, T.G.; Hall, J.E.; Appel, L.J.; Falkner, B.E.; Graves, J.W.; Hill, M.N.; Jones, D.H.; Kurtz, T.; Sheps, S.G.; Roccella, F.J.; et al. Recommendations for blood pressure measurement in humans: An AHA scientific statement from the Council on High Blood Pressure Research Professional and Public Education Subcommittee. *J. Clin. Hypertens* **2005**, *7*, 102–109. [CrossRef] [PubMed]
23. Hypertension in pregnancy. Report of the American College of Obstetricians and Gynecologists' Task Force on Hypertension in Pregnancy. *Obstet. Gynecol.* **2013**, *122*, 1122–1131. [CrossRef]
24. Basu, R.; Poglitsch, M.; Yogasundaram, H.; Thomas, J.; Rowe, B.H.; Oudit, G.Y. Roles of Angiotensin Peptides and Recombinant Human ACE2 in Heart Failure. *J. Am. Coll. Cardiol.* **2017**, *69*, 805–819. [CrossRef]
25. Pavo, N.; Wurm, R.; Goliasch, G.; Novak, J.F.; Strunk, G.; Gyongyosi, M.; Poglitsch, M.; Saemann, M.D.; Hulsmann, M. Renin-Angiotensin System Fingerprints of Heart Failure with Reduced Ejection Fraction. *J. Am. Coll. Cardiol.* **2016**, *68*, 2912–2914. [CrossRef]
26. Burrello, J.; Buffolo, F.; Domenig, O.; Tetti, M.; Pecori, A.; Monticone, S.; Poglitsch, M.; Mulatero, P. Renin-Angiotensin-Aldosterone System Triple-A Analysis for the Screening of Primary Aldosteronism. *Hypertension* **2020**, *75*, 163–172. [CrossRef]
27. Guo, Z.; Poglitsch, M.; McWhinney, B.C.; Ungerer, J.P.J.; Ahmed, A.H.; Gordon, R.D.; Wolley, M.; Stowasser, M. Measurement of Equilibrium Angiotensin II in the Diagnosis of Primary Aldosteronism. *Clin. Chem.* **2020**, *66*, 483–492. [CrossRef]
28. Van Rooyen, J.M.; Poglitsch, M.; Huisman, H.W.; Mels, C.; Kruger, R.; Malan, L.; Botha, S.; Lammertyn, L.; Gafane, L.; Schutte, A.E. Quantification of systemic renin-angiotensin system peptides of hypertensive black and white African men established from the RAS-Fingerprint(R). *J. Renin Angiotensin Aldosterone Syst.* **2016**, *17*, 1470320316669880. [CrossRef]

29. Arisido, M.W.; Foco, L.; Shoemaker, R.; Melotti, R.; Delles, C.; Gogele, M.; Barolo, S.; Baron, S.; Azizi, M.; Dominiczak, A.F.; et al. Cluster analysis of angiotensin biomarkers to identify antihypertensive drug treatment in population studies. *BMC Med. Res. Methodol.* **2023**, *23*, 131. [CrossRef]
30. Binder, C.; Poglitsch, M.; Duca, F.; Rettl, R.; Dachs, T.M.; Dalos, D.; Schrutka, L.; Seirer, B.; Ligios, L.C.; Capelle, C.; et al. Renin Feedback Is an Independent Predictor of Outcome in HFpEF. *J. Pers. Med.* **2021**, *11*, 370. [CrossRef]
31. Reindl-Schwaighofer, R.; Hödlmoser, S.; Eskandary, F.; Poglitsch, M.; Bonderman, D.; Strassl, R.; Aberle, J.H.; Oberbauer, R.; Zoufaly, A.; Hecking, M. Angiotensin-Converting Enzyme 2 (ACE2) Elevation in Severe COVID-19. *Am. J. Respir. Crit. Care Med.* **2021**, *203*, 1191–1196. [CrossRef] [PubMed]
32. Brown, M.A.; Zammit, V.C.; Mitar, D.A.; Whitworth, J.A. Renin-aldosterone relationships in pregnancy-induced hypertension. *Am. J. Hypertens.* **1992**, *5*, 366–371. [CrossRef] [PubMed]
33. Zitouni, H.; Raguema, N.; Gannoun, M.B.A.; Hebert-Stutter, M.; Zouari, I.; Maleh, W.; Faleh, R.; Letaifa, D.B.; Almawi, W.Y.; Fournier, T.; et al. Impact of obesity on the association of active renin and plasma aldosterone concentrations, and aldosterone-to-renin ratio with preeclampsia. *Pregnancy Hypertens.* **2018**, *14*, 139–144. [CrossRef] [PubMed]
34. Gant, N.F.; Daley, G.L.; Chand, S.; Whalley, P.J.; MacDonald, P.C. A study of angiotensin II pressor response throughout primigravid pregnancy. *J. Clin. Investig.* **1973**, *52*, 2682–2689. [CrossRef] [PubMed]
35. Jensen, E.; Wood, C.; Keller-Wood, M. The normal increase in adrenal secretion during pregnancy contributes to maternal volume expansion and fetal homeostasis. *J. Soc. Gynecol. Investig.* **2002**, *9*, 362–371. [CrossRef] [PubMed]
36. Todkar, A.; Di Chiara, M.; Loffing-Cueni, D.; Bettoni, C.; Mohaupt, M.; Loffing, J.; Wagner, C.A. Aldosterone deficiency adversely affects pregnancy outcome in mice. *Pflugers Arch.* **2012**, *464*, 331–343. [CrossRef] [PubMed]
37. Shukri, M.Z.; Tan, J.W.; Manosroi, W.; Pojoga, L.H.; Rivera, A.; Williams, J.S.; Seely, E.W.; Adler, G.K.; Jaffe, I.Z.; Karas, R.H.; et al. Biological Sex Modulates the Adrenal and Blood Pressure Responses to Angiotensin II. *Hypertension* **2018**, *71*, 1083–1090. [CrossRef]
38. Monori-Kiss, A.; Antal, P.; Szekeres, M.; Varbiro, S.; Fees, A.; Szekacs, B.; Nadasy, G.L. Morphological remodeling of the intramural coronary resistance artery network geometry in chronically Angiotensin II infused hypertensive female rats. *Heliyon* **2020**, *6*, e03807. [CrossRef]
39. Vasan, R.S.; Evans, J.C.; Benjamin, E.J.; Levy, D.; Larson, M.G.; Sundstrom, J.; Murabito, J.M.; Sam, F.; Colucci, W.S.; Wilson, P.W. Relations of serum aldosterone to cardiac structure: Gender-related differences in the Framingham Heart Study. *Hypertension* **2004**, *43*, 957–962. [CrossRef]
40. Faulkner, J.L.; Wright, D.; Antonova, G.; Jaffe, I.Z.; Kennard, S.; Belin de Chantemele, E.J. Midgestation Leptin Infusion Induces Characteristics of Clinical Preeclampsia in Mice, Which Is Ablated by Endothelial Mineralocorticoid Receptor Deletion. *Hypertension* **2022**, *79*, 1536–1547. [CrossRef]
41. Jia, G.; Habibi, J.; Aroor, A.R.; Martinez-Lemus, L.A.; DeMarco, V.G.; Ramirez-Perez, F.I.; Sun, Z.; Hayden, M.R.; Meininger, G.A.; Mueller, K.B.; et al. Endothelial Mineralocorticoid Receptor Mediates Diet-Induced Aortic Stiffness in Females. *Circ. Res.* **2016**, *118*, 935–943. [CrossRef]
42. Bentley-Lewis, R.; Adler, G.K.; Perlstein, T.; Seely, E.W.; Hopkins, P.N.; Williams, G.H.; Garg, R. Body mass index predicts aldosterone production in normotensive adults on a high-salt diet. *J. Clin. Endocrinol. Metab.* **2007**, *92*, 4472–4475. [CrossRef] [PubMed]
43. Wilson, M.; Morganti, A.A.; Zervoudakis, I.; Letcher, R.L.; Romney, B.M.; Von Oeyon, P.; Papera, S.; Sealey, J.E.; Laragh, J.H. Blood pressure, the renin-aldosterone system and sex steroids throughout normal pregnancy. *Am. J. Med.* **1980**, *68*, 97–104. [CrossRef] [PubMed]
44. Lewandowski, K.C.; Tadros-Zins, M.; Horzelski, W.; Grzesiak, M.; Lewinski, A. Establishing Reference Ranges for Aldosterone, Renin and Aldosterone-to-Renin Ratio for Women in the Third-Trimester of Pregnancy. *Exp. Clin. Endocrinol. Diabetes* **2022**, *130*, 210–216. [CrossRef] [PubMed]
45. Pasten, V.; Tapia-Castillo, A.; Fardella, C.E.; Leiva, A.; Carvajal, C.A. Aldosterone and renin concentrations were abnormally elevated in a cohort of normotensive pregnant women. *Endocrine* **2022**, *75*, 899–906. [CrossRef]
46. Salas, S.P.; Marshall, G.; Gutierrez, B.L.; Rosso, P. Time course of maternal plasma volume and hormonal changes in women with preeclampsia or fetal growth restriction. *Hypertension* **2006**, *47*, 203–208. [CrossRef]
47. Rehan, M.; Raizman, J.E.; Cavalier, E.; Don-Wauchope, A.C.; Holmes, D.T. Laboratory challenges in primary aldosteronism screening and diagnosis. *Clin. Biochem.* **2015**, *48*, 377–387. [CrossRef]
48. Sealey, J.E.; Gordon, R.D.; Mantero, F. Plasma renin and aldosterone measurements in low renin hypertensive states. *Trends Endocrinol. Metab.* **2005**, *16*, 86–91. [CrossRef]
49. Omar, M.B.; Kogler, W.; Maharaj, S.; Aung, W. Renal artery stenosis presenting as preeclampsia. *Clin. Hypertens.* **2020**, *26*, 6. [CrossRef]
50. Armanini, D.; Zennaro, C.M.; Martella, L.; Scali, M.; Pratesi, C.; Grella, P.V.; Mantero, F. Mineralocorticoid effector mechanism in preeclampsia. *J. Clin. Endocrinol. Metab.* **1992**, *74*, 946–949. [CrossRef]
51. Armanini, D.; Zennaro, C.M.; Martella, L.; Pratesi, C.; Scali, M.; Zampollo, V. Regulation of aldosterone receptors in hypertension. *Steroids* **1993**, *58*, 611–613. [CrossRef] [PubMed]
52. Szmuilowicz, E.D.; Adler, G.K.; Williams, J.S.; Green, D.E.; Yao, T.M.; Hopkins, P.N.; Seely, E.W. Relationship between aldosterone and progesterone in the human menstrual cycle. *J. Clin. Endocrinol. Metab.* **2006**, *91*, 3981–3987. [CrossRef] [PubMed]

53. Quinkler, M.; Meyer, B.; Bumke-Vogt, C.; Grossmann, C.; Gruber, U.; Oelkers, W.; Diederich, S.; Bahr, V. Agonistic and antagonistic properties of progesterone metabolites at the human mineralocorticoid receptor. *Eur. J. Endocrinol.* **2002**, *146*, 789–799. [CrossRef]
54. Tita, A.T.; Szychowski, J.M.; Boggess, K.; Dugoff, L.; Sibai, B.; Lawrence, K.; Hughes, B.L.; Bell, J.; Aagaard, K.; Edwards, R.K.; et al. Treatment for Mild Chronic Hypertension during Pregnancy. *N. Engl. J. Med.* **2022**, *386*, 1781–1792. [CrossRef] [PubMed]
55. Mulatero, P.; Bertello, C.; Veglio, F.; Monticone, S. Approach to the Patient on Antihypertensive Therapy: Screen for Primary Aldosteronism. *J. Clin. Endocrinol. Metab.* **2022**, *107*, 3175–3181. [CrossRef]
56. Piani, F.; Agnoletti, D.; Baracchi, A.; Scarduelli, S.; Verde, C.; Tossetta, G.; Montaguti, E.; Simonazzi, G.; Degli Esposti, D.; Borghi, C. Serum uric acid to creatinine ratio and risk of preeclampsia and adverse pregnancy outcomes. *J. Hypertens.* **2023**, *41*, 1333–1338. [CrossRef]
57. Mule, G.; Castiglia, A.; Morreale, M.; Geraci, G.; Cusumano, C.; Guarino, L.; Altieri, D.; Panzica, M.; Vaccaro, F.; Cottone, S. Serum uric acid is not independently associated with plasma renin activity and plasma aldosterone in hypertensive adults. *Nutr. Metab. Cardiovasc. Dis.* **2017**, *27*, 350–359. [CrossRef]
58. Guo, M.; Zhou, C.; Xu, G.; Tang, L.; Ruan, Y.; Yu, Y.; Lin, X.; Wu, D.; Chen, H.; Yu, P.; et al. An alternative splicing variant of mineralocorticoid receptor discovered in preeclampsia tissues and its effect on endothelial dysfunction. *Sci. China Life Sci.* **2020**, *63*, 388–400. [CrossRef]
59. Manosroi, W.; Atthakomol, P. High body fat percentage is associated with primary aldosteronism: A cross-sectional study. *BMC Endocr. Disord.* **2020**, *20*, 175. [CrossRef]
60. Verdonk, K.; Saleh, L.; Lankhorst, S.; Smilde, J.E.; van Ingen, M.M.; Garrelds, I.M.; Friesema, E.C.; Russcher, H.; van den Meiracker, A.H.; Visser, W.; et al. Association studies suggest a key role for endothelin-1 in the pathogenesis of preeclampsia and the accompanying renin-angiotensin-aldosterone system suppression. *Hypertension* **2015**, *65*, 1316–1323. [CrossRef]
61. Bentley-Lewis, R.; Graves, S.W.; Seely, E.W. The renin-aldosterone response to stimulation and suppression during normal pregnancy. *Hypertens. Pregnancy* **2005**, *24*, 1–16. [CrossRef] [PubMed]
62. Gennari-Moser, C.; Escher, G.; Kramer, S.; Dick, B.; Eisele, N.; Baumann, M.; Raio, L.; Frey, F.J.; Surbek, D.; Mohaupt, M.G. Normotensive blood pressure in pregnancy: The role of salt and aldosterone. *Hypertension* **2014**, *63*, 362–368. [CrossRef] [PubMed]
63. Erdelyi, L.S.; Hunyady, L.; Balla, A. V2 vasopressin receptor mutations: Future personalized therapy based on individual molecular biology. *Front. Endocrinol.* **2023**, *14*, 1173601. [CrossRef]
64. Sandgren, J.A.; Deng, G.; Linggonegoro, D.W.; Scroggins, S.M.; Perschbacher, K.J.; Nair, A.R.; Nishimura, T.E.; Zhang, S.Y.; Agbor, L.N.; Wu, J.; et al. Arginine vasopressin infusion is sufficient to model clinical features of preeclampsia in mice. *JCI Insight* **2018**, *3*, e99403. [CrossRef] [PubMed]

Disclaimer/Publisher's Note: The statements, opinions and data contained in all publications are solely those of the individual author(s) and contributor(s) and not of MDPI and/or the editor(s). MDPI and/or the editor(s) disclaim responsibility for any injury to people or property resulting from any ideas, methods, instructions or products referred to in the content.

Article

Physical Training vs. Perindopril Treatment on Arterial Stiffening of Spontaneously Hypertensive Rats: A Proteomic Analysis and Possible Mechanisms

Danyelle Siqueira Miotto [1], Francine Duchatsch [1], Aline Dionizio [2], Marília Afonso Rabelo Buzalaf [2] and Sandra Lia Amaral [1,3,*]

1. Joint Graduate Program in Physiological Sciences (PIPGCF), Federal University of Sao Carlos and São Paulo State University, UFSCar/UNESP, São Carlos 14801-903, Brazil; dany.miotto21@gmail.com (D.S.M.); francine.duchatsch@gmail.com (F.D.)
2. Department of Biological Sciences, Bauru School of Dentistry, University of São Paulo—USP, Bauru 17012-901, Brazil; stars_line@hotmail.com (A.D.)
3. Department of Physical Education, School of Sciences, São Paulo State University—UNESP, Bauru 17033-360, Brazil
* Correspondence: amaral.cardoso@unesp.br

Abstract: (1) Background: Arterial stiffness is an important predictor of cardiovascular events. Perindopril and physical exercise are important in controlling hypertension and arterial stiffness, but the mechanisms are unclear. (2) Methods: Thirty-two spontaneously hypertensive rats (SHR) were evaluated for eight weeks: SHR_C (sedentary); SHR_P (sedentary treated with perindopril—3 mg/kg) and SHR_T (trained). Pulse wave velocity (PWV) analysis was performed, and the aorta was collected for proteomic analysis. (3) Results: Both treatments determined a similar reduction in PWV (−33% for SHR_P and −23% for SHR_T) vs. SHR_C, as well as in BP. Among the altered proteins, the proteomic analysis identified an upregulation of the EH domain-containing 2 (EHD2) protein in the SHR_P group, required for nitric oxide-dependent vessel relaxation. The SHR_T group showed downregulation of collagen-1 (COL1). Accordingly, SHR_P showed an increase (+69%) in the e-NOS protein level and SHR_T showed a lower COL1 protein level (−46%) compared with SHR_C. (4) Conclusions: Both perindopril and aerobic training reduced arterial stiffness in SHR; however, the results suggest that the mechanisms can be distinct. While treatment with perindopril increased EHD2, a protein involved in vessel relaxation, aerobic training decreased COL1 protein level, an important protein of the extracellular matrix (ECM) that normally enhances vessel rigidity.

Keywords: SHR; pulse wave velocity; perindopril; collagen and aerobic training

Citation: Miotto, D.S.; Duchatsch, F.; Dionizio, A.; Buzalaf, M.A.R.; Amaral, S.L. Physical Training vs. Perindopril Treatment on Arterial Stiffening of Spontaneously Hypertensive Rats: A Proteomic Analysis and Possible Mechanisms. *Biomedicines* **2023**, *11*, 1381. https://doi.org/10.3390/biomedicines11051381

Academic Editors: Mária Szekeres, György L. Nádasy and András Balla

Received: 27 March 2023
Revised: 21 April 2023
Accepted: 2 May 2023
Published: 6 May 2023

Copyright: © 2023 by the authors. Licensee MDPI, Basel, Switzerland. This article is an open access article distributed under the terms and conditions of the Creative Commons Attribution (CC BY) license (https:// creativecommons.org/licenses/by/ 4.0/).

1. Introduction

An important predictor of cardiovascular events is arterial stiffness, which is assessed by pulse wave velocity (PWV). It is present in hypertension and has been considered an index of vascular aging [1–4]. In this sense, the imbalance between collagen and elastin, the rupture of elastic fibers, and alterations in several proteins of the vascular components contribute to increased arterial stiffening [5–9].

Both pharmacological and non-pharmacological treatments are promising; however, the mechanisms are still uncertain. Among antihypertensive drugs, it has been shown that perindopril, a renin–angiotensin system (RAS) inhibitor, has higher effectiveness, lower mortality rate among others from the same class, and prevents vascular remodeling [10–12], which helps to reduce arterial stiffening when compared to others drugs [13,14]. Recently, our group [15] demonstrated that perindopril reduces blood pressure and does improvements in the RhoA/Rho-kinase/LIMK/Cofilin-1 pathway.

Likewise, physical exercise is a globally accepted tool to control hypertension and PWV [16–18], mainly by adjusting the structural components of the vessel wall such as subendothelial matrix proteins and elastic fibers [19].

This study used proteomic analysis to provide a deeper understanding of vascular wall modulations under perindopril treatment and exercise training. Our hypothesis was that both treatments would be effective in controlling arterial stiffening, but probably through different mechanisms.

2. Materials and Methods

Thirty-two spontaneously hypertensive rats (SHR, 200–250 g, 2 months old) were obtained from the Animal Facility of the Institute of Biomedical Sciences, University of São Paulo, (USP) Brazil. All rats were housed in the animal facility maintenance at the School of Sciences, São Paulo State University—UNESP, campus of Bauru. All rats received water and food (Biobase, Águas Frias, SC, Brazil) ad libitum and were maintained in a dark–light cycle (12–12 h) and temperature-controlled room (22 ± 2 °C). The animal study protocol was approved by the Committee for Ethical Use of Animals (CEUA) of School of Sciences (UNESP, Bauru, #1320/2019 Vol. 1) and are in accordance with the Brazilian Ethical Principles in Animal Research.

2.1. Pharmacological Protocol

All rats were separated into 3 groups with similar body weight (BW) and randomly assigned to undergo an experimental protocol through 8 weeks: SHR_C (n = 12): sedentary SHR, daily treated with filtered water; SHR_P (n = 10): sedentary SHR, daily treated with perindopril (Conversyl, 3 mg/kg of BW, via gavage); SHR_T (n = 10): aerobic-trained SHR throughout the experimental protocol.

The dose of perindopril was chosen based on previous publications [17]. In order to test the effectiveness of the pharmacological treatment, a bolus of Angiotensin I was infused after the treatment period (100 µL, at a dose of 1 µg/µL, i.v.) in 2 treated and 2 control rats and the blood pressure (BP) response was evaluated.

2.2. Physical Training

After 7–10 days adaptation period on the treadmill (5–10 min, at 0.3–0.5 km/h), a maximum physical capacity test (Tmax) was performed on all rats, as published [20]. Then, the trained group was subjected to a physical training program on a treadmill (1 h/day, 5 days/week, at 50–60% of their maximum capacity). The Tmax was repeated after 4 weeks for the adjustment of the speed and at the end to evaluate the effects of training [17].

2.3. Pulse Wave Velocity

At the end of the experimental protocol, PWV was assessed, as previously published [8,15]. In summary, each rat was anesthetized with xylazine hydrochloride (Anasedan®, 10 mg/kg, i.p) and ketamine hydrochloride (Dopalen®, 50 mg/kg, i.p), and two pOpet® probes (Axelife SAS, Saint Nicolas de Redon, France) were positioned on the right forelimb (close to elbow) and hindlimb (close to knee). After stabilization of the signal (in a quiet room), the transit time (TT, ms) was recorded for 10 s and registered using pOpet 1.0 software. Taking together the distance between probes (D, cm) and TT, PWV was calculated using the following formula, as previously published [8]:

$$PWV\ (m/s) = D\ (m)\ /TT\ (s) \tag{1}$$

For PWV values, 10 measurements of each rat were obtained, and the average was calculated.

2.4. Arterial Pressure

Twenty-four hours after the PWV measurement, all rats were anesthetized with xylazine hydrochloride (Anasedan®, 10 mg/mL, i.p) and ketamine hydrochloride (Dopalen®, 50 mg/kg, i.p) and the carotid artery was catheterized, as previously published [21]. On the next day, the blood pressure of each awake animal was continuously recorded for at least 1 h, in a quiet room, using a pressure transducer (DPT100, Utah Medical Products Inc., Midvale, UT, USA) connected to the artery cannula, that sent the signal to an amplifier (Quad Bridge Amp, ADInstruments, Colorado Springs, CO, USA) and then to an acquisition board (Powerlab 4/35, ADInstruments, New South Wales, Australia), as published [22]. Systolic blood pressure (SBP) was derived from pulsatile AP recordings using computer software (Labchart pro v7.1, ADInstruments New South Wales, Australia).

2.5. Proteomic Analysis

2.5.1. Protein Extraction

After the cardiovascular parameter measurements, all rats were deeply anesthetized by an overload of xylazine hydrochloride and ketamine hydrochloride (Anasedan®, 20 mg/kg and Dopalen®, 160 mg/kg, i.v., respectively) and euthanized by guillotine. The thoracic aorta was excised, cleaned with saline solution, and homogenized in liquid nitrogen to prevent protein degradation. For the extraction, a total of 50 mg of tissue was homogenized in 500 µL of lysis buffer (7 M urea, 2 M thiourea, and 40 mM Dithiothreitol, all diluted in 50 mM of AMBIC solution) for 2 h in the refrigerator with continuous shaking and, in the end, centrifuged at $20,817 \times g$ for 30 min at 4 °C, after which the supernatant was taken. Total protein was quantified using the Quick Start ™ Bradford Protein Assay kit (Bio-Rad, Hercules, CA, USA) in duplicate, as described in the literature [23].

2.5.2. Proteomic Analysis of the Aorta

The proteomic analysis was performed as previously described [15,24,25]. A pooled sample of the aorta from 2 rats was analyzed and the proteomic analysis was performed in biological triplicates. They were subdivided into 50 µL aliquots containing 50 µg of proteins (1 µg/µL) and then 25 µL of a 0.2% RapiGest SF solution (Waters Corporation, Milford, MA, USA) was added, followed by agitation and another addition of 10 µL 50 mM of AMBIC. The samples were incubated at 37 °C for 30 min, after which the samples were reduced using 5 mM of dithiothreitol (DTT, Merck KGaA, Darmstadt, Germany), incubated at 37 °C for 40 min and alkylated with 10 mM of iodoacetamide (IAA, Sigma-Aldrich, Darmstadt, Germany), agitated and incubated in the dark at room temperature for 30 min. The samples were digested with the addition of 2% (w/w) trypsin (Thermo Scientific, Santa Clara, CA, USA) at 37 °C overnight. After the digestion, 10 µL of 5% trifluoroacetic acid (TFA) was added, followed by agitation and incubation at 37 °C for 90 min. Subsequently, the samples were centrifuged at $20,817 \times g$ at 6 °C for 30 min. The supernatants were purified and desalinated using a Pierce C18 Spin column (Thermo Scientific, Santa Clara, CA, USA). The supernatant was resuspended in 3% acetonitrile and 0.1% formic acid as standard. The peptide identification was performed on a nanoAcquity UPLC-Xevo QTof MS system (Waters Corporation, Manchester, United Kingdom), as previously described [26]. Protein identification and quantification were obtained using ProteinLynx Global Server (PLGS) version 3.0, using the ion-counting algorithm incorporated into the software. The data obtained were searched in the database of the species Rattus Norvegicus (UniProtKB/Swiss-Prot). The protein profile was obtained using the CYTOSCAPE® software v.3.7.0 (Java® 1.8.0_162) and the plugins ClusterMarker and ClueGO. All proteins identified by the mass spectrometer were inserted into the software, using their access number, and can also be seen in the UniProt database, free of charge available on the virtual platform (Uniprot 2022). After confirming the proteins in the Uniprot_acession database, the first network was created in STRING CONSORTIUM 2022 (STRING version 11.5). Then, it was necessary to make a filter with the taxonomy used in this study (Rattus norvegicus; 10116).

Within this classification, proteins were separated with a ratio value greater than 1 for those found to be upregulated, or with a ratio less than 1 for downregulated. Different numbers were assigned to identify the proteins specific to each group in the comparison.

2.6. Protein Analysis of the Aorta

From the aorta, 30 μg of protein was electrophoretically size-separated by using a polyacrylamide gel system (12%) in a running buffer solution for 55 min at 200 V/500 mA/150 W and then transferred to a nitrocellulose membrane at 120 V/500 mA/150 W for 90 min in a buffer solution. The membranes were stained with Ponceau for verification of the protein bands obtained by electrophoresis and washed in a Tris-buffered saline solution with tween-20 (TBS-T). Membranes were incubated within nonfat dry milk for 15 min in TBS-T solution for 10 min. Using the SNAP i.d. 2.0 Protein Detection system (Merck Millipore, Darmstadt, Germany), the membranes were incubated for 10–30 min in their respective primary antibodies (1% albumin bovine serum, BSA): e-Nos (Anti-eNOS/NOS, BD Transduction Laboratories (biosciences), cat #610297, 1:800), cofilin-1 (cofilin-1 (Ab-3), SAB Signalway Antibody cat#21164, 1:500), p-cofilin-1 (cofilin phosphoSer3, cat#ABP54967, 1:500), Collagen-1 (COL1A1, Antibody, Cell Signaling #84336S, 1:800) and Glyceraldehyde-3-phosphate dehydrogenase (GAPDH FL-335, Santa Cruz Biotechnology, INC #sc-25778, 1:800). Membranes were washed with TBS-T and incubated with their respective secondary antibody: anti-mouse (Polyclonal Peroxidase AffiniPure Goat Anti-Mouse IgG, Jackson ImmunoResearch®, #115035003, 1:1000) or anti-rabbit (Polyclonal Peroxidase AffiniPure goat anti-rabbit IgG, Jackson ImmunoResearch®, #111035003, 1:1000 or 1:800) according to each source of the primary antibody (diluted in 1% BSA). Then, membranes were washed again with TBS-T. The secondary antibodies were detected using a chemical reaction with enhanced luminescence (Immobilon® Crescendo western HRP substrate, Millipore cat#WBLUR0500), and the blots were visualized on C- Digit, Blot Scanner, Li-Cor. The bands were analyzed by using the software Image Studio Digits v. 5.2. The values were normalized by the amount of GAPDH and presented as % of the control group.

2.7. Statistical Analysis

All values were presented as mean ± standard error of the mean (SEM). Shapiro–Wilk test was used to test data for normality. For the samples with normal distribution, a one-way analysis of variance (ANOVA) was performed. When the data failed the normality test, we used the transform command in the SigmaStat software to adjust them to meet the normality requirements. All data were analyzed using Sigma Stat software (v4.0.0.37). Tukey's test was used for the necessary post hoc analysis ($p < 0.05$).

For the proteomic analysis, the comparison between groups was obtained using the Student *t*-test in the PLGS software, considering $p < 0.05$ for the significantly expressed proteins.

3. Results

All groups presented similar BW at the beginning of the experimental protocol (252 ± 22, 258 ± 28, and 245 ± 16 g for SHR_C, SHR_P, and SHR_T, respectively; $p > 0.05$) and similar gain during the eight weeks, since the final BW was similar (307 ± 32, 311 ± 44, and 309 ± 21 g for SHR_C, SHR_P, and SHR_T, respectively; $p > 0.05$). The maximal physical capacity (seconds during Tmax) was similar between the groups at the beginning (819 ± 244, 789 ± 288, and 726 ± 333 s for SHR_C, SHR_P, and SHR_T, respectively; $p > 0.05$). The trained rats ran during the first four weeks at a speed of 1.02 km/h (60% of max). After the second Tmax, the treadmill speed was increased by 1.65 km/h in order to maintain the intensity. At the end of eight weeks, the trained group had higher Tmax compared with the control and perindopril groups (577 ± 189, 743 ± 252, and 1511 ± 432 s for SHR_C, SHR_P, and SHR_T, respectively; $p < 0.0001$). After eight weeks of training or perindopril treatment, the values of SBP (206 ± 10, 131 ± 5, and 150 ± 6 mmHg for SHR_C, SHR_P, and SHR_T, respectively; $p < 0.0001$), MBP (184 ± 11, 115 ± 4 and 130 ± 7 mmHg for SHR_C, SHR_P, and SHR_T, respectively; $p < 0.0001$) and DBP (175 ± 12, 108 ± 5 and 120 ± 8 mmHg for SHR_C, SHR_P, and SHR_T, respectively; $p = 0.0001$) were lower than the control group.

Arterial Stiffness

As shown in Figure 1, both SHR$_P$ (−33%) and SHR$_T$ (−23%) groups presented lower PWV values, compared with SHR$_C$. There was no difference between the perindopril-treated and trained rats.

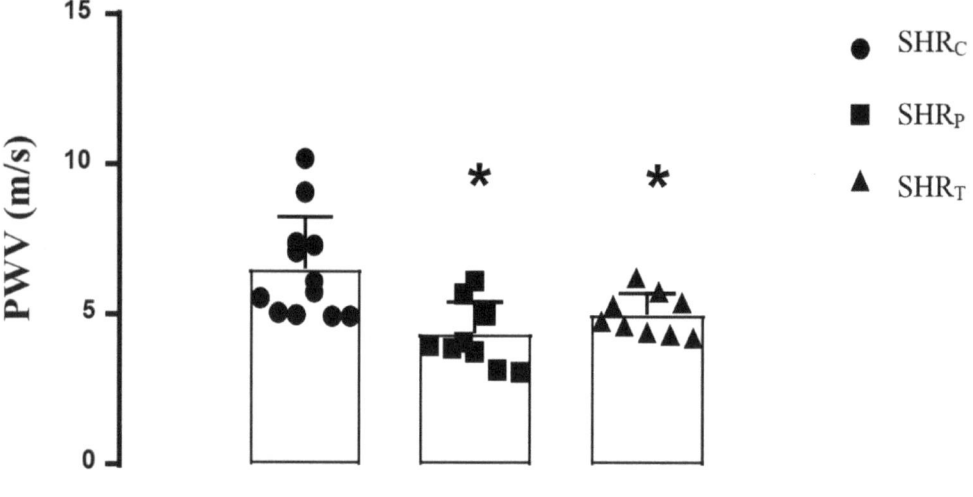

Figure 1. Pulse wave velocity (PWV) values of all SHR groups: sedentary control (SHR$_C$, n = 12), perindopril-treated sedentary (SHR$_P$, n = 10), and trained control (SHR$_T$, n = 10). Significance: * vs. SHR$_C$, p < 0.05.

The ClueGo® analysis, comparing SHR$_P$ vs. SHR$_C$ (Figure 2), demonstrates that 38 subcategories of the cellular component category were modulated. Among them, some subcategories, such as supramolecular fiber, actin cytoskeleton, supramolecular polymer, and membrane raft had the highest modulation. As shown in the Supplementary Materials, the process biologic category included 67 modulated subcategories, and the most modulated were the structural constituent of the cytoskeleton, energy derivation by the oxidation of organic compounds, supramolecular fiber organization and response to reactive oxygen species (Table S1, Supplementary Materials). Finally, in the immune system category (Figure S1, Supplementary Materials), only five subcategories were modulated: mature B cell differentiation involved in immune response, T-helper 1 cell differentiation, regulation of T cell-mediated cytotoxicity, negative regulation of myeloid leukocyte mediated immunity and dendritic cell chemotaxis.

On the other hand, the ClueGo® analysis, comparing SHR$_T$ vs. SHR$_C$ (Figure 3), demonstrated that 16 subcategories were modulated in the cellular component category and the highest modulated subcategory was collagen-containing extracellular matrix, followed by lamellipodium, actin filament bundle and cortical cytoskeleton. In the process biologic category (Table S2, Supplementary Materials), there were 48 modulated subcategories, such as actomyosin structure organization, response to heat, regulation of reactive oxygen species metabolic process, and myofibril assembly. In the immune system category, only four subcategories were modulated: positive regulation of leukocyte-mediated cytotoxicity, myeloid dendritic cell differentiation, regulation of T cell-mediated cytotoxicity, and negative regulation of regulatory T cell differentiation (Figure S2, Supplementary Materials).

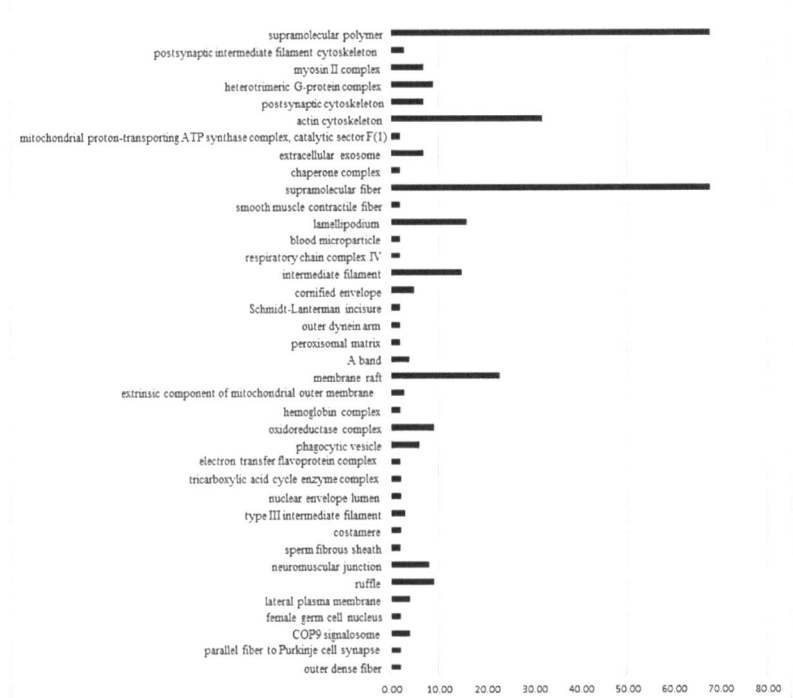

Figure 2. The ClueGo® analysis, comparing SHR$_P$ vs. SHR$_C$ (Figure 2), demonstrates that 38 subcategories of the cellular component category were modulated. Some categories, such as supramolecular fiber, actin cytoskeleton, supramolecular polymer, and membrane raft, had the highest modulation. The categories are presented and based on the gene ontology according to the cellular component in which they participate, provided by the Cytoscape® software v.3.7.0. Only significant terms were used, and the distribution was made according to the percentage of genes associated with each category. The protein access numbers were made available by UniProt.

Table S3 (Supplementary Materials) shows all the 138 expressed proteins under the effects of perindopril on hypertension (SHR$_P$ vs. SHR$_C$). Among them, 73 were upregulated and only 2 were downregulated. Figure 4 illustrates the network performed by the CYTOSCAPE® software using the proteins up- and downregulated present in Table S3 (Supplementary Materials) showing the results between the perindopril-treated rats compared with the control group (SHR$_P$ vs. SHR$_C$). The upregulated proteins are in green color: (P36201, Crip2) Cysteine-rich protein 2; (Q6AY56, Tuba8) Tubulin alpha-8 chain; (Q62736, Cald1) Caldesmon 1; (Q4QRB4, Tubb3) Tubulin beta-3 chain; (P31000, Vim) Vimentins; (Q3KRE8, Tubb2b) Tubulin beta-2B chain; (Q00715, Hist1h2bh) Histone cluster 1; (P04636, Mdh2) Malate dehydrogenase, mitochondrial; (P12346, Tf) Serotransferrin; (P47875, Csrp1) Cysteine and glycine-rich protein 1; (P07150, Anxa1) Annexin A1; (Q6P6Q2, Krt5) Keratin, type II cytoskeletal 5;(Q6AYZ1, Tuba1c) Tubulin alpha-1C chain; (P55063, Hspa1l) Heat shock 70 kDa protein 1-like; (P62963, Pfn1) Profilin-1; (P15999, Atp5a1) ATP synthase subunit alpha_ mitochondrial; (P69897, Tubb5) Tubulin beta-5 chain; (Q07936, Anxa2) Annexin A2; (P08010, Gstm2) Glutathione S-transferase Mu 2; (P05065, Aldoa) Fructose–bisphosphate aldolase A; (P63102, Ywhaz) 14-3-3 protein zeta/delta; (Q7M0E3, Dstn) Destrin; (P02454, Col1a1) Collagen alpha-1(I) chain; (Q9Z1P2, Actn1) Alpha-actinin-1; (P63269, Actg2) Actin, gamma-enteric smooth muscle; (Q5XI73, Arhgdia) Rho GDP-dissociation inhibitor 1; (P15650 Acadl) Long-chain specific acyl-CoA dehydrogenase, mitochondrial; (Q9ER34, Aco2) Aconitate hydratase, mitochondrial; (Q5XIF6, Tuba4a) Tubulin alpha-4A chain; (P31232,Tagln)

Transgelin; (P18666, Myl12b) Myosin regulatory light chain 12B; (P63018 Hspa8) Heat shock cognate 71 kDa protein; (P85108, Tubb2a) Tubulin beta-2A chain; (P02600, Myl1) Myosin light chain 1/3, skeletal muscle isoform; (Q6P9T8, Tubb4b) Tubulin, beta 4B chain; (P06399, Fga) Fibrinogen alpha chain; (Hspa5) 78 kDa glucose-regulated protein; (P70623, Fabp4) Fatty acid-binding protein 4, adipocyte; (Q68FR8, Tuba3b) Tubulin, alpha 3b; (P21807, Prph) Peripherin; (Q10758, Krt8) Keratin, type II cytoskeletal 8; (Q5RKI0, Wdr1) WD repeat-containing protein 1; (P62630, Eef1a1) Elongation factor 1-alpha 1; (Q9QXQ0, Actn4) Alpha-actinin-4; (P20760, Igg-2a) Ig gamma-2A chain C region; (P68136, Acta1) Actin, alpha skeletal muscle; (Q62812, Myh9) Myosin, heavy chain 9, non-muscle-like 1; (P68035, Acta2) Actin, alpha cardiac muscle 1; (Ptrf) Polymerase 1 and transcript release factor; (P02770, Alb) Serum albumin; (P60711, Actb) Actin, cytoplasmic 1; (Q6IG12, Krt7) Keratin; (P13832, Myl12a) Myosin regulatory light chain RLC-A; (P48675, Des) Desmin; (Q4V8H8, Ehd2) EH domain-containing protein 2; (P47853, Bgn) Biglycan; (Q9WVH8, Fbln5) Fibulin-5; (P14659, Hspa2) Heat shock-related 70 kDa protein 2; (P11762, Lgals1) Galectin-1; (Q9JLT0, Myh10) Myosin heavy chain 9/10/11/14; (P42930, Hspb1) Heat shock protein family b (small) member 1; (P85973, Pnp) Purine-nucleoside phosphorylase; (P68370, Tuba1a) Tubulin alpha-1A chain; (Q64122, Myl9) Myosin regulatory light chain 9; (P10111, Ppia) Peptidyl-prolyl cis-trans isomerase A; (P42930, Hspa1a) Heat shock 70kd protein 1b (mapped); (Q01129, Dcn) Decorin; (P16409, Myl3) Myosin light chain 3. On the other side, only two proteins were downregulated which are in red: (P01946, Hba1) Hemoglobin subunit alpha-1/2 and (P02091, Hbb) Hemoglobin subunit beta-1 after perindopril treatment (Figure 4).

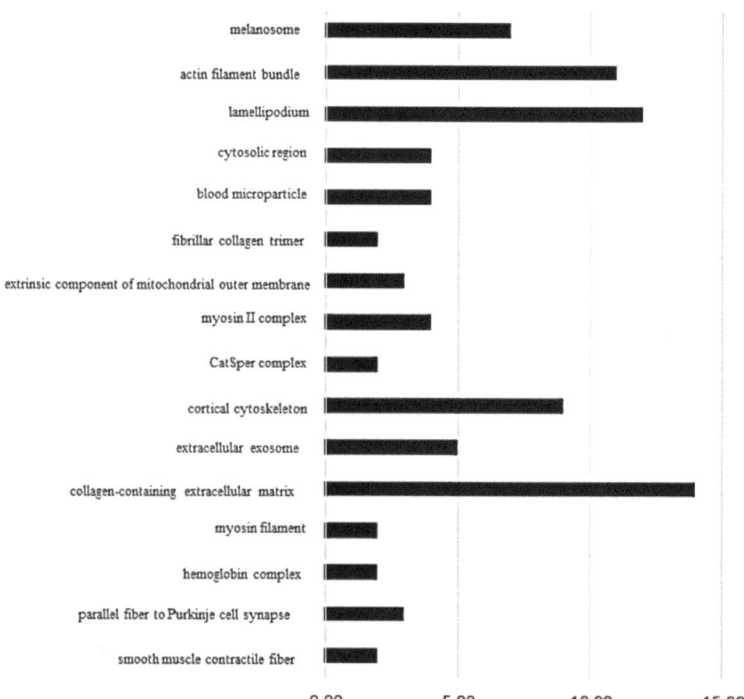

Figure 3. The ClueGo® analysis, comparing SHR$_T$ vs. SHR$_C$ (Figure 3), demonstrated that 16 subcategories were modulated in the cellular component category and the highest modulated subcategory was collagen-containing extracellular matrix, followed by lamellipodium, actin filament bundle, and cortical cytoskeleton. The categories are presented and based on the gene ontology according to the cellular component in which they participate, provided by the Cytoscape® software v.3.7.0. Only significant terms were used, and the distribution was made according to the percentage of genes associated with each category. The protein access numbers were made available by UniProt.

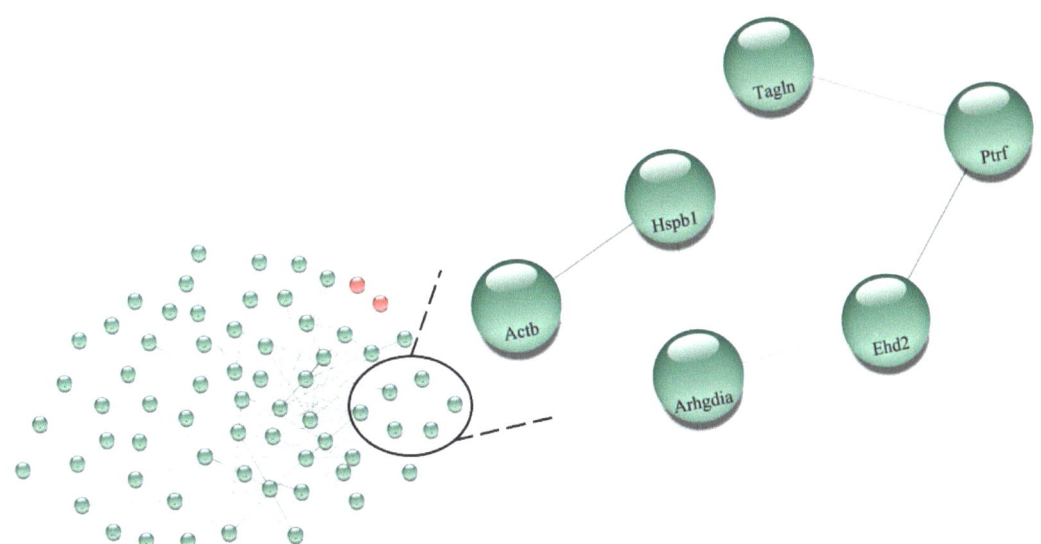

Figure 4. Network illustrating the results between the perindopril-treated rats compared with the control group (SHR$_P$ vs. SHR$_C$), performed by the CYTOSCAPE® software using the proteins up- and downregulated present in Table S3. The upregulated proteins are in green color: (P36201, Crip2) Cysteine-rich protein 2; (Q6AY56, Tuba8) Tubulin alpha-8 chain; (Q62736, Cald1) Caldesmon 1; (Q4QRB4, Tubb3) Tubulin beta-3 chain; (P31000, Vim) Vimentins; (Q3KRE8, Tubb2b) Tubulin beta-2B chain; (Q00715, Hist1h2bh) Histone cluster 1; (P04636, Mdh2) Malate dehydrogenase, mitochondrial; (P12346, Tf) Serotransferrin; (P47875, Csrp1) Cysteine and glycine-rich protein 1; (P07150, Anxa1) Annexin A1; (Q6P6Q2, Krt5) Keratin, type II cytoskeletal 5;(Q6AYZ1, Tuba1c) Tubulin alpha-1C chain; (P55063, Hspa1l) Heat shock 70 kDa protein 1-like; (P62963, Pfn1) Profilin-1; (P15999, Atp5a1) ATP synthase subunit alpha_ mitochondrial; (P69897, Tubb5) Tubulin beta-5 chain; (Q07936, Anxa2) Annexin A2; (P08010, Gstm2) Glutathione S-transferase Mu 2; (P05065, Aldoa) Fructose-bisphosphate aldolase A; (P63102, Ywhaz) 14-3-3 protein zeta/delta; (Q7M0E3, Dstn) Destrin; (P02454, Col1a1) Collagen alpha-1(I) chain; (Q9Z1P2, Actn1) Alpha-actinin-1; (P63269, Actg2) Actin, gamma-enteric smooth muscle; (Q5XI73, Arhgdia) Rho GDP-dissociation inhibitor 1; (P15650 Acadl) Long-chain specific acyl-CoA dehydrogenase, mitochondrial; (Q9ER34, Aco2) Aconitate hydratase, mitochondrial; (Q5XIF6, Tuba4a) Tubulin alpha-4A chain; (P31232,Tagln) Transgelin; (P18666, Myl12b) Myosin regulatory light chain 12B; (P63018 Hspa8) Heat shock cognate 71 kDa protein; (P85108, Tubb2a) Tubulin beta 2A chain; (P02600, Myl1) Myosin light chain 1/3, skeletal muscle isoform; (Q6P9T8, Tubb4b) Tubulin, beta 4B chain; (P06399, Fga) Fibrinogen alpha chain; (Hspa5) 78 kDa glucose-regulated protein; (P70623, Fabp4) Fatty acid-binding protein 4, adipocyte; (Q68FR8, Tuba3b) Tubulin, alpha 3b; (P21807, Prph) Peripherin; (Q10758, Krt8) Keratin, type II cytoskeletal 8; (Q5RKI0, Wdr1) WD repeat-containing protein 1; (P62630, Eef1a1) Elongation factor 1-alpha 1; (Q9QXQ0, Actn4) Alpha-actinin-4; (P20760, Igg-2a) Ig gamma-2A chain C region; (P68136, Acta1) Actin, alpha skeletal muscle; (Q62812, Myh9) Myosin, heavy chain 9, non-muscle-like 1; (P68035, Acta2) Actin, alpha cardiac muscle 1; (Ptrf) Polymerase 1 and transcript release factor; (P02770, Alb) Serum albumin; (P60711, Actb) Actin, cytoplasmic 1; (Q6IG12, Krt7) Keratin; (P13832, Myl12a) Myosin regulatory light chain RLC-A; (P48675, Des) Desmin; (Q4V8H8, Ehd2) EH domain-containing protein 2; (P47853, Bgn) Biglycan; (Q9WVH8, Fbln5) Fibulin-5; (P14659, Hspa2) Heat shock-related 70 kDa protein 2; (P11762, Lgals1) Galectin-1; (Q9JLT0, Myh10) Myosin heavy chain 9/10/11/14; (P42930, Hspb1) Heat shock protein family b (small) member 1; (P85973, Pnp) Purine-nucleoside phosphorylase; (P68370, Tuba1a) Tubulin alpha-1A chain; (Q64122, Myl9) Myosin regulatory light chain 9; (P10111, Ppia) Peptidyl-prolyl cis-trans isomerase A; (P42930, Hspa1a) Heat shock 70kd protein 1b (mapped); (Q01129, Dcn) Decorin; (P16409, Myl3) Myosin light chain 3. On the other hand, only two proteins

were downregulated which are in red: (P01946, Hba1) Hemoglobin subunit alpha-1/2 and (P02091, Hbb) Hemoglobin subunit beta-1 after perindopril treatment. Highlighted: (P60711, Actb) Actin, cytoplasmic 1; (P42930, Hspb1) Heat shock protein family b (small) member 1; (Q5XI73, Arhgdia) Rho GDP-dissociation inhibitor 1; (Q4V8H8, EHD2) EH domain-containing protein 2; (Ptrf) Polymerase 1 and transcript release factor; (P31232, Tagln) Transgelin.

The comparison between SHR_T and SHR_C, regarding the effects of training on hypertension as shown in Table S4 (Supplementary Materials), showed that 123 proteins were differently expressed. While 7 of them were upregulated, 22 were downregulated. The network made with the proteins in Table S4 (Supplementary Materials) is illustrated in Figure 5 (SHR_T vs. SHR_C). As shown, the upregulated proteins are in green color: (P68035, Acta2) Actin_ alpha cardiac muscle 1; (P63269, Actg2) Actin_ gamma-enteric smooth muscle; (P47853, Bgn) Biglycan; (P06761, Hspa5) Endoplasmic reticulum chaperone BiP; (P70490, Mfge8) Lactadherin; (Q6AY56, Tuba8) Tubulin alpha-8 chain; (Q9JLT0, Myh10) Myosin-10. The downregulated proteins are shown in red: (P47875, Csrp1) Cysteine and glycine-rich protein 1; (Q7M0E3, Dstn) Destrin; (P02454, Col1a1) Collagen alpha-1(I) chain; (P06866, Hp) Haptoglobin; (P50399, Gdi2) Rab GDP dissociation inhibitor beta; (P68136, Acta1) Actin_ alpha skeletal muscle; (P62738) Actin_ aortic smooth muscle; (P60711, Actb) Actin_ cytoplasmic 1; (P63259) Actin_ cytoplasmic 2; (Q9Z1P2, Actn1) Alpha-actinin-1; (P36201, Crip2) Cysteine-rich protein 2; (P04797, Gapdh) Glyceraldehyde-3-phosphate dehydrogenase; (P42930, Hspb1) Heat shock protein beta-1; (P01946, Hba1) Hemoglobin subunit alpha-1/2; (P02091, Hbb) Hemoglobin subunit beta-1; (P11517, ENSRNOP00000048546) Hemoglobin subunit beta-2; (P20760, Igg-2a) Ig gamma-2A chain C region; (P51886, Lum) Lumican; (Q64119, Myl6) Myosin light polypeptide 6; (Q64122, Myl9) Myosin regulatory light polypeptide 9; (P10111, Ppia) Peptidyl-prolyl cis-trans isomerase A; (P02770, Alb) Serum albumin (Figure 5).

Figure 6 illustrates the densitometric analysis of the e-NOS (Figure 6A) and COL1 (Figure 6B) protein levels in the aorta of all rats. As shown, the SHR_P group had higher values of aortic e-NOS protein level (+69%) when compared with the control group. Thus, it can be said that only perindopril treatment in SHR was able to increase e-NOS expression, while training did not significantly increase it when compared to the control group.

On the other hand, aortic COL1 level was lower in the SHR_T group, compared with the control group (−46%, Figure 6B), suggesting that training was able to reduce COL1 expression in SHR, but treatment with perindopril did not significantly reduce it.

The values of aortic cofilin-1 (Figure 7A), p-cofilin-1 (Figure 7B), and the ratio p-cofilin/cofilin-1 (Figure 7C) were similar between the groups, as shown in Figure 7. Therefore, neither perindopril treatment nor aerobic physical training was able to significantly modulate the total and/or phosphorylated cofilin-1 behavior in these SHR animals.

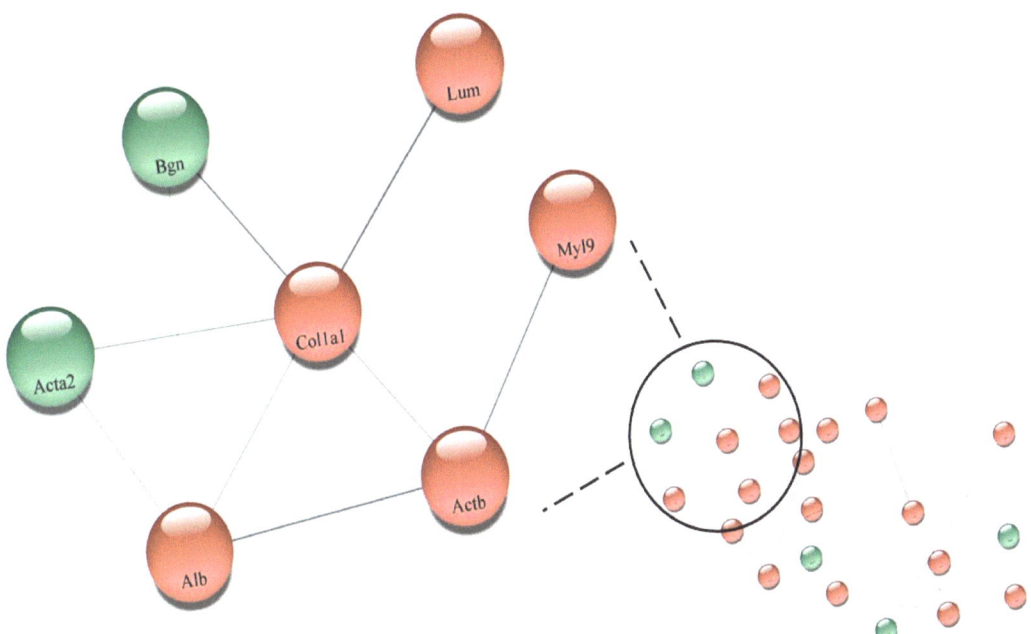

Figure 5. Network illustrating the effects of training on hypertension showed that 123 proteins were differently expressed (Table S4). While 7 of them were upregulated, 22 were downregulated (SHR$_T$ vs. SHR$_C$), performed by the CYTOSCAPE® software using the proteins up- and downregulated. The upregulated proteins are in green color: (P68035, Acta2) Actin_ alpha cardiac muscle 1; (P63269, Actg2) Actin_ gamma-enteric smooth muscle; (P47853, Bgn) Biglycan; (P06761, Hspa5) Endoplasmic reticulum chaperone BiP; (P70490, Mfge8) Lactadherin; (Q6AY56, Tuba8) Tubulin alpha-8 chain; (Q9JLT0, Myh10) Myosin-10. The downregulated proteins are shown in red: (P47875, Csrp1) Cysteine and glycine-rich protein 1; (Q7M0E3, Dstn) Destrin; (P02454, Col1a1) Collagen alpha-1(I) chain; (P06866, Hp) Haptoglobin; (P50399, Gdi2) Rab GDP dissociation inhibitor beta; (P68136, Acta1) Actin_ alpha skeletal muscle; (P62738) Actin_ aortic smooth muscle; (P60711, Actb) Actin_ cytoplasmic 1; (P63259) Actin_ cytoplasmic 2; (Q9Z1P2, Actn1) Alpha-actinin-1; (P36201, Crip2) Cysteine-rich protein 2; (P04797, Gapdh) Glyceraldehyde-3-phosphate dehydrogenase; (P42930, Hspb1) Heat shock protein beta-1; (P01946, Hba1) Hemoglobin subunit alpha-1/2; (P02091, Hbb) Hemoglobin subunit beta-1; (P11517, ENSRNOP00000048546) Hemoglobin subunit beta-2; (P20760, Igg-2a) Ig gamma-2A chain C region; (P51886, Lum) Lumican; (Q64119, Myl6) Myosin light polypeptide 6; (Q64122, Myl9) Myosin regulatory light polypeptide 9; (P10111, Ppia) Peptidyl-prolyl cis-trans isomerase A; (P02770, Alb) Serum albumin. Highlighted: (Q64122, Myl9) Myosin regulatory light polypeptide 9; (P02454, Col1a1) Collagen alpha-1(I) chain; (P60711, Actb) Actin_ cytoplasmic 1; (P51886, Lum) Lumican; (P02770, Alb) Serum albumin; (P68035, Acta2) Actin_ alpha cardiac muscle 1; (P47853, Bgn) Biglycan.

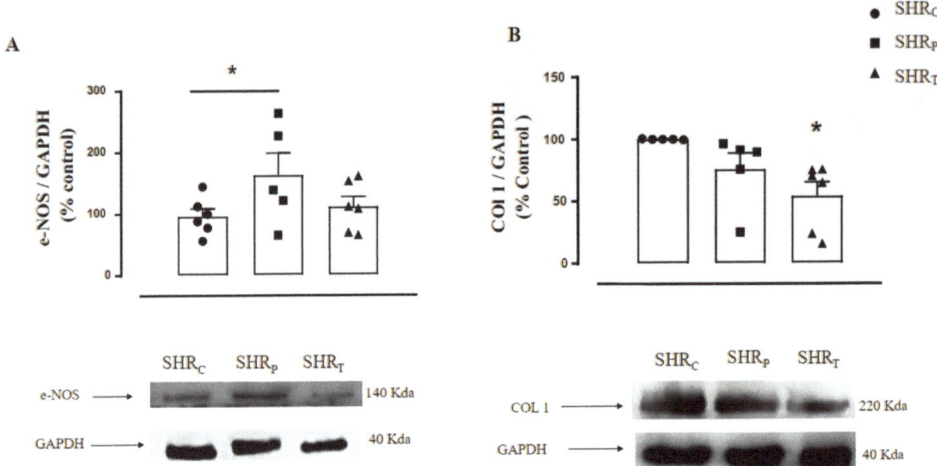

Figure 6. Illustration of the protein level in aortic endothelial nitric oxide synthase protein (e-NOS (**A**)) and collagen 1 (COL1, (**B**)) protein in all groups: sedentary control SHR (SHR$_C$, $n = 6$), sedentary treated with perindopril (SHR$_P$, $n = 6$) and trained control (SHR$_T$, $n = 6$). Figure 6 (bottom panel) also illustrates the representative Gel Blot of e-NOS and COL1 levels in the aorta of all groups, namely SHR$_C$, SHR$_P$, and SHR$_T$, respectively. Significance: * vs. SHR$_C$, $p < 0.05$.

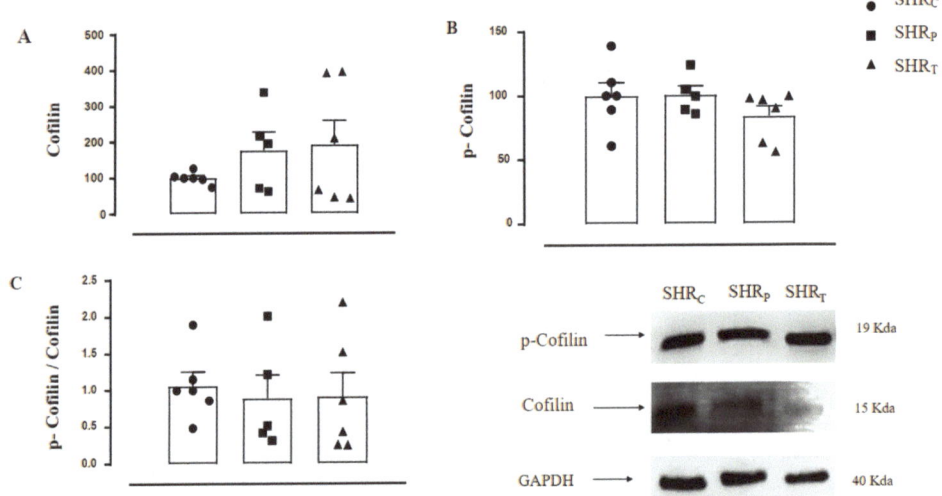

Figure 7. Values of aortic cofilin-1 protein (**A**), p-cofilin-1 protein (**B**), and the p-cofilin-1/cofilin-1 ratio (**C**) in all groups: sedentary control (SHR$_C$, $n = 6$), sedentary treated with perindopril (SHR$_P$, $n = 6$) and trained control (SHR$_T$, $n = 6$). Figure 7 also illustrates the representative Gel Blot of Cofilin-1, p-Cofilin-1, and GAPDH of the aorta in all groups, namely, SHR$_C$, SHR$_P$, and SHR$_T$, respectively.

4. Discussion

The main results of the present study were that either perindopril or physical training significantly reduced the pulse wave velocity of hypertensive rats. On the other hand, the proteomic analysis indicated that pharmacological and non-pharmacological treatment regulated distinct proteins in the aorta, suggesting that the mechanisms may be different

Since arterial stiffness has been considered a marker of vessel aging and a predictor for cardiovascular diseases and future events [2,27], there is a strong recommendation to include this measure in clinical practice [28], sometimes even for pediatric routine [29]. Although it is not clear if arterial stiffness precedes hypertension or vice versa [30,31], several studies clearly demonstrate that hypertensive individuals have higher PWV [8,15,32,33]. Therefore, the maintenance of normal values for blood pressure and PWV are the goals suggested by worldwide guidelines for the management of hypertension [34–36]. It has been shown that an increase of 1 m/s in PWV induces an increase of 15% in cardiovascular risk [37].

It is well-known that increased activity of the renin–angiotensin system (RAS) increases BP and causes vessel remodeling [13]. Therefore, angiotensin-converting enzyme (ACE) inhibitors are highly recommended, mainly because they alter the structure of vessels beyond BP lowering. Ong et al. [38] compared different antihypertensive drug classes, such as diuretic, beta-blocker, calcium antagonist, and ACEi, on arterial stiffness and BP improvement and concluded that the reduction in arterial stiffness is higher under ACEi than under calcium antagonist in a four-week treatment, while all classes had similar responses after four weeks of treatment. In addition, ACEi allows the circulating bradykinin bioavailability, which contributes to the formation of nitric oxide [39] and induces vasodilation.

Recently, our group has shown that SHR rats had higher BP and PWV, compared with normotensive rats, and eight weeks of perindopril treatment reduced both BP and PWV [17]. Likewise, the results of this present study (Figure 1) confirmed our previous results and showed that perindopril-treated SHR had lower BP and PWV compared with the control SHR.

Physical training has been considered an important adjunct to pharmacological treatment to control hypertension and is highly recommended by hypertension societies around the world [34–36], and the mechanism involves a better control of cardiac output and peripheral vascular resistance [40–43]. In addition, exercise training significantly decreases PWV, and the clinical relevance of different types of exercise on PWV reduction has been shown in several pathologies and hypertension [44–48]. In agreement, this present study showed that eight weeks of aerobic exercise training also reduced BP and PWV, and, interesting to note, both groups, the perindopril, and the trained SHR, presented similar values of BP and PWV compared with the control SHR.

From human studies, most of the mechanisms shown to be involved with PWV reduction are systemic, such as increases in plasma nitrite concentration and plasma NOx [48–50] and decreases in plasma levels of endothelin-1 and noradrenaline [48,49]. On the other hand, animal studies have shown important alterations both in the aortic extracellular matrix proteins and in the hypertrophy of vascular smooth muscle cells (VSMC), which contribute to altering vessel remodeling, but not all animal studies evaluate PWV [51]. Therefore, the present study used a non-invasive technique, previously validated by our group [8], which allows measuring PWV and performing histological and molecular analyses in the vessel of the same animal for a better understanding of the possible mechanisms.

The present study carried out a proteomic analysis in the aorta, which allowed the identification of differently expressed proteins of both groups, trained and perindopril-treated SHR. Previously, our group has identified an upregulation of GDP dissociation inhibitor protein (GDIs) in the aorta of perindopril-treated SHR, which is an internal regulator of RhoA pathway activation, suggesting that treated SHR had an inhibition of the RhoA/ROCK/LIMK/Cofilin-1 pathway [15]. Accordingly, Morales-Quinones et al. [52] showed that LIMK inhibition reduces p-Cofilin/Cofilin, which was followed by a reduction in arterial stiffness. Although the results of this present study also showed an upregulation of GDIs after treatment with perindopril followed by a reduction in arterial stiffness, the results of aortic p-Cofilin/Cofilin protein level were only slightly reduced. Probably the higher variability between rats interfered with the results, and this is a limitation of this study.

Additionally, the proteomic analysis showed an interaction between GDIs and EHD2 protein, which was also upregulated in the SHR$_P$ group. Cellular homeostasis is maintained due to an organized process of internalization of nutrients and molecules that move along a series of tubular membranes, and this process is known as the endocytic trafficking system [53]. Likewise, this system is necessary for the product's degradation to return to the membrane surface. Several proteins are recruited to orchestrate this endocytic transport, and among these are the C terminal Eps15 homology domain (EHD)-containing proteins. EHD2 is one of these proteins (family of 4 EHD) that is highly expressed in many tissues including fat, skeletal muscle, lung, spleen, kidney, heart, and tissues rich in caveolae like blood vessels [53–55]. It has been shown that this protein is important for the eNOS-NO-dependent vessel relaxation since EHD2 knockout mice present lower NO abundance in the vascular endothelium and impaired acetylcholine-induced relation in mesenteric arteries [54]. Moreover, the proteomic analysis also demonstrated an interaction of the EHD2 protein with the caveolae-associated protein 1 (Ptrf), which is an important protein involved in the formation of caveolae. Ptrf is essential for caveolae recruitment in the presence of caveolin-1. Matthaeus et al. [54] demonstrated that EHD2 knockout mice showed a decrease in NO production, regardless of eNOS levels, which did not change. Furthermore, they showed that, in these mice lacking EHD2, the caveola was detached from the membrane, which resulted in the redistribution of eNOS into the cytoplasm. Indeed, EHD2 knockdown HUVECs showed that detached caveolae still contained eNOS; however, they observed reduced phosphorylation of eNOS Ser1177 in EHD2 knockdown endothelial cells, which was indicative of reduced eNOS activity. Therefore, these authors concluded that EHD2 in the caveolae neck is required for correct eNOS localization and signaling, and therefore for proper endothelial function. Based on this proposition, and on the results of this present study, we hypothesized that the up-regulated EHD2 observed in the proteomic results could be contributing to maintaining the stabilization of the caveolae at the plasma membrane and, in turn, to the correct location and function of eNOS, which could be modulating vessel relaxation in the perindopril group, as demonstrated by reduced PWV. Although the level of aortic eNOS protein was increased in the present study, we did not evaluate its activity. We also did not evaluate NO formation. However, we have previously shown [15] that perindopril treatment increased plasma nitrite concentration (indicative of NO formation) by 83% in SHR, and this response was negatively correlated with PWV. Therefore, we believe that the correct stabilization of caveolae in the plasma membrane, modulated by the level of the EHD2 protein, could orchestrate the localization and activity of eNOS. We may assume that the lower PWV observed in the group treated with perindopril was induced by the eNOS/NO pathway, which was allowed by a correct stabilization of the caveolae in the membrane, induced by the upregulation of EHD2.

Unlike the effects of perindopril, proteomic analysis revealed that aerobic training downregulated the COL1a1 protein in the aorta of SHR. The elastic characteristic of arteries depends on the balance between structural proteins responsible for determining contraction and relaxation, such as collagen and elastin. Any change in these components, such as increased collagen synthesis and deposition, elastin degradation, as well as disruption of elastic fibers, can lead to vessel remodeling and increased stiffness [7,8,32,56]. Collagen (COL1 and COL3) along with elastin are the major extracellular matrix structural proteins of the cardiovascular system. It is widely distributed extracellularly in most tissues. Both collagen types are predominantly secreted by fibroblasts and smooth muscle cells. While COL1 is the main collagen type present in bone, tendons, dermis, ligaments, and connective tissue, COL3 is distributed mostly in the skin, vessel walls and reticular fibers of most tissues [57]. Collagen and elastin have a key role in modulating the tight balance of elasticity, resilience, and rigidity, which is necessary for physiological functions. Since the elastic fiber network is the most distensible component of the arterial wall, and the collagen fiber network provides rigidity and strength of the arterial wall, the vascular balance of COL and elastin is necessary for vessel physiological function [56,58]. In addition, both fibrillar collagens have similar physiological functions, but COL1 is stiffer and provides structural rigidity over COL 3, which is thinner. In this sense, recently, Witting and Szulcek [58] have proposed that the normal physiologic range of the aortic COL1/COL3 ratio is around 2.04 to

3.83, which is in agreement with a recent study from our group [59] and that COL1 increases to collagen-III in all non-physiologic cases, including hypertension and atherosclerosis.

To confirm the proteomic finding, we evaluated the COL1 protein level on the aorta of the trained SHR and observed that the trained group had a 46% lower level of COL1 when compared with the control SHR group. In addition, less aortic collagen level has been demonstrated after aerobic training [19,60] which contributes to decreased arterial stiffness. The mechanism induced by aerobic training to reduce the level of aortic COL1 protein may involve a lower sympathetic drive to the vessel [17,51] since the synthesis of collagen is mediated by increased sympathetic nerve activity through the beta receptor [61].

Additionally, the network performed in this present work indicated that the downregulated COL1a1 protein directly interacted with the protein Lumican, which was also downregulated. Lumican is a proteoglycan of the extracellular matrix involved in collagen fibrillogenesis and changes in its content may affect collagen organization and, consequently, blood vessels' elastic properties [62]. In this sense, higher expressions of lumican have been found in the aorta of patients with chronic renal failure [63] and in patients with aortic dissection [64]. The reduced regulation of COL1 and Lumican in the aorta of the trained SHR could contribute to decreased vessel rigidity observed in hypertension.

5. Conclusions

In conclusion, the present study indicated that both perindopril and aerobic training similarly reduced arterial stiffness in SHR; however, the proteomic analysis on the aorta revealed that the mechanisms can be distinct. While treatment with perindopril increased the EHD2, a protein involved in the vessel relaxation induced by the e-NOS-NO pathway, aerobic training decreased the aortic COL1 protein level, an important protein of the ECM that normally enhances vessel rigidity.

Supplementary Materials: The following supporting information can be downloaded at: https://www.mdpi.com/article/10.3390/biomedicines11051381/s1, Figure S1: Protein analysis in ClueGO plugins and the number of genes involved within the system immune category between SHR_P vs. SHR_C groups; Figure S2: Protein analysis in ClueGO plug-ins and the number of genes involved within the sistem immune category between SHR_T vs. SHR_C groups; Table S1: Protein analysis in ClueGO plugins and the number of genes involved within the process biologic category between SHR_P vs. SHR_C groups; Table S2: Protein analysis in ClueGO plug-ins and the number of genes involved within the process biological category between SHR_T vs SHR_C groups; Table S3: The protein of comparison SHR_P vs. SHR_C; Table S4: The protein of comparison SHR_T vs. SHR_C.

Author Contributions: D.S.M.: Conceptualization, Methodology, Formal analysis, Investigation, Writing—Original Draft, Writing—Review and Editing, Project administration, Funding acquisition; F.D.: Methodology, Investigation; A.D.: Methodology, Investigation, Resources; M.A.R.B.: Methodology, Investigation, Resources, Visualization; S.L.A.: Conceptualization, Resources Writing—Original Draft, Writing—Review and Editing, Supervision, Project administration, Funding acquisition. All authors have read and agreed to the published version of the manuscript.

Funding: This study was supported by São Paulo Research Foundation (FAPESP #2019/25603-6). SLA was a fellow of the National Council for Scientific and Technological Development (CNPq), grant #311071/2020-1. This study was funded in part by the Coordination for the Improvement of Higher Education Personnel—Brazil (CAPES)—Finance Code 001. DM was the recipient of a scholarship from Coordination for the Improvement of Higher Education Personnel (CAPES #88887.634301/2021-00). FD was the recipient of a scholarship from Coordination for the Improvement of Higher Education Personnel (CAPES #88887.634304/2021-00). AD was the recipient of a scholarship from (FAPESP 2019/26117-8) and an MB thematic project from (FAPESP 2019/26070-1).

Institutional Review Board Statement: The animal study protocol was approved by the Committee for Ethical Use of Animals (CEUA) of School of Sciences (UNESP, Bauru, #1320/2019 Vol. 1) and are in accordance with the Brazilian Ethical Principles in Animal Research.

Informed Consent Statement: Not applicable.

Data Availability Statement: The data presented in this study are available on request from the corresponding author.

Conflicts of Interest: The authors declare no conflict of interest.

References

1. Wadstrom, B.N.; Fatehali, A.H.; Engstrom, G.; Nilsson, P.M. A Vascular Aging Index as Independent Predictor of Cardiovascular Events and Total Mortality in an Elderly Urban Population. *Angiology* **2019**, *70*, 929–937. [CrossRef] [PubMed]
2. Rovella, V.; Gabriele, M.; Sali, E.; Barnett, O.; Scuteri, A.; Di Daniele, N. Is Arterial Stiffness a Determinant of Hypotension? *High Blood Press. Cardiovasc. Prev.* **2020**, *27*, 315–320. [CrossRef] [PubMed]
3. Valencia-Hernandez, C.A.; Lindbohm, J.V.; Shipley, M.J.; Wilkinson, I.B.; McEniery, C.M.; Ahmadi-Abhari, S.; Singh-Manoux, A.; Kivimaki, M.; Brunner, E.J. Aortic Pulse Wave Velocity as Adjunct Risk Marker for Assessing Cardiovascular Disease Risk: Prospective Study. *Hypertension* **2022**, *79*, 836–843. [CrossRef] [PubMed]
4. Vlachopoulos, C.; Aznaouridis, K.; Stefanadis, C. Prediction of cardiovascular events and all-cause mortality with arterial stiffness: A systematic review and meta-analysis. *J. Am. Coll. Cardiol.* **2010**, *55*, 1318–1327. [CrossRef]
5. Morgan, E.E.; Casabianca, A.B.; Khouri, S.J.; Kalinoski, A.L. In vivo assessment of arterial stiffness in the isoflurane anesthetized spontaneously hypertensive rat. *Cardiovasc. Ultrasound* **2014**, *12*, 37. [CrossRef]
6. Lindesay, G.; Ragonnet, C.; Chimenti, S.; Villeneuve, N.; Vayssettes-Courchay, C. Age and hypertension strongly induce aortic stiffening in rats at basal and matched blood pressure levels. *Physiol. Rep.* **2016**, *4*, e12805. [CrossRef]
7. Lindesay, G.; Bezie, Y.; Ragonnet, C.; Duchatelle, V.; Dharmasena, C.; Villeneuve, N.; Vayssettes-Courchay, C. Differential Stiffening between the Abdominal and Thoracic Aorta: Effect of Salt Loading in Stroke-Prone Hypertensive Rats. *J. Vasc. Res.* **2018**, *55*, 144–158. [CrossRef]
8. Fabricio, M.F.; Jordao, M.T.; Miotto, D.S.; Ruiz, T.F.R.; Vicentini, C.A.; Lacchini, S.; Santos, C.F.; Michelini, L.C.; Amaral, S.L. Standardization of a new non-invasive device for assessment of arterial stiffness in rats: Correlation with age-related arteries' structure. *MethodsX* **2020**, *7*, 100901. [CrossRef]
9. Steppan, J.; Jandu, S.; Savage, W.; Wang, H.; Kang, S.; Narayanan, R.; Nyhan, D.; Santhanam, L. Restoring Blood Pressure in Hypertensive Mice Fails to Fully Reverse Vascular Stiffness. *Front. Physiol.* **2020**, *11*, 824. [CrossRef]
10. Pilote, L.; Abrahamowicz, M.; Eisenberg, M.; Humphries, K.; Behlouli, H.; Tu, J.V. Effect of different angiotensin-converting-enzyme inhibitors on mortality among elderly patients with congestive heart failure. *CMAJ* **2008**, *178*, 1303–1311. [CrossRef]
11. Dinicolantonio, J.J.; Lavie, C.J.; O'Keefe, J.H. Not all angiotensin-converting enzyme inhibitors are equal: Focus on ramipril and perindopril. *Postgrad. Med.* **2013**, *125*, 154–168. [CrossRef] [PubMed]
12. Ahimastos, A.A.; Aggarwal, A.; D'Orsa, K.M.; Formosa, M.F.; White, A.J.; Savarirayan, R.; Dart, A.M.; Kingwell, B.A. Effect of perindopril on large artery stiffness and aortic root diameter in patients with Marfan syndrome: A randomized controlled trial. *JAMA* **2007**, *298*, 1539–1547. [CrossRef] [PubMed]
13. Laurent, S.; Agabiti-Rosei, C.; Bruno, R.M.; Rizzoni, D. Microcirculation and Macrocirculation in Hypertension: A Dangerous Cross-Link? *Hypertension* **2022**, *79*, 479–490. [CrossRef] [PubMed]
14. Sonawane, K.B.; Deshmukh, A.A.; Segal, M.S. Pulse pressure, arterial stiffening, and the efficacy of renin-angiotensin system inhibitor combinations. *J. Hum. Hypertens.* **2018**, *32*, 165–166. [CrossRef] [PubMed]
15. Miotto, D.S.; Dionizio, A.; Jacomini, A.M.; Zago, A.S.; Buzalaf, M.A.R.; Amaral, S.L. Identification of Aortic Proteins Involved in Arterial Stiffness in Spontaneously Hypertensive Rats Treated With Perindopril:A Proteomic Approach. *Front. Physiol.* **2021**, *12*, 624515. [CrossRef]
16. Lopes, S.; Mesquita-Bastos, J.; Garcia, C.; Leitao, C.; Bertoquini, S.; Ribau, V.; Carvalho, P.; Oliveira, J.; Viana, J.; Figueiredo, D.; et al. Physical Activity is Associated with Lower Arterial Stiffness in Patients with Resistant Hypertension. *Heart Lung Circ.* **2021**, *30*, 1762–1768. [CrossRef]
17. Miotto, D.S.; Duchatsch, F.; Macedo, A.G.; Ruiz, T.F.R.; Vicentini, C.A.; Amaral, S.L. Perindopril Reduces Arterial Pressure and Does Not Inhibit Exercise-Induced Angiogenesis in Spontaneously Hypertensive Rats. *J. Cardiovasc. Pharmacol.* **2021**, *77*, 519–528. [CrossRef]
18. Zhou, H.; Wang, S.; Zhao, C.; He, H. Effect of exercise on vascular function in hypertension patients: A meta-analysis of randomized controlled trials. *Front. Cardiovasc. Med.* **2022**, *9*, 1013490. [CrossRef]
19. Kohn, J.C.; Bordeleau, F.; Miller, J.; Watkins, H.C.; Modi, S.; Ma, J.; Azar, J.; Putnam, D.; Reinhart-King, C.A. Beneficial Effects of Exercise on Subendothelial Matrix Stiffness are Short-Lived. *J. Biomech. Eng.* **2018**, *140*, 0745011–0745015. [CrossRef]
20. Herrera, N.A.; Jesus, I.; Dionisio, E.J.; Dionisio, T.J.; Santos, C.F.; Amaral, S.L. Exercise Training Prevents Dexamethasone-induced Rarefaction. *J. Cardiovasc. Pharmacol.* **2017**, *70*, 194–201. [CrossRef]
21. Amaral, S.L.; Zorn, T.M.; Michelini, L.C. Exercise training normalizes wall-to-lumen ratio of the gracilis muscle arterioles and reduces pressure in spontaneously hypertensive rats. *J. Hypertens.* **2000**, *18*, 1563–1572. [CrossRef] [PubMed]
22. Duchatsch, F.; Constantino, P.B.; Herrera, N.A.; Fabricio, M.F.; Tardelli, L.P.; Martuscelli, A.M.; Dionisio, T.J.; Santos, C.F.; Amaral, S.L. Short-term exposure to dexamethasone promotes autonomic imbalance to the heart before hypertension. *J. Am. Soc. Hypertens.* **2018**, *12*, 605–613. [CrossRef] [PubMed]

23. Bradford, M.M. A rapid and sensitive method for the quantitation of microgram quantities of protein utilizing the principle of protein-dye binding. *Anal. Biochem.* **1976**, *72*, 248–254. [CrossRef] [PubMed]
24. Batista, T.B.D.; Chaiben, C.L.; Penteado, C.A.S.; Nascimento, J.M.C.; Ventura, T.M.O.; Dionizio, A.; Rosa, E.A.R.; Buzalaf, M.A.R.; Azevedo-Alanis, L.R. Salivary proteome characterization of alcohol and tobacco dependents. *Drug Alcohol Depend.* **2019**, *204*, 107510. [CrossRef]
25. Dionizio, A.; Melo, C.G.S.; Sabino-Arias, I.T.; Araujo, T.T.; Ventura, T.M.O.; Leite, A.L.; Souza, S.R.G.; Santos, E.X.; Heubel, A.D.; Souza, J.G.; et al. Effects of acute fluoride exposure on the jejunum and ileum of rats: Insights from proteomic and enteric innervation analysis. *Sci. Total Environ.* **2020**, *741*, 140419. [CrossRef]
26. Leite, A.L.; Lobo, J.G.V.M.; Pereira, H.A.B.S.; Fernandes, M.S.; Martini, T.; Zucki, F.; Sumida, D.H.; Rigalli, A.; Buzalaf, M.A. Proteomic analysis of gastrocnemius muscle in rats with streptozotocin-induced diabetes and chronically exposed to fluoride. *PLoS ONE* **2014**, *9*, e106646. [CrossRef]
27. Vasan, R.S.; Pan, S.; Xanthakis, V.; Beiser, A.; Larson, M.G.; Seshadri, S.; Mitchell, G.F. Arterial Stiffness and Long-Term Risk of Health Outcomes: The Framingham Heart Study. *Hypertension* **2022**, *79*, 1045–1056. [CrossRef]
28. Sang, D.S.; Zhang, Q.; Song, D.; Tao, J.; Wu, S.L.; Li, Y.J. Association between brachial-ankle pulse wave velocity and cardiovascular and cerebrovascular disease in different age groups. *Clin. Cardiol.* **2022**, *45*, 315–323. [CrossRef]
29. Agbaje, A.O. To prevent hypertension in Africans: Do we need to eat more vegetables? *Eur. J. Prev. Cardiol.* **2022**, *29*, 2333–2335. [CrossRef]
30. Oh, Y.S. Arterial stiffness and hypertension. *Clin. Hypertens.* **2018**, *24*, 17. [CrossRef]
31. Loboz-Rudnicka, M.; Jaroch, J.; Kruszynska, E.; Bociaga, Z.; Rzyczkowska, B.; Dudek, K.; Szuba, A.; Loboz-Grudzien, K. Gender-related differences in the progression of carotid stiffness with age and in the influence of risk factors on carotid stiffness. *Clin. Interv. Aging* **2018**, *13*, 1183–1191. [CrossRef] [PubMed]
32. Tardelli, L.P.; Duchatsch, F.; Herrera, N.A.; Vicentini, C.A.; Okoshi, K.; Amaral, S.L. Differential effects of dexamethasone on arterial stiffness, myocardial remodeling and blood pressure between normotensive and spontaneously hypertensive rats. *J. Appl. Toxicol.* **2021**, *41*, 1673–1686. [CrossRef]
33. Tardelli, L.P.; Duchatsch, F.; Herrera, N.A.; Ruiz, T.F.R.; Pagan, L.U.; Vicentini, C.A.; Okoshi, K.; Amaral, S.L. Benefits of combined exercise training on arterial stiffness and blood pressure in spontaneously hypertensive rats treated or not with dexamethasone. *Front. Physiol.* **2022**, *13*, 916179. [CrossRef] [PubMed]
34. Barroso, W.K.S.; Rodrigues, C.I.S.; Bortolotto, L.A.; Mota-Gomes, M.A.; Brandao, A.A.; Feitosa, A.D.M.; Machado, C.A.; Poli-de-Figueiredo, C.E.; Amodeo, C.; Mion Junior, D.; et al. Brazilian Guidelines of Hypertension—2020. *Arq. Bras. De Cardiol.* **2021**, *116*, 516–658. [CrossRef] [PubMed]
35. Pelliccia, A.; Sharma, S.; Gati, S.; Back, M.; Borjesson, M.; Caselli, S.; Collet, J.P.; Corrado, D.; Drezner, J.A.; Halle, M.; et al. 2020 ESC Guidelines on sports cardiology and exercise in patients with cardiovascular disease. *Eur. Heart J.* **2021**, *42*, 17–96. [CrossRef]
36. Bull, F.C.; Al-Ansari, S.S.; Biddle, S.; Borodulin, K.; Buman, M.P.; Cardon, G.; Carty, C.; Chaput, J.P.; Chastin, S.; Chou, R.; et al. World Health Organization 2020 guidelines on physical activity and sedentary behaviour. *Br. J. Sport. Med.* **2020**, *54*, 1451–1462. [CrossRef] [PubMed]
37. Vlachopoulos, C.; Terentes-Printzios, D.; Laurent, S.; Nilsson, P.M.; Protogerou, A.D.; Aznaouridis, K.; Xaplanteris, P.; Koutagiar, I.; Tomiyama, H.; Yamashina, A.; et al. Association of Estimated Pulse Wave Velocity with Survival: A Secondary Analysis of SPRINT. *JAMA Netw. Open* **2019**, *2*, e1912831. [CrossRef] [PubMed]
38. Ong, K.T.; Delerme, S.; Pannier, B.; Safar, M.E.; Benetos, A.; Laurent, S.; Boutouyrie, P.; Investigators. Aortic stiffness is reduced beyond blood pressure lowering by short-term and long-term antihypertensive treatment: A meta-analysis of individual data in 294 patients. *J. Hypertens.* **2011**, *29*, 1034–1042. [CrossRef]
39. Ferrari, R.; Pasanisi, G.; Notarstefano, P.; Campo, G.; Gardini, E.; Ceconi, C. Specific properties and effect of perindopril in controlling the renin-angiotensin system. *Am. J. Hypertens.* **2005**, *18*, 142S–154S. [CrossRef]
40. Amaral, S.L.; Michelini, L.C. Effect of gender on training-induced vascular remodeling in SHR. *Braz. J. Med. Biol. Res.* **2011**, *44*, 814–826. [CrossRef]
41. Fernandes, T.; Magalhaes, F.C.; Roque, F.R.; Phillips, M.I.; Oliveira, E.M. Exercise training prevents the microvascular rarefaction in hypertension balancing angiogenic and apoptotic factors: Role of microRNAs-16, -21, and -126. *Hypertension* **2012**, *59*, 513–520. [CrossRef] [PubMed]
42. Herrera, N.A.; Jesus, I.; Shinohara, A.L.; Dionisio, T.J.; Santos, C.F.; Amaral, S.L. Exercise training attenuates dexamethasone-induced hypertension by improving autonomic balance to the heart, sympathetic vascular modulation and skeletal muscle microcirculation. *J. Hypertens.* **2016**, *34*, 1967–1976. [CrossRef] [PubMed]
43. Constantino, P.B.; Dionisio, T.J.; Duchatsch, F.; Herrera, N.A.; Duarte, J.O.; Santos, C.F.; Crestani, C.C.; Amaral, S.L. Exercise attenuates dexamethasone-induced hypertension through an improvement of baroreflex activity independently of the renin-angiotensin system. *Steroids* **2017**, *128*, 147–154. [CrossRef]
44. Lopes, S.; Afreixo, V.; Teixeira, M.; Garcia, C.; Leitao, C.; Gouveia, M.; Figueiredo, D.; Alves, A.J.; Polonia, J.; Oliveira, J.; et al. Exercise training reduces arterial stiffness in adults with hypertension: A systematic review and meta-analysis. *J. Hypertens.* **2021**, *39*, 214–222. [CrossRef]
45. Xin, C.; Ye, M.; Zhang, Q.; He, H. Effect of Exercise on Vascular Function and Blood Lipids in Postmenopausal Women: A Systematic Review and Network Meta-Analysis. *Int. J. Environ. Res. Public Health* **2022**, *19*, 12074. [CrossRef] [PubMed]

46. Saz-Lara, A.; Cavero-Redondo, I.; Alvarez-Bueno, C.; Notario-Pacheco, B.; Reina-Gutierrez, S.; Sequi-Dominguez, I.; Ruiz, J.R.; Martinez-Vizcaino, V. What type of physical exercise should be recommended for improving arterial stiffness on adult population? A network meta-analysis. *Eur. J. Cardiovasc. Nurs.* **2021**, *20*, 696–716. [CrossRef] [PubMed]
47. Li, G.; Lv, Y.; Su, Q.; You, Q.; Yu, L. The effect of aerobic exercise on pulse wave velocity in middle-aged and elderly people: A systematic review and meta-analysis of randomized controlled trials. *Front. Cardiovasc. Med.* **2022**, *9*, 960096. [CrossRef]
48. Mynard, J.P.; Clarke, M.M. Arterial Stiffness, Exercise Capacity and Cardiovascular Risk. *Heart Lung Circ.* **2019**, *28*, 1609–1611. [CrossRef]
49. Son, W.M.; Sung, K.D.; Bharath, L.P.; Choi, K.J.; Park, S.Y. Combined exercise training reduces blood pressure, arterial stiffness, and insulin resistance in obese prehypertensive adolescent girls. *Clin. Exp. Hypertens.* **2017**, *39*, 546–552. [CrossRef]
50. Otsuki, T.; Namatame, H.; Yoshikawa, T.; Zempo-Miyaki, A. Combined aerobic and low-intensity resistance exercise training increases basal nitric oxide production and decreases arterial stiffness in healthy older adults. *J. Clin. Biochem. Nutr.* **2020**, *66*, 62–66. [CrossRef]
51. Jordao, M.T.; Ladd, F.V.; Coppi, A.A.; Chopard, R.P.; Michelini, L.C. Exercise training restores hypertension-induced changes in the elastic tissue of the thoracic aorta. *J. Vasc. Res.* **2011**, *48*, 513–524. [CrossRef] [PubMed]
52. Morales-Quinones, M.; Ramirez-Perez, F.I.; Foote, C.A.; Ghiarone, T.; Ferreira-Santos, L.; Bloksgaard, M.; Spencer, N.; Kimchi, E.T.; Manrique-Acevedo, C.; Padilla, J.; et al. LIMK (LIM Kinase) Inhibition Prevents Vasoconstriction- and Hypertension-Induced Arterial Stiffening and Remodeling. *Hypertension* **2020**, *76*, 393–403. [CrossRef] [PubMed]
53. Simone, L.C.; Naslavsky, N.; Caplan, S. Scratching the surface: Actin' and other roles for the C-terminal Eps15 homology domain protein, EHD2. *Histol. Histopathol.* **2014**, *29*, 285–292. [CrossRef]
54. Matthaeus, C.; Lian, X.; Kunz, S.; Lehmann, M.; Zhong, C.; Bernert, C.; Lahmann, I.; Muller, D.N.; Gollasch, M.; Daumke, O. eNOS-NO-induced small blood vessel relaxation requires EHD2-dependent caveolae stabilization. *PLoS ONE* **2019**, *14*, e0223620. [CrossRef] [PubMed]
55. Torrino, S.; Shen, W.W.; Blouin, C.M.; Mani, S.K.; Viaris de Lesegno, C.; Bost, P.; Grassart, A.; Koster, D.; Valades-Cruz, C.A.; Chambon, V.; et al. EHD2 is a mechanotransducer connecting caveolae dynamics with gene transcription. *J. Cell Biol.* **2018**, *217*, 4092–4105. [CrossRef] [PubMed]
56. Lacolley, P.; Regnault, V.; Segers, P.; Laurent, S. Vascular Smooth Muscle Cells and Arterial Stiffening: Relevance in Development, Aging, and Disease. *Physiol. Rev.* **2017**, *97*, 1555–1617. [CrossRef]
57. Goel, S.A.; Guo, L.W.; Shi, X.D.; Kundi, R.; Sovinski, G.; Seedial, S.; Liu, B.; Kent, K.C. Preferential secretion of collagen type 3 versus type 1 from adventitial fibroblasts stimulated by TGF-beta/Smad3-treated medial smooth muscle cells. *Cell Signal* **2013**, *25*, 955–960. [CrossRef]
58. Wittig, C.; Szulcek, R. Extracellular Matrix Protein Ratios in the Human Heart and Vessels: How to Distinguish Pathological from Physiological Changes? *Front. Physiol.* **2021**, *12*, 708656. [CrossRef]
59. de Paula, V.F.; Tardelli, L.P.; Amaral, S.L. Dexamethasone-Induced Arterial Stiffening Is Attenuated by Training due to a Better Balance Between Aortic Collagen and Elastin Levels. *Cardiovasc. Drugs Ther.* **2023**. [CrossRef]
60. Guers, J.J.; Farquhar, W.B.; Edwards, D.G.; Lennon, S.L. Voluntary Wheel Running Attenuates Salt-Induced Vascular Stiffness Independent of Blood Pressure. *Am. J. Hypertens.* **2019**, *32*, 1162–1169. [CrossRef]
61. Dab, H.; Kacem, K.; Hachani, R.; Dhaouadi, N.; Hodroj, W.; Sakly, M.; Randon, J.; Bricca, G. Physiological regulation of extracellular matrix collagen and elastin in the arterial wall of rats by noradrenergic tone and angiotensin II. *J. Renin-Angiotensin-Aldosterone Syst.* **2012**, *13*, 19–28. [CrossRef] [PubMed]
62. He, F.; Chu, J.F.; Chen, H.W.; Lin, W.; Lin, S.; Chen, Y.Q.; Peng, J.; Chen, K.J. Qingxuan Jiangya Decoction (清眩降压汤) Prevents Blood Pressure Elevation and Ameliorates Vascular Structural Remodeling via Modulating TGF-beta 1/Smad Pathway in Spontaneously Hypertensive Rats. *Chin. J. Integr. Med.* **2020**, *26*, 180–187. [CrossRef] [PubMed]
63. Fassot, C.; Briet, M.; Rostagno, P.; Barbry, P.; Perret, C.; Laude, D.; Boutouyrie, P.; Bozec, E.; Bruneval, P.; Latremouille, C.; et al. Accelerated arterial stiffening and gene expression profile of the aorta in patients with coronary artery disease. *J. Hypertens.* **2008**, *26*, 747–757. [CrossRef] [PubMed]
64. Deng, T.; Liu, Y.; Gael, A.; Fu, X.; Deng, X.; Liu, Y.; Wu, Y.; Wu, Y.; Wang, H.; Deng, Y.; et al. Study on Proteomics-Based Aortic Dissection Molecular Markers Using iTRAQ Combined with Label Free Techniques. *Front. Physiol.* **2022**, *13*, 862732. [CrossRef] [PubMed]

Disclaimer/Publisher's Note: The statements, opinions and data contained in all publications are solely those of the individual author(s) and contributor(s) and not of MDPI and/or the editor(s). MDPI and/or the editor(s) disclaim responsibility for any injury to people or property resulting from any ideas, methods, instructions or products referred to in the content.

Article

Blood ACE Phenotyping for Personalized Medicine: Revelation of Patients with Conformationally Altered ACE

Sergei M. Danilov [1,2,3,*], Mark S. Jain [3], Pavel A. Petukhov [4], Olga V. Kurilova [3], Valery V. Ilinsky [5], Pavel E. Trakhtman [6], Elena L. Dadali [7], Larisa M. Samokhodskaya [3], Armais A. Kamalov [3] and Olga A. Kost [8]

1. Department of Medicine, Division of Pulmonary, Critical Care, Sleep and Allergy, University of Illinois, Chicago, IL 60607, USA
2. Department of Medicine, University of Arizona Health Sciences, Tucson, AZ 85721, USA
3. Medical Center, Lomonosov Moscow State University, 119992 Moscow, Russia
4. Department of Pharmaceutical Sciences, College of Pharmacy, University of Illinois, Chicago, IL 60612, USA
5. Genotek, Ltd., 105120 Moscow, Russia
6. Dmitry Rogachev National Medical Research Center of Pediatric Hematology, Oncology and Immunology, 117997 Moscow, Russia
7. Medico-Genetic Center, 115478 Moscow, Russia
8. Chemistry Faculty, Lomonosov Moscow State University, 119991 Moscow, Russia
* Correspondence: danilov@uic.edu

Abstract: Background: The angiotensin-converting enzyme (ACE) metabolizes a number of important peptides participating in blood pressure regulation and vascular remodeling. Elevated blood ACE is a marker for granulomatous diseases and elevated ACE expression in tissues is associated with increased risk of cardiovascular diseases. **Objective and Methodology:** We applied a novel approach —ACE phenotyping—to find a reason for conformationally impaired ACE in the blood of one particular donor. Similar conformationally altered ACEs were detected previously in 2–4% of the healthy population and in up to 20% of patients with uremia, and were characterized by significant increase in the rate of angiotensin I hydrolysis. **Principal findings:** This donor has (1) significantly increased level of endogenous ACE inhibitor in plasma with MW less than 1000; (2) increased activity toward angiotensin I; (3) M71V mutation in *ABCG2* (membrane transporter for more than 200 compounds, including bilirubin). We hypothesize that this patient may also have the decreased level of free bilirubin in plasma, which normally binds to the N domain of ACE. Analysis of the local conformation of ACE in plasma of patients with Gilbert and Crigler-Najjar syndromes allowed us to speculate that binding of mAbs 1G12 and 6A12 to plasma ACE could be a natural sensor for estimation of free bilirubin level in plasma. Totally, 235 human plasma/sera samples were screened for conformational changes in soluble ACE. **Conclusions/Significance:** ACE phenotyping of plasma samples allows us to identify individuals with conformationally altered ACE. This type of screening has clinical significance because this conformationally altered ACE could not only result in the enhancement of the level of angiotensin II but could also serve as an indicator of free bilirubin levels.

Keywords: angiotensin I-converting enzyme; blood; conformational changes; plasma ACE; screening; bilirubin; Gilbert syndrome; *ABCG2* mutations

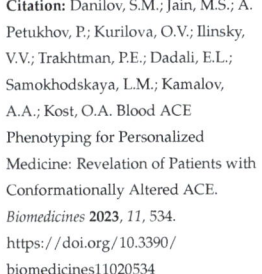

Citation: Danilov, S.M.; Jain, M.S.; A. Petukhov, P.; Kurilova, O.V.; Ilinsky, V.V.; Trakhtman, P.E.; Dadali, E.L.; Samokhodskaya, L.M.; Kamalov, A.A.; Kost, O.A. Blood ACE Phenotyping for Personalized Medicine: Revelation of Patients with Conformationally Altered ACE. *Biomedicines* **2023**, *11*, 534. https://doi.org/10.3390/biomedicines11020534

Academic Editors: Mária Szekeres, György L. Nádasy and András Balla

Received: 26 January 2023
Revised: 4 February 2023
Accepted: 8 February 2023
Published: 13 February 2023

Copyright: © 2023 by the authors. Licensee MDPI, Basel, Switzerland. This article is an open access article distributed under the terms and conditions of the Creative Commons Attribution (CC BY) license (https://creativecommons.org/licenses/by/4.0/).

1. Introduction

Angiotensin I-converting enzyme (ACE, CD143) is a Zn^{2+} carboxydipeptidase, which plays key roles in the regulation of blood pressure and in the development of vascular pathology. ACE is constitutively expressed on the endothelial cell surface, absorptive epithelial, and neuroepithelial cells and cells of the immune system (macrophages, dendritic cells), as reviewed in [1,2]. Blood ACE likely originates from endothelial cells [3], primarily lung capillary endothelium [4], by proteolytic cleavage [5–7]. In healthy individuals, blood

ACE levels are very stable over their lifetime [8], whereas in granulomatous diseases (e.g., sarcoidosis and Gaucher's disease), blood ACE activity is significantly increased [9–11].

Due to the increased frequency of sarcoidosis [12], we can expect that correct and quantitative determination of ACE in the blood becomes increasingly necessary. Additionally, changes in mentality in clinical medicine toward personalized/precision medicine [13–15] aroused the need for more accurate determination of ACE levels, as well as ACE status (ACE phenotype) [16–21].

We established a novel approach, blood ACE phenotyping, for the purpose of full characterization of ACE in plasma or serum [17–21]. The kinetic and conformational aspects of ACE phenotyping allowed identification of patients with conformationally changed ACE in the blood of patients with uremia (and in a small part of the blood samples from healthy donors). This kind of ACE was characterized by an increased activity toward angiotensin I [17]. Potential clinical relevance of this finding is that the presence of such ACE levels in the blood of patients can contribute to ACE inhibitor resistance in non-responders and lead to continuous local increase in angiotensin II formation. Moreover, such ACEs could be considered to be an indicator of a risk of the End Stage Kidney Disease (ESKD), which is especially important for African Americans [22].

During routine testing of plasma from different categories of donors and patients, we found an apparently healthy donor, donating his blood for transfusion procedures, who seemed to possess conformationally changed ACE. As we had access to a large volume of plasma from this donor (plasma was expired for transfusion) we had a possibility to study the putative reasons for these changes in his ACE conformation.

We did not find ACE mutation to be the reason for these conformational changes in ACE in this donor, but we found that the changes in ACE conformation in this given patient could be explained by the presence of some plasma components, or at least increased level of endogenous ACE inhibitors. Moreover, we came to conclusion that the results of really precise determination of ACE status in the blood of any given patient are associated with the bilirubin status of this patient, because likely the level of free bilirubin in this patient not only determines an apparent conformation of ACE surface topography but also influence the rate of ACE shedding, i.e., blood ACE level. We hypothesize that the parameter we introduced, the ratios of the binding of two monoclonal antibodies (mAbs) 1G12/9B9 and 6A12/9B9, reflects the level of free bilirubin, which could be especially useful in perinatology for the detection of patients with high levels of free bilirubin, i.e., patients with high risk of toxic brain injury [23].

2. Materials and Methods

Chemicals. ACE substrates, benzyloxycarbonyl-L-phenylalanyl-L-histidyl-L-leucine (Z-Phe-His-Leu) and hippuryl-L-histidyl-L-leucine (Hip-His-Leu) were purchased from Bachem Bioscience Inc. (King of Prussia, PA, USA) and Sigma (St. Louis, MO, USA). Other reagents (unless otherwise indicated) were obtained from Sigma (St. Louis, MO, USA).

Antibodies. Antibodies used in this study include a set of 17 (mAbs) to human ACE, recognizing native conformation of the N and C domains of human ACE [24,25].

Study participants. The study was approved by the Ethics Committee of the Medical Center of Moscow University (protocol # 9, 26 November 2018). All corresponding procedures were carried out in accordance with institutional guidelines and the Code of Ethics of the World Medical Association (Declaration of Helsinki). All patients provided written informed consent to have serum and citrated plasma for ACE characterization.

ACE activity assay. ACE activity in serum, plasma or purified lung ACE preparations was measured using a fluorimetric assay with two ACE substrates, 2 mM Z-Phe-His-Leu or 5 mM Hip-His-Leu [26,27]. Briefly, 20 µL aliquots of serum or plasma (diluted 1/5 in PBS) or aliquots of purified lung ACE preparations with corresponding specific ACE activity, were added to 100 µL of ACE substrate and incubated for the appropriate time at 37 °C. The His-Leu product was quantified fluorometrically. ACE activity in individual patients was expressed as % from pooled citrated plasma (control) collected from plasma of healthy donors and purchased from Interstate Blood Bank, Inc. (Memphis, TN). Before pooling, each plasma sample was preliminary tested for the presence of ACE inhibitors or conformationally changed ACEs, as described in this study. ACE activity in serum/plasma was also determined with 0.3 mM angiotensin I as a substrate in PBS, pH 7.5, also using fluorimetric assay as described above.

Calculation of ZPHL/HHL ratio [27] was performed by dividing fluorescence of the sample with ZPHL to that with HHL. Human lung ACE was purified using lisinopril-affinity chromatography exactly as described before [28].

Immunological characterization of the blood ACE. Microtiter (96-well) plates (Corning, Corning, NY) were coated with anti-ACE mAbs via goat anti-mouse IgG (Pierce, Rockford, IL) bridge and incubated with plasma/serum/lung ACE samples. After washing of unbound ACE, plate-bound ACE activity was measured by adding a substrate for ACE (Z-Phe-His-Leu) directly into wells [26]. The level of ACE immunoreactive protein, using strong mAb 9B9, was quantified as described previously [26]. Conformational fingerprinting of blood ACE with mAbs to ACE was performed and presented as described previously [19,24].

Bilirubin determination. Plasma bilirubin levels were measured using AU480 analyzer (Beckman Coulter, USA).

Sequencing (Sanger) and genotyping. Genomic DNA was obtained from the whole blood of donor 2D (and his sister) by QIAamp DNA Mini Kit (Qiagen, Valencia, CA, USA), and six exons of ACE gene (*ACE*), 7–11th and 13th, were amplified and sequenced as in [29]. Two exons of the lysozyme gene (*LYZL1*), 2th and 4th, coding most of the amyloidogenic mutations [30], were amplified and sequenced using primers, kindly provided by Dr. T. Prokaeva (Boston University, Boston, MA). Genotyping of *UGT1A1* gene for polymorphism of TATAA repeats (UGT1A1 * 28) was performed as in [31].

Next-Generation sequencing of exomes. The sequencing of 6000 + clinically relevant genes in the proband and his sister was performed by Genotek Ltd. (Moscow, Russia), after Ethics committee approved the study (07/2018). The proband and his sister gave written informed consent for studies and publication of their clinical information, images, and sequencing data. DNA libraries were constructed using the NEBNext Ultra DNA Library Prep Kit for Illumina (New England Biolabs, Ipswich, MA, USA) with adapters for sequencing on Illumina platform according to manufacturer's protocol. For target enrichment, we used SureSelect XT2 (Agilent Technologies, Santa Clara, CA, USA). Enriched samples were sequenced using an Illumina HiSeq 2500 system (Illumina, San Diego, CA, USA) in paired-end mode (100 bp reads) and analyzed as described exactly in [32].

Modeling of bilirubin binding to ACE. Coordinates of X-ray model of human ACE (PDB: 3NXQ, [33]) were downloaded from the PDB. All molecular modeling studies were performed in Molecular Operating Environment [34]. Two residues of N-acetylglucosamine were attached to Asn289 similar to that found for Asn45 in 3NXQ. The proteins were subjected to the "structure preparation" procedure. Hydrogen atoms were added using the Protonate 3D algorithm. The energy of the resulting structure was minimized using AMBER14EHT force-field implemented in MOE [35,36]. The proteins were minimized until the root mean square (RMS) gradient was less than 0.001 kcal/mol/Å^2. Bilirubin was assigned MMFF94x charges and minimized using the MMFF94x force-field until the RMS gradient was less than 0.001 kcal/mol/Å^2. The MOE docking module "Dock" was used for docking/scoring using the default parameters and settings. The approximate location of the docking site was assigned based on the outcomes of the analysis with antibodies [37,38]. Docking was performed using the "induced fit" algorithm, "Triangle Matcher" for placement, London dG for scoring of the binding poses after placement, and GBVI/WSA dG [34] for rescoring of the resulting poses. A total of 10 poses were stored after the refinement step. The docking poses that did not meet antibodies epitope mapping [37,38] were discarded, and the top remaining poses were considered for further analysis. The molecular surface colored to electrostatic potential was generated for the resulting structure except Asn45, Asn289, and Asn416 which were colored in green. All the post-translational modifications with sugars were rendered with "space fill" and colored cyan. In addition, the carboxyl groups in the top pose of bilirubin were glucuronidated, and the resulting structure was co-minimized with ACE using the same procedure as above.

Statistical analysis. All experiments were conducted independently in duplicate or triplicate, the results were expressed as mean value ± standard deviation, SD. Statistica for Windows (version 10.0, Stat.Soft. Inc., Tulsa, OK, USA) was used for statistical analysis. Significance was analyzed by Mann–Whitney test with $p \leq 0.05$ considered statistically significant and $p \leq 0.01$ considered highly statistically significant.

3. Results and Discussion

Blood ACE phenotyping and identification of patient with conformationally changed ACE. Previously, we developed a new approach to characterize blood ACE in individual patients-blood ACE phenotyping [17–21]. This approach includes not just determination of ACE activity (with two substrates, ZPHL and HHL), but also determination of a novel kinetic parameter, the ratio of the rates of the hydrolysis of these two substrates (ZPHL/HHL ratio), which is able to control the native state of N and C domains of ACE active centers and to reveal the presence of ACE inhibitors [17–21,27]. The third parameter is the concentration of ACE immunoreactive protein [24], and, finally, the fourth and most sensitive approach is conformational fingerprinting of ACE using anti-ACE mAbs showing subtle conformational changes in ACE surface topography [17–21,24,39].

ACE phenotyping was performed on 10 citrated plasma samples from six patients obtained after therapeutic apheresis (marked as ##A) and four healthy donors (marked as ##D), for whom the plasma for transfusion was already expired. The results are presented in Figure 1 in comparison with corresponding results for previously obtained control pooled serum from 83 healthy donors. Quantification of ACE activity (with ZPHL as a substrate, Figure 1A) and the levels of ACE immunoreactive protein determined with strong mAb 9B9 (Figure 1B) demonstrated excellent correlation (R = 0.962).

We already performed ACE phenotyping in 300 apparently healthy individuals [21] and found that standard deviation (SD) from mean of ACE activity or level of ACE immunoreactive protein (with mAb 9B9) for the population was about 25% for both methods. It means that normal blood ACE values (for 95% of population) are within range of mean ± 2SD (50–150% of the mean), i.e., inter-individual variations in ACE level are significant, at least three-fold, which confirmed previous estimations [8,26]. However, one sample from 10 plasmas phenotyped in this study (from donor 2D) had ACE activity and immunoreactive ACE protein level more than 150% from the mean value (brown colored in Figure 1A,B). We also determined the ratio of the rates of the hydrolysis of two substrates, ZPHL/HHL ratio, which elevated values serves as an indicator of the presence of commercial ACE inhibitors [27], but did not find any significant elevation of ZPHL/HHL ratio (Figure 1C), indicating that all these 10 plasmas do not contain exogenous (commercial) ACE inhibitors [19,27].

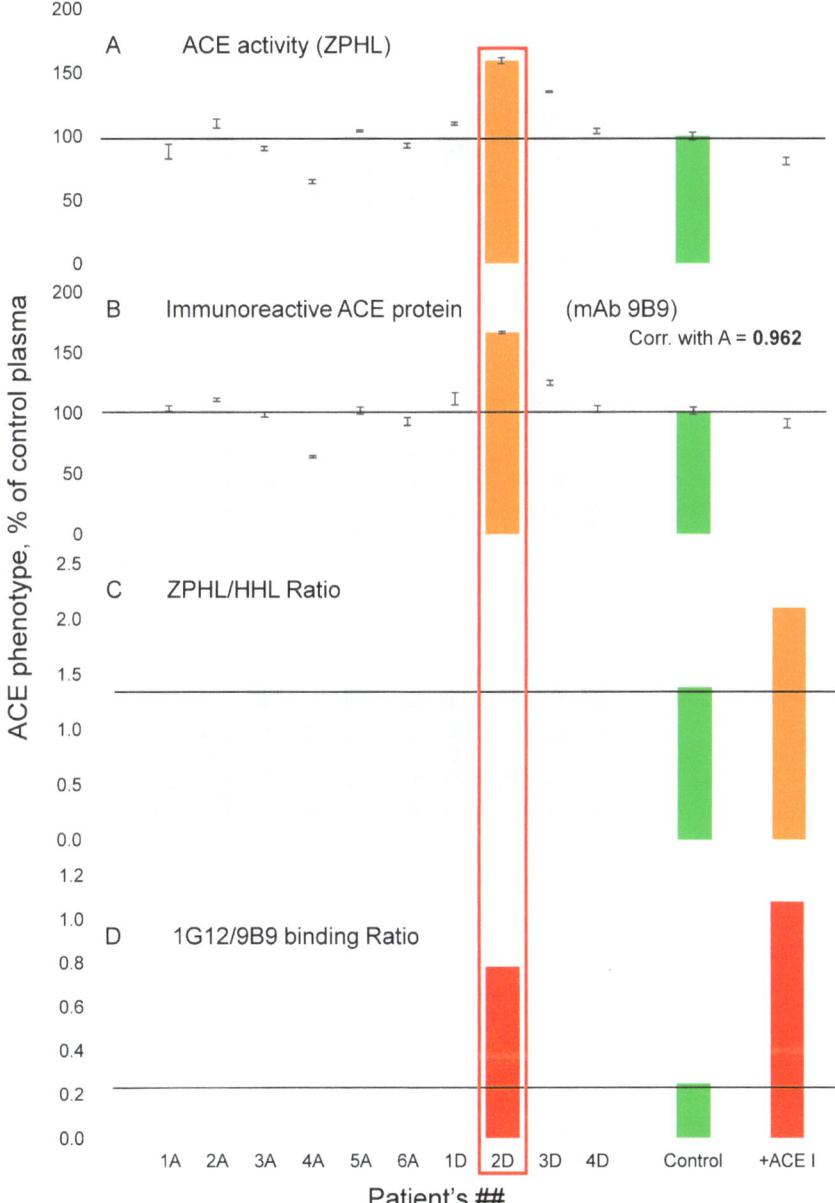

Figure 1. ACE phenotyping in citrated plasma samples. (**A**) ACE activity was measured by a spectrofluorometric assay with ZPHL (2 mM as a substrate) (**B**) The immunoreactive ACE protein was quantified by precipitation of ACE activity from plasma samples by mAb 9B9. (**C**) Ratio of the rates of the hydrolysis of two substrates, 2 mM ZPHL and 5 mM HHL (ZPHL/HHL ratio) (**D**) Ratio of ACE activity precipitation from plasma by mAbs 1G12 and 9B9 (1G12/9B9 ratio). Data were expressed as % of parameters of ACE phenotype from corresponding values for control pooled plasma samples from healthy controls (green bars). The same pooled plasma with 5 nM of ACE inhibitor enalaprilat was used as a positive control for the putative presence of ACE inhibitor in plasma samples. Bars highlighted with brown and red-samples with values of ACE parameters higher than 150% and 200% of controls. Bars for patient 2D are red boxed.

The amount of ACE immunoreactive protein (Figure 1B) was estimated using the strongest mAb to ACE, clone 9B9 [38,40,41]. A great advantage of this approach is a possibility to measure ACE levels in plasma taken with EDTA or in plasma containing ACE inhibitors, because EDTA or ACE inhibitors are washed out during washing step with distilled water with Tween-20 while ACE is still bound to this mAb [19,26]. However, precipitation of ACE activity from tested plasmas was also performed with mAb 1G12, which binding to blood ACE is extremely sensitive to the presence of ACE inhibitors in the blood [17–21,37]. Patient 2D demonstrated dramatically increased 1G12/9B9 ratio (Figure 1D), as if this patient had ACE inhibitors in his blood (for comparison, see an effect of 5 nM ACE inhibitor enalaprilat on ACE activity precipitation from control plasma, red bars in Figure 1D) but without increase in ZPHL/HHL ratio (Figure 1C). On the base of these results, we formed a pool of citrated plasma from three plasma samples from healthy donors (not including plasma from patient 2D) which was used as a control for further experiments.

Previously, we already found several persons (3 from tested 48, i.e., approximately 6%) among healthy donors, which plasma ACE demonstrated dramatic increase in 1G12/9B9 binding ratio without concomitant increase in ZPHL/HHL ratio [17]. The proportion of patients with such conformationally changed ACE significantly increased (at least up to at least 20%) among patients with uremia [17]. Pathophysiological effects of conformationally changed ACE from such patients were the following: (1) elevated rate of the hydrolysis of angiotensin I (AI) [17], which could theoretically increase the local concentrations of AII, which in turn, has numerous deleterious cardiovascular and inflammatory effects [42,43]; (2) decreased efficacy of ACE inhibitors toward these ACEs; (3) elevated blood pressure [17]. Therefore, we could not exclude the possibility that the detection of patients with such conformationally changed ACE could be clinically relevant as a screening of patients with high risk factor for ESRD (End Stage Renal Disease). As the volume of plasma from donor 2D was rather big, it gave us an opportunity to try to find the biochemical reasons for such conformational changes in blood ACE (including possible heritability of such phenotype). It is noteworthy that the ACE phenotype in donor 2D's blood (i.e., ACE activity, the level of immunoreactive ACE protein, and enhanced 1G12/9B9 ratio) was the same at first and second blood donations which were held at an annual interval, indicating that this phenotype is not accidental but rather an intrinsic characteristic of donor 2D.

3.1. Characterization of ACE in Donor 2D (with Conformationally Changed ACE)

When donor 2D plasma and control plasma were equilibrated by ACE activity with ZPHL, we found that ACE in donor 2D plasma appeared to be twice as active with angiotensin I as a substrate (205 ± 25, $p < 0.05$). Thus, we confirmed our previous finding [17] that conformationally changed ACE with increased 1G12/9B9 ratio can possess enhanced ACE activity toward this substrate (which undoubtedly is of clinical importance).

Efficacy of the inhibition of plasma ACE activity with HHL or ZPHL as substrates by common ACE inhibitor enalaprilat was reproducibly lower for plasma 2D compared to control plasma, but the difference was quite small, 5–10% (Figure S1A–C).

The effect of the presence of enalaprilat on several mAbs binding is shown in Figure S1D–F. While enalaprilat did not affect the binding of mAbs 9B9 to both control ACE and ACE from donor 2D (Figure S1D) and rather equally diminished the binding of mAb 4E3 (Figure S1E), the effect of the inhibitor on mAb 1G12 binding to control and donor 2D ACEs was strikingly different. The presence of enalaprilat dramatically increased binding of mAb 1G12 to blood ACE from control plasma, whereas it almost did not affect the binding of this mAb to blood ACE from donor 2D (Figure S1F) which is already high (Figure 1D). Previously, we found that bilirubin binds to ACE exactly in the region for epitopes for mAbs 1G12/6A12 and binding of ACE inhibitors induces dissociation of bilirubin from ACE, which leads to dramatic increase in 1G12/6A12 binding. Moreover, we demonstrated that mutation in ACE (R532W) abolished bilirubin binding to ACE and caused significant increase in ACE shedding and, therefore, increase in ACE levels in the blood [29]. Thus, increased level of ACE in the 2D plasma (Figure 1A,B), high 1G12/9B9 ratio (Figure 1D) and the absence of the effect of enalaprilat on 1G12 binding to ACE from 2D plasma (Figure S1F) indicates that donor 2D could have a mutation in ACE. This mutation could be similar, but not identical to R532W, because increase in blood ACE level in donor 2D was about 160%, in contrast to 450% for patient with mutation R532W [29]. Alternatively, there could be some non-canonical bilirubin in patient 2D, which binds less to ACE, i.e., more conjugated, or optical stereoisomer [44].

We compared conformational fingerprint of ACE from donor 2D plasma with control plasma (Figure 2A) in order to detect the regions of possible local changes in ACE conformation in the blood of patient 2D, and to find the region of possible ACE mutation. The results for each mAb are shown as the ratio of the effectiveness of the binding of this particular mAb with ACE from 2D plasma to that with control ACE. Dramatically increased binding of two mAbs, 1G12 and 6A12, to ACE from 2D plasma closely resembles an effect of plasma dilution, filtration, dialysis, or addition of enalaprilat on mAbs binding to control plasma ACE (Figure 2C,D and Figures S12 and S13 in [29]). The similarity of Figure 2A with these figures in the cited paper allowed us to suggest that low molecular weight (LMW) blood components, which binds to ACE in normal plasma and dissociate from ACE as a result of dilution, filtration, or dialysis (or action of common ACE inhibitors inducing this dissociation) may not be able to bind to ACE in 2D plasma. It was shown [29] that this LMW blood component is bilirubin, but the possibility exists that there could be other compounds in the blood able to bind to ACE. Thus, the results of conformational fingerprinting of ACE could indicate that conformational change in ACE surface topography observed in donor 2D (Figure 2A) could be caused both by changes in ACE structure in this donor due to mutations, post-translational modifications, or changes in blood components that bind (or not bind) to ACE.

The results presented in Figure 2A as well as the experiments in Figure 2C,D and Figures S12 and S13 in [29] were obtained by washing the plates with distilled water containing 0.05% Tween-20. Strikingly, washing with PBS/Tween-20 (Figure 2B) showed fewer differences in the binding of mAbs 1G12 and 6A12 to donor 2D and control ACEs than in the case of water/Tween-20 (Figure 2A). In control plasma, washing with water/Tween-20 left a lower proportion of ACE bound to mAbs 1G12, 6A12 and, to a lesser extent, mAb i2H5 than washing with PBS/Tween-20 as evidenced by lower ACE activity (Figure 2C). A similar difference after washing with PBS or distilled water was observed with ACE purified from lung, heart, and seminal fluid (not shown). The effect appears to be focused on the same area of ACE protein as the epitopes for these very mAbs are overlapping [37,45]. Charged amino acid residues were estimated to account for 44% of the total amount of amino acid residues in the epitope area for mAb 1G12, and hydrophobic residues represent approximately 25%. For mAb 6A12, charged amino acid residues constitute 51% of amino acid residues in its epitope area, whereas hydrophobic residues account for 16% [37]. Since water/Tween-20 solution has lower ionic strength compared to that of PBS/Tween-20, it is expected to be more effective at disrupting hydrophobic interactions. Overall, these findings suggest that disruption of the hydrophobic interactions is more critical for 1G12, 6A12 and, to a lesser extent, mAb i2H5, than for other mAbs and their corresponding epitopes. The opposite is true for mAb i1A8, which showed higher remaining ACE activity after washing with water/Tween-20 compared to washing with PBS/Tween-20 (Figure 2C). Regardless of the exact reasons for this phenomenon, we stopped using distilled water for washing unbound ACE off the plate (as we did before) because very low ACE activity still bound to mAbs 1G12 and 6A12 (and, therefore, enhanced determination errors) and started using only PBS.

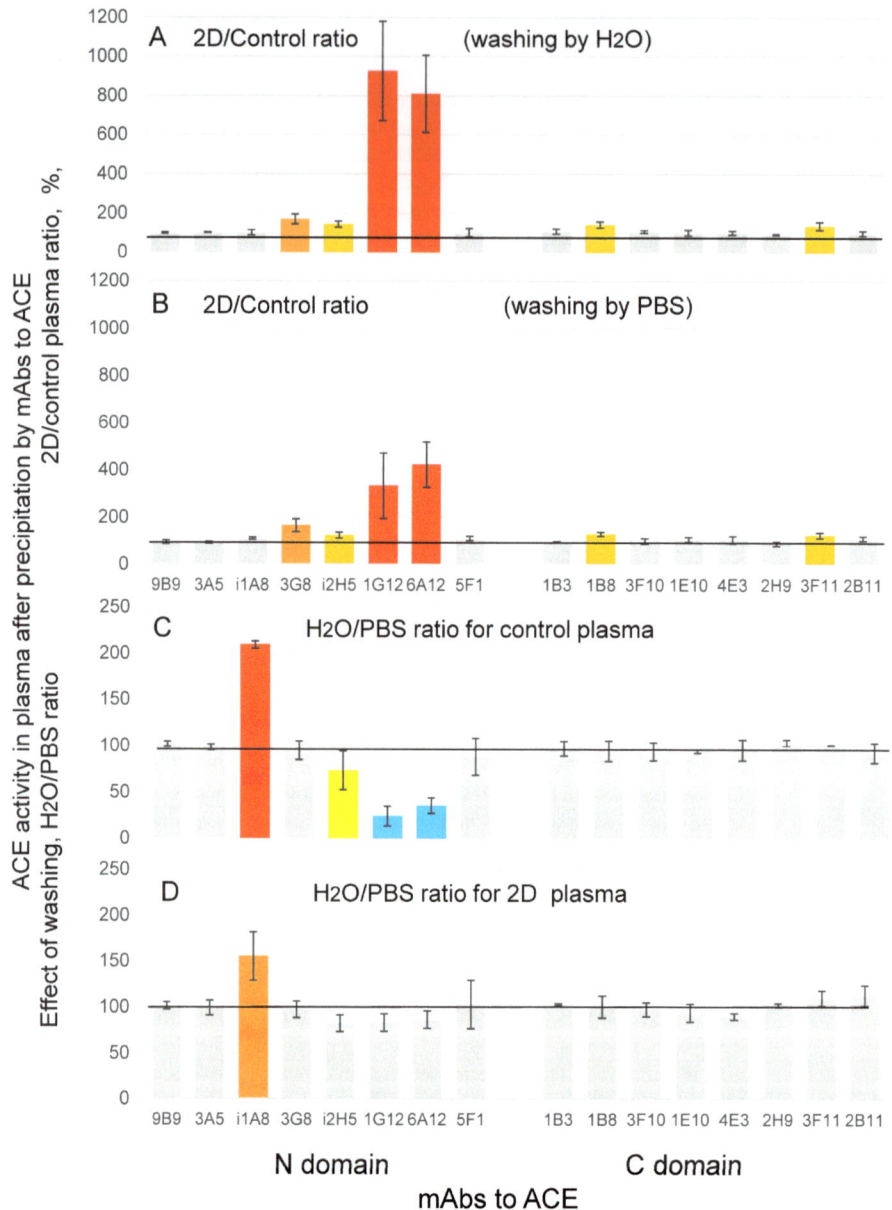

Figure 2. Conformational fingerprinting of blood ACE (using mAbs to ACE). (A–D) ACE activity was precipitated from plasma of donor 2D (boxed in Figure 1) and control plasma (diluted 1/5 with PBS) with 16 mAbs to different epitopes on the N and C domains of human ACE. (**A,B**) Data were expressed as a % of ACE activity from plasma of donor D precipitated by different mAbs from that for control plasma. Plates were washed from unbound ACE with distilled water (**A**) or with PBS (**B**). (**C,D**) Data are expressed as a % of ACE activity from control plasma (**C**) or plasma from donor 2D (**D**) with washing by water from that with washing by PBS. Orange bars-increase of ACE precipitation more than 20%, brown bars-more than 50%, red-more than 2-fold. Yellow bars-decrease of ACE precipitation more than 20%, blue bars-more than 50%.

Unlike the control, 2D plasma showed no buffer-dependent differences in the binding of these mAbs to ACE (Figure 2D). Many factors, including mutations, post-translational modifications (PTMs), and binding of endogenous ligands, may make this region on 2D ACE surface insensitive to changes in ionic strength. Since only a few mAbs displayed the buffer-dependent binding to ACE and binding of i1A8 remained to be buffer-dependent in 2D plasma (Figure 2D), the phenomenon responsible for switching between buffer-dependent and independent states of ACE is likely to be located on a relatively small portion of the ACE protein surface.

We tested the effects of serial dilutions of control and 2D plasma and found that while the relative binding of mAbs 1G12 and 6A12 (normalized to binding of mAb 9B9) to control ACE remarkably increased upon dilution, twice more for mAb 1G12 than for 6A12 (Figure S2B,C), the binding of mAb 1G12 to ACE from 2D plasma did not depend on the dilution and relative binding of 6A12 even slightly decreased (Figure S2B,C). This result may indicate on ACE mutation/PTM in patient 2D or (alternatively) on the significant lack (or even absence) of blood components able to bind to ACE in 2D plasma, but present in the control plasma or in favor of putative endogenous ACE inhibitor/effector in 2D blood of different nature than in control plasma, which is tightly bound to ACE and does not dissociate upon dilution.

We tested the effect of 2D and control plasma filtration through filters of different pore size 3–100 kDa on four mAbs binding to plasma ACE. The binding of mAbs 1G12 and 6A12 to control ACE (in contrast to mAb 9B9 to the N domain and mAb 1E10 to the C domain) significantly increased after filtration (Figure 3A–D) in accordance with previous results [29] that filtration helps remove LMW ACE effector, likely bilirubin. However, filtration of 2D plasma did not increase neither 1G12 nor 6A12 binding (Figure 3C,D) which was already significantly elevated (Figures 1 and 2) in comparison to control ACE, indicating that possibly bilirubin in the blood of donor 2D less binds to ACE due to some peculiarities of this ACE or decreased concentration of bilirubin (or its modifications) able to bind to this ACE. or there are other blood components in the blood of this donor that also bind to this region on N domain of ACE.

Then, we tested an effect of the filtrates from these two plasmas on the precipitation of model ACE (purified ACE from human lung) by mAbs to ACE and found that all 3–100 kDa filtrates from control plasma similarly and much more effectively decreased mAbs 1G12 and 6A12 binding to purified ACE than filtrates from donor 2D (Figure 3E–H). The effect of filtration and filtrates on ACE precipitation by 9 tested mAbs (Figure S3) confirmed reciprocal (mirrored) effect of filtration and filtrates on mAbs 1G12 and 6A12 binding. These results indicated that local conformational changes in ACE surface topography in plasma of donor 2D could be due to at least two factors—changes in ACE structure preventing effective binding of bilirubin to the epitopes of these 2 mAbs to ACE or the lack of LMW ACE-binding components in 2D plasma.

To clarify this, we purified ACE from both plasmas by affinity chromatography (batch procedure) on Lisinopril-Sepharose (Figure S4A,B). It appeared that long (24–48 h) incubation of plasma with Lisinopril-Sepharose during a batch procedure likely influenced local conformation of ACE surface topography in the region of overlapping epitopes of mAb 1G12 and 6A12, as the binding of these two mAbs was not increased (due to elimination of bilirubin after passing of plasma through Lisinopril-Sepharose column, as in Figure 10B in [39]) but significantly decreased after purification (Figure S4A,B). The only difference between the effects of purification on ACE from 2D and control plasma was less binding of mAb 6A12 to purified ACE from 2D plasma compared to control (red box). Nevertheless, conformational fingerprint of ACE partially purified from plasma of donor 2D by affinity chromatography was almost identical to that for ACE purified from control plasma (Figure S4C), in a sharp contrast to that in whole plasmas (Figure S4D), thus ruling out genuine changes in ACE conformation in donor 2D. The purification procedure dramatically changed binding of few mAbs to purified ACEs from both plasmas, and especially dramatically increased binding of mAb 5F1 (red bars in Figure S4A,B), localized in the interface of dimerization [20,46]. This fact may indicate that purification on Lisinopril-Sepharose can decrease the extent of dimerization of ACE from plasma, unmasking the epitope for mAb 5F1.

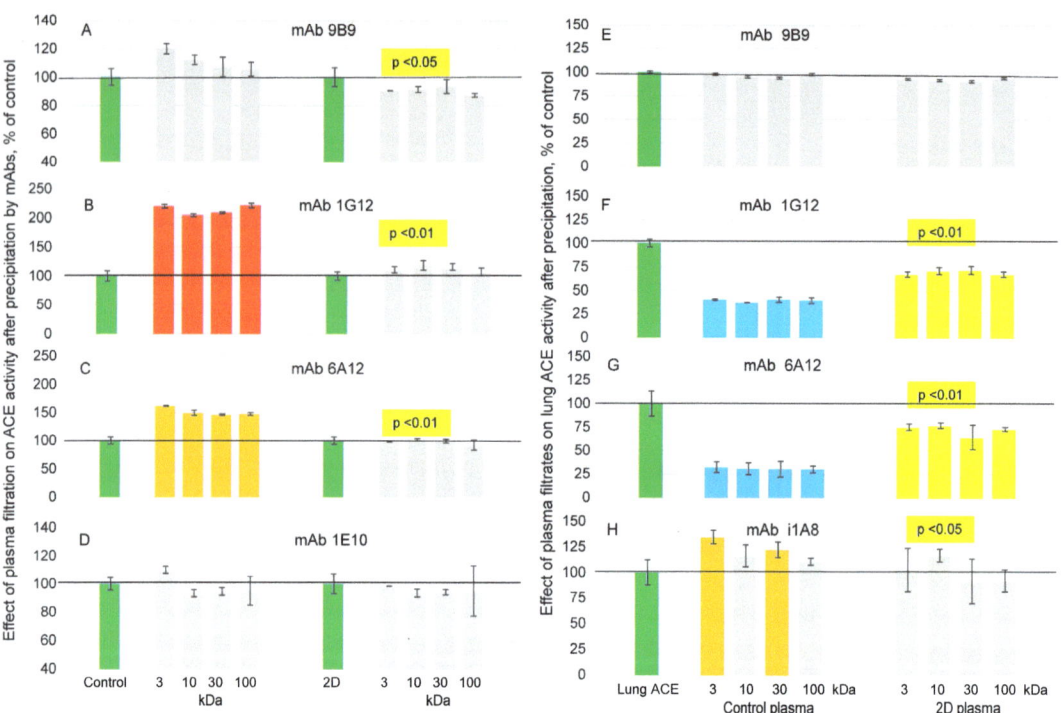

Figure 3. Effect of plasma filtration and filtrates on mAbs binding to ACEs. Plasma samples from donor 2D and control plasma (both 2 mL) were filtered by centrifugation through filters with 3, 10, 30, and 100 kD pores. (**A–D**) Plasmas were concentrated on filters 10-fold and then diluted 10-fold to return to initial volume. ACE activity precipitated from recovered plasmas by different mAbs was measured as in Figures 1 and 2. Data were expressed as a % of precipitated ACE activity from that for initial non-filtrated plasmas (mean ± SD of triplicates) by each mAb. (**E–H**) Filtrates (at 90% concentration) were added to purified lung ACE (final ACE activity about 10 mU/mL). ACE activity precipitation by different mAbs was measured as in Figures 1 and 2. Data were expressed as % (mean ± SD) of precipitated ACE activity in the presence of filtrates from that for controls (PBS instead of filtrates—green bars).

Sequencing of five exons (from 7th to 11th) coding the surface of the N domain of ACE, where overlapping region of mAbs 1G12 and 6A12 were localized [37], did not reveal any ACE mutation (Figure S5), confirming that the amino acid replacements in ACE molecule are not responsible for an apparent changed surface topography of plasma ACE of donor 2D. Nevertheless, confirming the power of personalized/precision medicine approach, we found polymorphic variant of ACE gene in this patient, heterozygous genotype CT of rs4613 in the 13th exon, coding Pro27 in the signal peptide of testicular ACE. (File S1). Homozygous TT genotype is associated with significant decrease in testicular ACE expression on spermatozoa and lower fertilization rates [47]. As we previously found that another blood component, lysozyme, binds to ACE [29], we also sequenced 2th and 4h exons of lysozyme (where most of the amyloidogenic mutations were localized [30]) and also did not found any mutation in these two exons of lysozyme gene of donor 2D.

3.2. ABCG2 Mutation in Donor 2D

After the experiments described above, the changes in bilirubin metabolism in patient 2D became an alternative hypothesis for the explanation of conformational changes in its blood ACE. There are numerous genes (and their products-proteins) that participate in bilirubin synthesis and metabolism, members of UTG family, biliverdin reductases, membrane transporters, for review see [48–50]. Therefore, in order to find causal mutation(s) which could be responsible for an apparent conformational changes in ACE surface topography in plasma of this donor, we performed exome sequencing of 6000+ clinically relevant genes in donor 2D and her sister who had no conformational changes in her plasma ACE (File S1, Figure S5).

We did not find any mutation in ACE and lysozyme genes of donor 2D nor any loss-of-function mutation in the *UGT1A1* gene [51], the product of which, bilirubin UDP-glucuronosyltransferase 1, is the only relevant enzyme responsible for bilirubin glucoronidation [52], but found several mutations (polymorphic variants) in some members of UGT family (Supplemental Materials File S1), and, most important, found mutation (M71V, rs148475733) in membrane transporter *ABCG2*, which participates in transport of more than 200 compounds (for review see [53]) including bilirubin [54]. The same mutation was described recently [55] and was shown to lead to decreased surface expression of ABCG2, but influence of this mutation on bilirubin metabolism was not analyzed.

Soon after that, we found plasma sample (from unrelated patient with chronic prostatitis) also with conformationally altered ACE (Index patient IP in Figure 4). Accidentally, this patient previously ordered exome sequence of 6000+ clinically relevant genes and we found that this patient has also mutation of *ABCG2*, but another one, N596S, implicated in protein trafficking and stability [53,56]. ACE phenotyping of citrated plasma from all patients with conformationally altered ACE (increased 1G12/9B9 ratio) that were found in 200 unrelated plasma samples and patients with two mutations in *ABCG2* are shown in Figure 4. Most of the patients with increased 1G12/9B9 ratio demonstrated also elevated level of blood ACE (Figure 4A,B), which is consistent with the hypothesis that these patients could have reduced level of bilirubin, which normally prevents an excessive ACE shedding and an appearance of extra ACE in the blood [29]. We expressed the 1G12/9B9 binding ratio in control citrated plasma as a percentage of this ratio in control EDTA plasma, where EDTA causes dissociation of zinc-ions from the active centers of ACE and, thus, reversibly inactivates the enzyme. We found that the epitope for 1G12 in citrated plasma of two control patients, 2DS and #66A, was unmasked by 20% in comparison to EDTA plasma (Figure 4D). The strikingly different results were obtained for donor 2D and Index patient: while in Index patient the epitope for mAb 1G12 in citrated plasma was unmasked by 50%, in donor 2D this epitope was completely unmasked in both citrate and EDTA plasma.

Thus, it is possible that some (but perhaps not all) mutations in *ABCG2* could lead to changes in bilirubin metabolism, which in turn, could influence an apparent local ACE surface conformation (1G12/9B9 binding ratio) in plasma. Therefore, we expected *ABCG2* null mutations to have more influence on bilirubin metabolism than other heterozygous mutations which we already analyzed. However, when we performed ACE phenotyping in plasma from three different *ABCG2* null mutations (Figure S6) in three patients. We did not find any effect of these null mutations on 1G12/9B9 binding ratio. The only reasonable hypothesis we could make is that the effect of tested *ABCG2* mutations (at least M71V and N596S) on bilirubin metabolism might be realized via homo- or even heterodimerization of *ABCG2* [53]. In the case of *ABCG2* null mutations, there could be no partners for homo- or heterodimerization of *ABCG2* and, as a result, no effect on bilirubin metabolism.

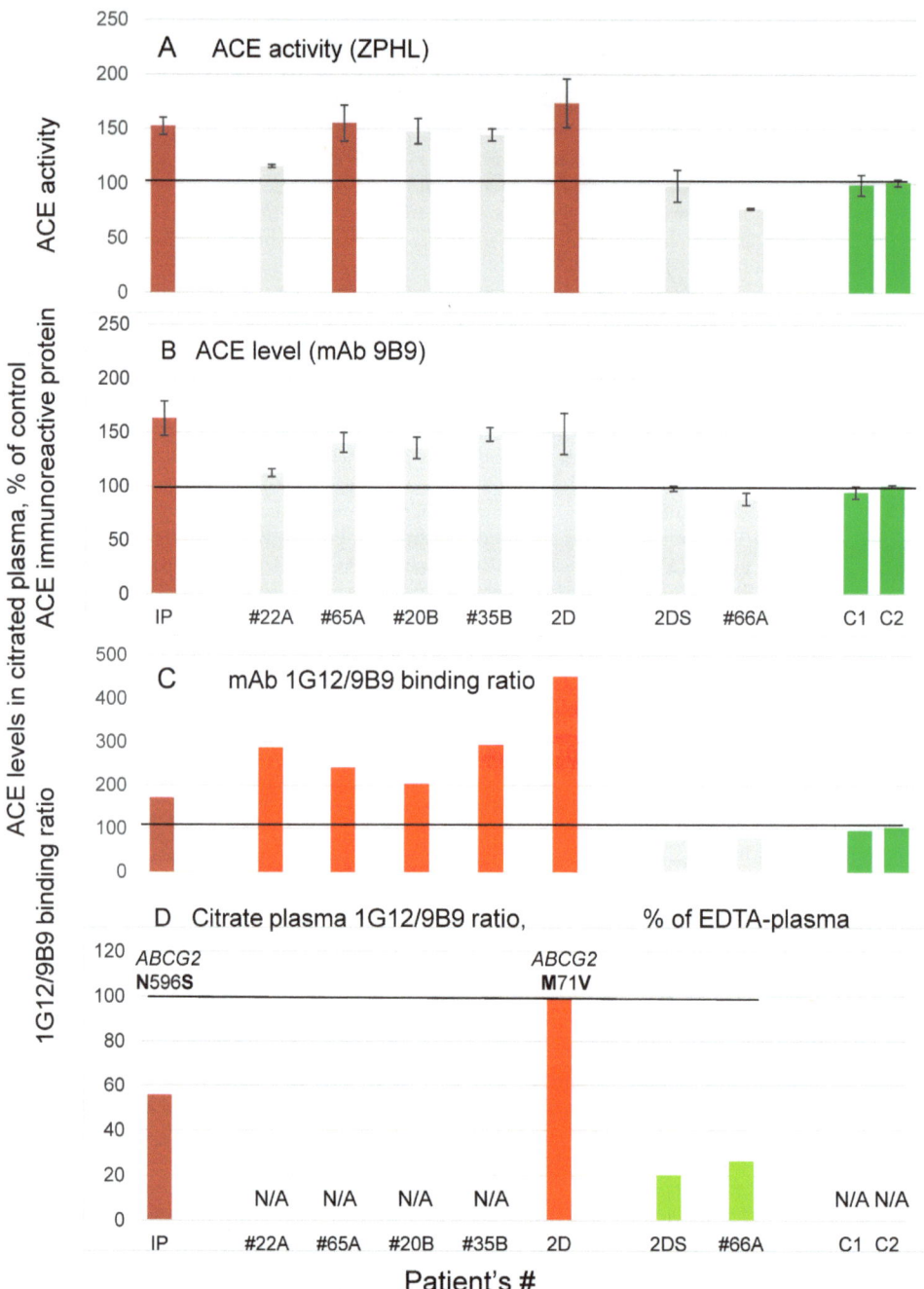

Figure 4. ACE phenotyping in patients with conformationally changed ACEs. (**A**) ACE activity; (**B**) amount of immunoreactive ACE protein determined with mAb 9B9: (**C**) and 1G12/9B9 binding ratio were quantified and presented as in Figure 1. (**D**) 1G12/9B9 binding ratio in citrated plasmas was calculated as a % from that in available EDTA plasma. As a control, we used pooled plasma samples from two different pools (1–2). Bars highlighted as in Figure 2.

3.3. Molecular Modeling of Bilirubin Binding to ACE

To gain additional insights on the possible binding mode of bilirubin to ACE, we docked it to the putative binding site previously identified that involved R532 on the N domain of ACE [29]. The top pose of bilirubin on the human N domain of ACE (PDB: 3NXQ) is shown in Figures 5 and 6. As shown in Figure 5A, bilirubin binds to a shallow hydrophobic pocket formed by the hydrophobic residues Pro308, Ile408, and Val296 and the hydrophobic portions of the side-chains of Lys407, Tyr531, Glu299, Thr302, Glu298, Arg295, Thr291, His292, and Trp299. The two negatively charged carboxyl groups of bilirubin participate in bidentate ionic interactions with positively charged Arg532 (yellow dotted lines). In addition, one of the carboxyl groups of bilirubin forms hydrogen bond with C=O of Gly409, making the resulting binding even stronger.

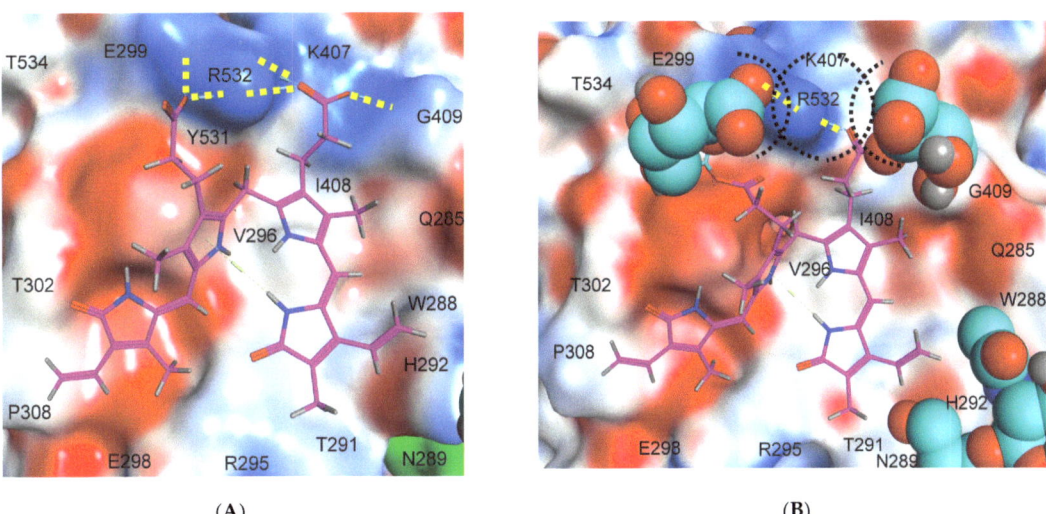

Figure 5. Protein-ligand interaction between ACE and bilirubins. N domain of human ACE (PDB: 3NXQ) with docked bilirubin (**A**) and bilirubin diglucuronide (**B**). Bilirubin scaffold is rendered by magenta. The hydrogen bonds with Arg532 are shown as yellow dotted lines. The steric interactions between the glucuronidate portions of bilirubin and Arg532 are marked by black dotted lines. The solvent accessible protein surface is mapped by electrostatic potential, red-negative, blue-positive. Gln289 is rendered as green surface. The sugar PTMs portions (Post-Translational Modifications) are rendered by cyan.

Mutation R532W is expected to disrupt these strong ionic interactions, resulting in a weakening bilirubin-ACE interaction as demonstrated in [29]. The docked molecule of bilirubin is only c.a. 10 Å away from Asn289, suggesting that glycan at potential glycosylation site Asn289 may potentially interfere with bilirubin binding to ACE (note proximity of the glycan rendered by cyan to bilirubin rendered by magenta (Figure 5A)).

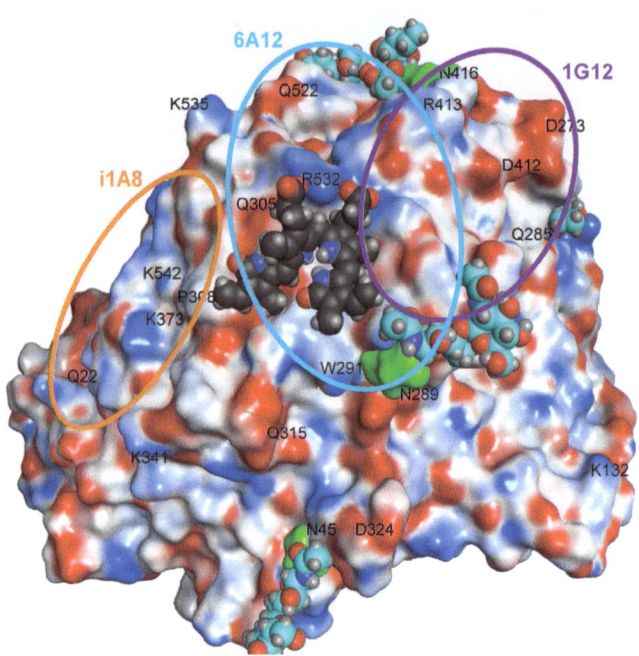

Figure 6. General view of bilirubin docked to ACE. N domain of human ACE (PDB: 3NXQ) near the epitopes for mAbs 6A12, 1G12 and i1A8. The surface of the protein is mapped by electrostatic potential, red-negative, blue-positive. Asn 45, Asn289 (Q in 3NXQ), and Asn416 are rendered by green surface. The sugar PTMs portions are rendered by cyan and bilirubin molecule -by dark gray.

Next, we analyzed how glucuronidation of bilirubin may impact its binding to ACE (Figure 5B). Glucuronidation of bilirubin leads to an increase in bulk of the resulting adduct in the region next to the newly formed ester bond between bilirubin and glucuronide portions. Moreover, in the glucuronidated bilirubin, the position of the negatively charged carboxyl groups are extended relatively to the heterocyclic core of bilirubin. Altogether, these changes in the structure of the glucuronidated bilirubin result in c.a. 2 Å shift in its position toward modeled Asn289 and loss of at least one strong electrostatic interaction with Arg532. A preliminary modeling shows that the carboxyl groups of the two glucuronidate groups may still form bidentate interactions with Arg532, but it would also require dissociation of the bilirubin core from the only hydrophobic pocket at the intersection of the epitopes for antibodies 6A12 and 1G12 (Figure 6). It is tempting to speculate that these changes would also disrupt the binding of glucuronidated bilirubin to ACE.

A comparison of bilirubin binding site on ACE with that on PPARα (peroxisome proliferator-activated receptor-α (docked structure, Figure 2A in [57]) and on albumin-PDB:2VUE [58]) shows that in all the cases the carboxylic acid groups of bilirubin form ionic interactions with the positively charged residues of the binding site whereas the remaining part of the ligand interacts with the hydrophobic portion of the binding site. For instance, ionic interactions between Arg117 and Arg186 on albumin with carboxylic groups of bilirubin are reminiscent to the interaction between bilirubin docked to ACE and Arg532.

Magnification of bilirubin docking model shows how bilirubin in complex with ACE maybe located in both epitopes for mAbs 1G12 and 6A12 (Figure 6). This picture also helps explain an unusual effect of filtrates of control plasma (i.e., mainly bilirubin in these filtrates) on an increase of mAb i1A8 binding to purified lung ACE (Figures S3B and S4D). Likely, binding of bilirubin to the area within epitopes for mAbs 1G12 and 6A12 changes local conformation of ACE remotely (in the region of epitope for mAb i1A8 (probably bumps formed by Q305, P308 and maybe K542)).

We compared the effects of free and conjugated (tartar) bilirubin on the binding of mAbs to purified lung ACE (Figure 7). Whereas bilirubin did not affect the binding of mAb 2H9 to the C domain and only slightly affected the binding of mAb 9B9 to the N domain (Figure 7C,D), both bilirubin remarkably decreased the binding of mAbs 6A12 and 1G12 to the N domain of ACE (Figure 7A,B). The ability of conjugated bilirubin to decrease binding of mAbs 6A12/1G12 was the same, while free bilirubin apparently better disrupted mAb 6A12 interactions with ACE, in accordance with the above modeling and results in [29].

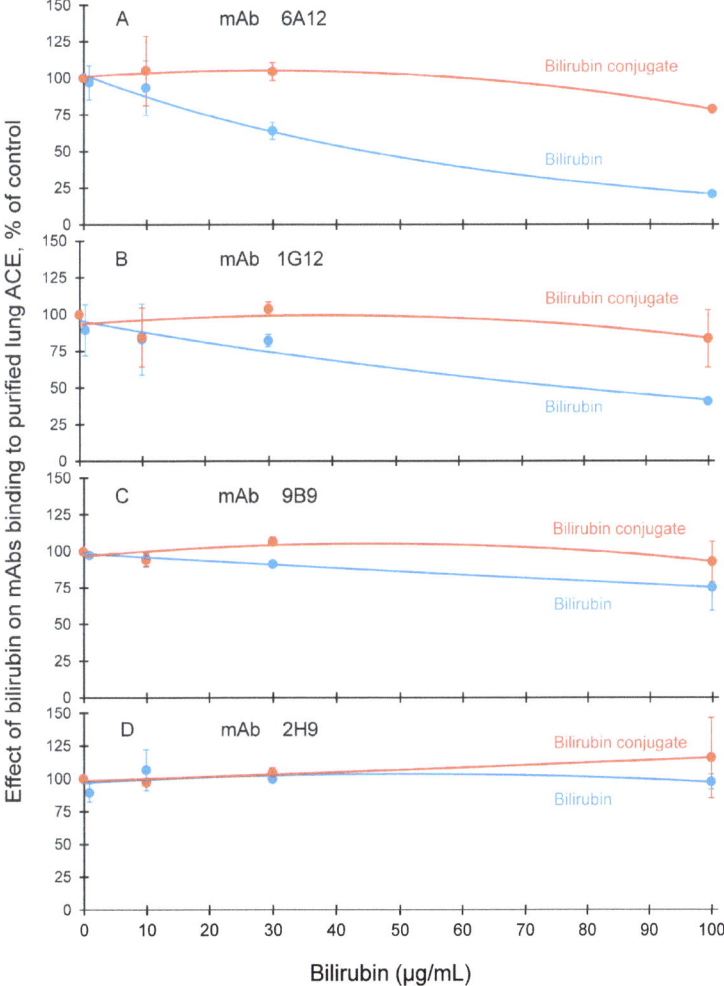

Figure 7. Effect of bilirubins on mAbs binding to purified human lung ACE. Bilirubin and bilirubin tartar conjugate in PBS were added to purified lung ACE (final ACE activity about 10 mU/mL). ACE activity precipitation by different mAbs was measured with as in Figures 1 and 2. (**A**) mAb 6A12; (**B**) mAb 1G12; (**C**) mAb 9B9; (**D**) mAb 2H9. Data were expressed as % (mean ± SD) of precipitated ACE activity in the presence of bilirubins from that for controls (PBS instead of bilirubins).

It should be emphasized that the effect of free bilirubin on mAb 6A12 (and 1G12) binding was much more pronounced than the effect of the same concentrations of conjugated bilirubin (Figure 7A,B) confirming the conclusions that free bilirubin better binds to ACE molecule.

Very rough estimation of the comparative efficacy of free and conjugated bilirubins (Figure 7A) toward 6A12/9B9 ratio showed that free bilirubin at least five-fold better binds to ACE than conjugated form. In normal conditions, however, the concentration of direct bilirubin is two orders higher than free unbound bilirubin. Therefore, there are much more complexes of ACE with direct bilirubin than complexes with free bilirubin. At the pathological level, however, increased or decreased concentration of free bilirubin might influence on the proportion of the complexes of ACE with both bilirubin.

Much evidence suggests that free (unbound and unconjugated) bilirubin concentration correlates more strongly with bilirubin toxicity than the total bilirubin [59]. These results justified further investigations into the clinical use of free bilirubin measurements [60]. Unfortunately, accurate measuring the free bilirubin concentration in the presence of much higher concentrations of protein-bound bilirubin is difficult [61]. However, at least two reliable methods for the measurement of free bilirubin (Bf) were established. Green fluorescent protein from eel was cloned, the fluorescence of which was significantly increased in the presence of free bilirubin [62]. In another approach, fluorescently labeled mutants of fatty acid binding proteins were used [63]. This sensor binds unconjugated bilirubin with high affinity (Kd = 16 nM) but binds conjugated bilirubin much worse, Kd > 300 nM [64]. Both methods started to be used for clinical determination of free bilirubin in human plasma [64,65]. This discrimination in free and conjugated bilirubin binding is similar to the effect of these bilirubins on ACE binding by mAbs 1G12 and 6A12 (Figure 7). Therefore, we may speculate that the 6A12/9B9 (or 1G12/9B9) ACE-binding ratio could be an additional (and natural) sensor for free bilirubin determination in the blood of patients.

3.4. Endogenous ACE Inhibitors in Plasma of Donor 2D

We analyzed an effect of dilution and filtration of plasma 2D and control plasma, as well as an effect of filtrates of plasmas on purified ACE, not only on mAbs binding (Figures 3 and 4), but also on ACE activity (Figure S7). Filtration of 2D and normal plasmas (Figure S7D,E) shows that this procedure resulted in a remarkable increase of ACE activity in 2D plasma, but only toward the substrate HHL, which in turn, indicates the presence in 2D plasma of endogenous ACE inhibitors preferably inhibiting the C domain of the enzyme. This conclusion is supported by the findings that simple 2D plasma dilution (Figure S7C) and filtration (Figure S7F) significantly decreased ZPHL/HHL ratio for plasma ACE, while filtrates of this plasma increased ZPHL/HHL ratio for purified ACE (Figure S7F). Therefore, while the balance of endogenous ACE inhibitors in normal plasma is shifted toward inhibitors preferably inhibiting the N domain [27], the balance of ACE inhibitors in 2D plasma is shifted toward inhibitors preferably inhibiting the C domain of the enzyme. The overall number of inhibitors in 2D plasma is apparently higher than in normal plasma as the effect of 2D filtrates on the activity of purified ACE is greater (Figure S7H).

In addition, we found that ACE activity in the serum was higher than that in citrated plasma for almost all tested normal blood samples by $11.4 \pm 7.7\%$ (mean for nine samples) (Figure S8A), indicating that some ACE inhibitor(-s) binds to the blood clot. However, this difference was much more pronounced for donor 2D ($+58.0 \pm 4.4\%$), while the amount of immunoreactive ACE protein was the same in his serum and plasma (Figure S8B). Therefore, we can speculate that concentration of this endogenous ACE inhibitor(s) is five-fold higher in 2D plasma than in plasma of nine healthy volunteers. Purification of ACE from normal and 2D plasma also revealed that ACE binding with Lisinopril-Sepharose was more effective for control plasma than ACE from 2D plasma as less ACE activity appeared under washing: $18.6\% \pm 10.8$ for control plasma versus $48.9\% \pm 19.5$ ($p < 0.001$) for plasma 2D. This fact could be attributed to the enhanced concentration of ACE inhibitors in 2D plasma competing with Lisinopril on the matrix.

Conformational fingerprinting of ACEs in serum versus plasma for donor 2D (and his unaffected sister) indicated that the elimination of the inhibitor(s) binding by blood clot somewhat changed the efficacy of binding of mAb 6A12 to the N domain and mAbs 1E10 and 4E3 to the overlapping region of their epitopes on the C domain (Figure S8C,D).

Therefore, in addition to decrease in bilirubin binding, the changes in content and concentration of endogenous ACE inhibitors can change an apparent local surface ACE conformation in plasma.

3.5. Blood ACE Phenotyping in Patients with Gilbert and Crigler-Najjar Syndromes

Bilirubin is the final product of heme catabolism, mainly originating from hemoglobin after degradation of old erythrocytes. In healthy individuals, the total level of blood bilirubin is under 17 μM [66]. Bilirubin is poorly soluble in water and in the blood is mostly bound to serum albumin [50], while water soluble (or direct) bilirubin results from conjugation of the initial molecule with one or two glucuronide groups by the enzyme UDP-glucuronosyl transferase 1A1 (UGT1A1) [52]. Plasma (serum) total and direct bilirubin concentrations are common laboratory criteria of bilirubin status in many diseases, but especially in jaundiced newborns [67]. Displacement of bilirubin from albumin binding sites or mutations in *UGT1A1* gene leading to a decreased expression of UGT1A1 are the reasons for enhanced levels of free, unbound bilirubin which usually presents in very small quantities in the blood and is highly neurotoxic. The aftermath health effects could be Gilbert syndrome (with a frequency of up to 15% in the Western population) with up to 90 μM total bilirubin [31], which is considered relatively harmless, but jaundice can be triggered by different types of stress. Rather rare but much more severe case is Crigler-Najjar syndrome with ≤10% UGT1A1 activity and a total bilirubin level in a range 100–750 μM (depending on I or II Type) with a risk of brain damage in infancy and teenage years, including encephalopathy and neurological impairment [23].

Previously, we showed that bilirubin is able to bind to ACE and thus cause the decrease of the 1G12/9B9 binding ratio [29]. The modeling of bilirubin-ACE interaction shown in the present work demonstrates that both direct conjugated and free unconjugated bilirubins can bind to ACE, but free bilirubin is able to form the most favorable interactions with the enzyme. Thus, we performed plasma ACE phenotyping in two patients with Gilbert syndrome (from Russia) and two patients with Crigler-Najjar syndrome (Type II, from Netherlands) and compared it with four patients with conformationally changed ACE, exhibiting increased 1G12/9B9 ratio, which we found in an independent study, along with 2D plasma (Figure 8, clinical details are in File S2). As negative controls, we used three different pools of citrated plasmas from patients with native ACE according to their ZPHL/HHL and 1G12/9B9 ratios [17] and plasma from the sister of donor 2D, also with naïve ACE conformation.

ACE activity was not measured quantitatively in the two patients with Crigler-Najjar syndrome (Figure 8A) due to very high concentrations of bilirubin in these two patients, 240 and 147 μM (Figure 8D), which interfered with fluorimetric assay of ACE activity, but the amount of immunoreactive ACE protein, measured with mAb 9B9, appeared to be very different (5-fold) in these two patients (Figure 8B). Both ACE activity and the amount of immunoreactive ACE protein in plasma of patients with Gilbert syndrome with mild hyperbilirubinemia −24.6 and 32 μM (Figure 8D) were rather similar and slightly higher than in control plasmas (Figure 8A,B).

The values of 1G12/9B9 ratio were significantly decreased compared to normal values for both patients with Gilbert syndrome (Figure 8C) in accordance with earlier observation [29] that bilirubin can bind to ACE and thus decrease 1G12/9B9 ratio. We can hypothesize that these patients may have elevated levels of free bilirubin. This conclusion is supported by increased conjugated/serum total bilirubin ratio in patients with Gilbert syndrome [66].

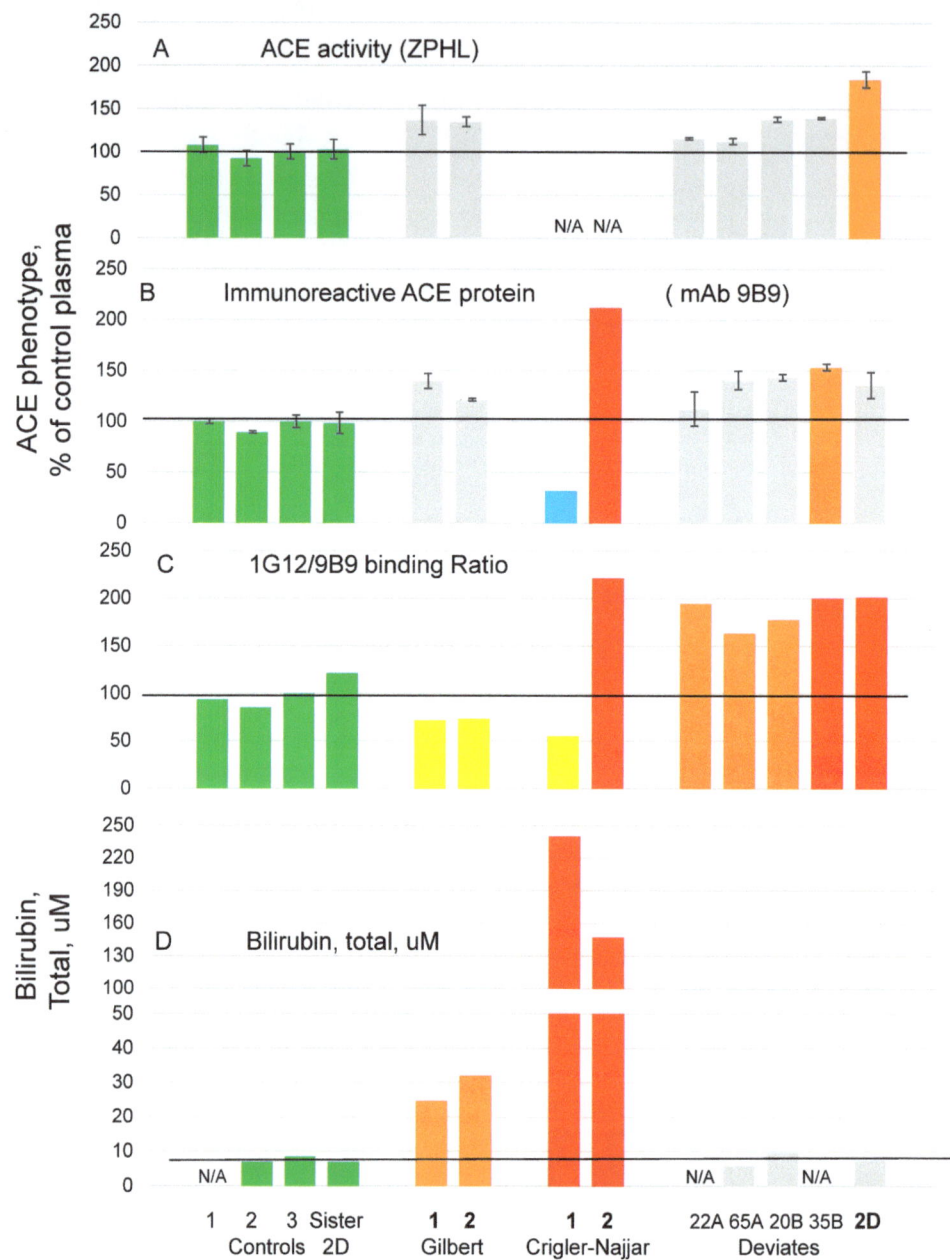

Figure 8. ACE phenotyping and bilirubin status in patients with Gilbert and Crigler-Najjar syndromes. ACE activity (**A**), amount of immunoreactive ACE protein determined with mAb 9B9 (**B**) and 1G12/9B9 binding ratio (**C**) were quantified and presented as in Figure 1. The available data on total bilirubin concentration are also presented (**D**). As controls, we used plasma samples from pools 1-2-3 and from sister of donor 2D. Several plasma samples with conformationally changed ACE, which we found in apparently healthy population, as well as plasma from donor 2D, are presented for comparison. Bars highlighted as in Figure 2.

For patients with Crigler-Najjar syndrome, however, the situation appeared to be equivocal. While ACE in the plasma of patient #1 with extremely high, total bilirubin was characterized by a low 1G12/9B9 ratio; the ACE in plasma of patient #2, also with high total bilirubin, unexpectedly, exhibited a high 1G12/9B9 ratio (Figure 8C,D). It is noteworthy that patients with Crigler-Najjar syndrome are characterized by much lower expression of UGT1A1 and higher extent of hyperbilirubinemia than Gilbert syndrome [48]. We could also expect higher amount of free bilirubin in plasma of these two patients than in normal plasma and plasma of patients with Gilbert syndrome and, therefore, even lower values of 1G12/9B9 ratio, as well as the decrease in blood ACE levels. It appeared to be true, but only for patient #1 (Figure 8B,C), while patient #2 demonstrated significantly elevated 1G12/9B9 ratio Figure 8C) and the level of ACE in plasma (Figure 8B). The primary hypothesis is that, despite dramatic hyperbilirubinemia, patient #2 has much lower levels of free bilirubin and thus this patient is not in a risk group for bilirubin-induced encephalopathy, because the concentration of free bilirubin in his plasma should not be high (and toxic). (Alternative explanation could be that patient #2 may have conformationally impaired ACE). The only way to clarify the reason for such differences in mAbs to ACE binding for these two patients with Crigler-Najjar syndrome is to measure free bilirubin by independent methods [64,65]. Unfortunately, none of these methods are available yet.

We also performed ACE phenotyping in 100 apparently healthy volunteers and calculated 1G12/9B9 ratio for those patients who had no ACE inhibitors in their blood (Figure S9). We found four patients with increased 1G12/9B9 ratio (orange bars) in accordance with previous results [17] demonstrating that apparently healthy donors could have conformationally altered ACE in their blood and, therefore, are at risk of hypertension and renal disease development. However, we also found nine patients with significantly decreased 1G12/9B9 ratio (yellow bars). Based on the results of 1G12/9B9 ratio calculation in two patients with Gilbert in Figure 8C we can hypothesize that these nine patients with 1G12/9B9 ratio less than 80% may have elevated levels of free bilirubin and, therefore, could be candidates for Gilbert syndrome. High frequency of Gilbert syndrome in Caucasians [31] fit with this hypothesis.

4. Conclusions

We can state that donor 2D (with conformationally impaired ACE) characterized by significantly increased activity toward angiotensin I, is also characterized by significantly increased level of endogenous ACE inhibitor (with MW less than 1000) which likely induced dissociation of bilirubin from its binding site near the epitope for mAb 6A12.

Moreover, this donor has a mutation in membrane transporter *ABCG2* participating in transport of more than 200 compounds, including bilirubin. We also demonstrated by modeling and confirmed by in vitro experiment that free bilirubin binds more effectively to its binding site on the N domain of ACE, than conjugated bilirubin.

Therefore, our data allowed us to hypothesize that we made a starting point for the indirect method that can reflect probable concentration of free bilirubin in plasma (serum). Estimation of (patho)physiologically active free bilirubin could be diagnostically extremely interesting, because an excess of free bilirubin is very toxic for the brain and clinical measurement of free bilirubin is limited to the small numbers of very big and well-equipped hospitals.

Supplementary Materials: The following supporting information can be downloaded at: https://www.mdpi.com/article/10.3390/biomedicines11020534/s1.

Author Contributions: Conceptualization, S.M.D.; methodology, P.A.P. and V.V.I.; formal analysis, P.A.P.; investigation, M.S.J., P.A.P., O.V.K. and V.V.I.; resources, P.E.T., E.L.D., L.M.S. and A.A.K.; data curation, S.M.D., M.S.J. and V.V.I.; writing—original draft preparation, S.M.D.; writing—review and editing, S.M.D. and O.A.K.; visualization, P.A.P.; supervision, S.M.D.; project administration, L.M.S.; funding acquisition, L.M.S., A.A.K. and O.A.K. All authors have read and agreed to the published version of the manuscript.

Funding: This research was funded partially by the project Scientific basis for national bank-depositary of living systems (RSF agreement #14-50-00029) and by M. V. Lomonosov Moscow State University (Registration Theme 121041500039-8-for OAK).

Institutional Review Board Statement: The study was approved by the Ethics Committee of the Medical Center of Moscow University (protocol # 9, 26 November 2018). All corresponding procedures were carried out in accordance with institutional guidelines and the Code of Ethics of the World Medical Association (Declaration of Helsinki).

Informed Consent Statement: Informed consent was obtained from all subjects involved in the study.

Data Availability Statement: All data are included in the manuscript.

Acknowledgments: We are grateful to Andrew B. Nesterovitch (then Rush University, Chicago, IL, USA) for help with PCR amplification and sequencing of several exons of ACE and lysozyme genes in patient 2D. We also are grateful to Piter Bosma (Tytgat Institute for Liver and Intestinal Research, University of Amsterdam Medical Center, The Netherlands) for providing blood samples from patients with Crigler-Najjar syndrome and to Connie Westhoff (New York Blood Center, New York, NY, USA) for providing blood samples from patients with null mutations of *ABCG2*. The study was conducted using biomaterials collected and preserved in the frame of the project Scientific basis for national bank-depositary of living systems (RSF agreement #14-50-00029) using the equipment purchased as a part of Lomonosov Moscow State University Program of Development and also in part by M. V. Lomonosov Moscow State University (Registration Theme 121041500039-8-for OAK).

Conflicts of Interest: The authors declare no conflict of interest.

References

1. Sturrock, E.D.; Anthony, C.S.; Danilov, S.M. Peptidyl-dipeptidase A/Angiotensin I-converting enzyme. In *Handbook of Proteolytic Enzymes*, 3rd ed.; Rawlings, N., Salvesen, G., Eds.; Academic Press: Oxford, UK, 2012; Chapter 98, pp. 480–494.
2. Bernstein, K.E.; Ong, F.S.; Blackwell, W.-L.B.; Shah, K.H.; Giani, J.F.; Gonzalez-Villalobos, R.A.; Shen, X.Z.; Fuchs, S. A Modern Understanding of the Traditional and Nontraditional Biological Functions of Angiotensin-Converting Enzyme. *Pharmacol. Rev.* **2012**, *65*, 1–46. [CrossRef] [PubMed]
3. Ching, S.F.; Hayes, L.W.; Slakey, L.L. Angiotensin-converting enzyme in cultured endothelial cells. Synthesis, degradation, and transfer to culture medium. *Arteriosclerosis* **1983**, *3*, 581–588. [CrossRef] [PubMed]
4. Metzger, R.; Franke, F.; Bohle, R.-M.; Alhenc-Gelas, F.; Danilov, S.M. Heterogeneous distribution of Angiotensin I-converting enzyme (CD143) in the human and rat vascular systems: Vessels, organs and species specificity. *Microvasc. Res.* **2011**, *82*, 206–215. [CrossRef] [PubMed]
5. Hooper, N.; Keen, J.; Pappin, D.; Turner, A. Pig kidney angiotensin converting enzyme. Purification and characterization of amphipatic and hydrophilic forms of the enzyme establishes C-terminal anchorage to the plasma membrane. *Biochem. J.* **1987**, *247*, 85–93. [CrossRef]
6. Wei, L.; Gelas, F.A.; Soubrier, F.; Michaud, A.; Corvol, P.; Clauser, E. Expression and characterization of recombinant human angiotensin I-converting enzyme. Evidence for a C-terminal transmembrane anchor and for a proteolytic processing of the secreted recombinant and plasma enzymes. *J. Biol. Chem.* **1991**, *266*, 5540–5546. [CrossRef]
7. Ehlers, M.; Gordon, K.; Schwager, S.; Sturrock, E. Shedding the load of hypertension: The proteolytic processing of angiotensin-converting enzyme. *S. Afr. Med. J.* **2017**, *102*, 461–464. [CrossRef]
8. Alhenc-Gelas, F.; Richard, J.; Courbon, D.; Warnet, J.M.; Corvol, P. Distribution of plasma angiotensin I-converting enzyme levels in healthy men: Relationship to environmental and hormonal parameters. *J. Lab. Clin. Med.* **1991**, *117*, 33–39.
9. Lieberman, J. Elevation of serum angiotensin-converting enzyme level in sarcoidosis. *Am. J. Med.* **1975**, *59*, 365–372. [CrossRef]
10. Lieberman, J.; Beutler, E. Elevation of angiotensin-converting enzyme in Gaucher's disease. *N. Engl. J. Med.* **1976**, *294*, 1442–1444. [CrossRef]
11. Romer, F. Clinical and biochemical aspects of sarcoidosis. With special reference to angiotensin-converting enzyme (ACE). *Acta Med. Scand. Suppl.* **1984**, *690*, 3–96.
12. Sawahata, M.; Sugiyama, Y.; Nakamura, Y.; Nakayama, M.; Mato, N.; Yamasawa, H.; Bando, M. Age-related and historical changes in the clinical characteristics of sarcoidosis in Japan. *Resp. Med.* **2015**, *109*, 272–278. [CrossRef] [PubMed]
13. Sznajder, J.; Ciechanover, A. Personalized Medicine. The Road Ahead. *Am. J. Respir. Crit. Care Med.* **2012**, *186*, 945–947. [CrossRef] [PubMed]
14. Beckmann, J.; Lew, D. Reconciling evidence-based medicine and precision medicine in the era of big data: Challenges and opportunities. *Genome Med.* **2016**, *8*, 134. [CrossRef]
15. König, I.R.; Fuchs, O.; Hansen, G.; von Mutius, E.; Kopp, M.V. What is precision medicine? *Eur. Respir. J.* **2017**, *50*, 1700391. [CrossRef]

16. Danser, A.J.; Batenburg, W.W.; Meiracker, A.H.V.D.; Danilov, S.M. ACE phenotyping as a first step toward personalized medicine for ACE inhibitors. Why does ACE genotyping not predict the therapeutic efficacy of ACE inhibition. *Pharmacol. Ther.* **2007**, *113*, 607–618. [CrossRef] [PubMed]
17. Petrov, M.N.; Shilo, V.Y.; Tarasov, A.V.; Schwartz, D.E.; Garcia, J.G.N.; Kost, O.A.; Danilov, S.M. Conformational changes of blood ACE in uremia. *PLoS ONE* **2012**, *7*, e49290. [CrossRef]
18. Danilov, S.M.; Tovsky, S.I.; Schwartz, D.E.; Dull, R. ACE Phenotyping as a guide toward personalized therapy with ACE inhibitors. *J. Cardiovasc. Pharmacol. Ther.* **2017**, *22*, 374–386. [CrossRef]
19. Danilov, S.M.; Tikhomirova, V.E.; Kryukova, O.V.; Balatsky, A.V.; Bulaeva, N.I.; Golukhova, E.Z.; Bokeria, L.A.; Samokhodskaya, L.M.; Kost, O.A. Conformational fingerprint of blood and tissue ACEs: Personalized approach. *PLoS ONE* **2018**, *13*, e0209861. [CrossRef]
20. Danilov, S.M.; Jain, M.S.; Petukhov, P.A.; Goldman, C.; DiSanto-Rose, M.; Vancavage, R.; Francuzevitch, L.Y.; Samokhodskaya, L.M.; Kamalov, A.A.; Arbieva, Z.H.; et al. Novel ACE mutations mimicking sarcoidosis by increasing blood ACE Levels. *Transl. Res.* **2021**, *230*, 5–20. [CrossRef]
21. Samokhodskaya, L.M.; Jain, M.S.; Kurilova, O.V.; Bobkov, A.P.; Kamalov, A.A.; Dudek, S.M.; Danilov, S.M. Blood ACE phenotyping: A necessary approach for precision medicine. *J. Appl. Lab. Med.* **2021**, *6*, 1179–1191. [CrossRef]
22. Umeukeje, E.M.; Young, B.A. Genetics and ESKD disparities in African-Americans. *Am. J. Kidney Dis.* **2019**, *74*, 811–821. [CrossRef] [PubMed]
23. Watchko, J.F.; Tiribelli, C. Bilirubin-induced neurologic damage-mechanisms and management approaches. *N. Engl. J. Med.* **2013**, *369*, 2021–2030. [CrossRef]
24. Danilov, S.M.; Balyasnikova, I.V.; Danilova, A.S.; Naperova, I.A.; Arablinskaya, N.E.; Borisov, S.E.; Metzger, R.; Franke, F.E.; Schwartz, D.E.; Gachok, I.V.; et al. Conformational fingerprinting of the angiotensin-converting enzyme (ACE): Application in sarcoidosis. *J. Proteome Res.* **2010**, *9*, 5782–5793. [CrossRef]
25. Kryukova, O.V.; Tikhomirova, V.E.; Golukhova, E.Z.; Evdokimov, V.V.; Kalantarov, G.F.; Trakht, I.N.; Schwartz, D.E.; Dull, R.O.; Gusakov, A.V.; Uporov, I.V.; et al. Tissue Specificity of human angiotensin I-converting enzyme. *PLoS ONE* **2015**, *10*, e0143455. [CrossRef] [PubMed]
26. Danilov, S.; Savoie, F.; Lenoir, B.; Jeunemaitre, X.; Azizi, M.; Tarnow, L.; Alhenc-Gelas, F. Development of enzyme-linked immunoassays for human angiotensin I converting enzyme suitable for large-scale studies. *J. Hypertens.* **1996**, *14*, 719–727. [CrossRef]
27. Danilov, S.M.; Balyasnikova, I.V.; Albrecht, R.F.; Kost, O.A. Simultaneous determination of ACE activity with two substrates provide an information on the nativity of somatic ACE and allow to detect ACE inhibitors in the human blood. *J. Cardiovasc. Pharmacol.* **2008**, *52*, 90–103. [CrossRef] [PubMed]
28. Kost, O.A.; Grinshtein, S.V.; Nikolskaya, I.; Shevchenko, A.; Binevski, P.V. Purification of soluble and membrane forms of somatic angiotensin-converting enzyme by cascade affinity chromatography. *Biochemistry* **1997**, *62*, 321–328.
29. Danilov, S.M.; Lünsdorf, H.; Akinbi, H.T.; Nesterovitch, A.B.; Epshtein, Y.; Letsiou, E.; Kryukova, O.V.; Piegeler, T.; Golukhova, E.Z.; Schwartz, D.E.; et al. Lysozyme and bilirubin bind to ACE and regulate ACE conformation and shedding. *Sci. Rep.* **2016**, *6*, 34913. [CrossRef]
30. Dumoulin, M.; Kumita, J.R.; Dobson, C.M. Normal and aberrant biological self-assembly: Insights from studies of human lysozyme and its amyloidogenic variants. *Acc. Chem. Res.* **2006**, *39*, 603–610. [CrossRef]
31. Bosma, P.J.; Chowdhury, J.R.; Bakker, C.; Gantla, S.; de Boer, A.; Oostra, B.A.; Lindhout, D.; Tytgat, G.N.; Jansen, P.L.; Elferink, R.P.; et al. The genetic basis of the reduced expression of bilirubin UDP-glucuronosyltransferase 1 in Gilbert's syndrome. *N. Engl. J. Med.* **1995**, *333*, 1171–1175. [CrossRef]
32. Okuneva, E.G.; Kozina, A.A.; Baryshnikova, N.V.; Krasnenko, A.Y.; Tsukanov, K.; Klimchuk, O.I.; Surkova, E.I.; Ilinsky, V.V. A novel elastin gene frameshift mutation in a Russian family with cutis laxa: A case report. *BMC Dermatol.* **2019**, *19*, 4. [CrossRef] [PubMed]
33. Anthony, C.S.; Corradi, H.R.; Schwager, S.L.; Redelinghuys, P.; Georgiadis, D.; Dive, V.; Acharya, K.R.; Sturrock, E.D. The N domain of human angiotensin-I-converting enzyme: The role of N-glycosylation and the crystal structure in complex with an N domain-specific phosphinic inhibitor, RXP407. *J. Biol. Chem.* **2010**, *285*, 35685–35693. [CrossRef] [PubMed]
34. *Molecular Operating Environment (MOE)*, 2019.0101; Chemical Computing Group Inc.: Montreal, QC, Canada, 2019.
35. Gerber, P.; Muller, K. MAB, a generally applicable molecular force field for structure modelling in medicinal chemistry. *J. Comput. Aided Mol. Des.* **2015**, *9*, 251–268. [CrossRef]
36. Maier, J.A.; Martinez, C.; Kasavajhala, K.; Wickstrom, L.; Hauser, K.E.; Simmerling, C. ff14SB: Improving the Accuracy of Protein Side Chain and Backbone Parameters from ff99SB. *J. Chem. Theor. Comput.* **2015**, *11*, 3696–3713. [CrossRef]
37. Balyasnikova, I.V.; Skirgello, O.E.; Binevski, P.V.; Nesterovitch, A.B.; Albrecht, R.F.; Kost, O.A.; Danilov, S.M. Monoclonal antibodies 1G12 and 6A12 to the N-domain of human angiotensin-converting enzyme: Fine epitope mapping and antibody-based method for revelation and quantification of ACE inhibitors in the human blood. *J. Proteome Res.* **2007**, *6*, 1580–1594. [CrossRef] [PubMed]
38. Gordon, K.; Balyasnikova, I.V.; Nesterovitch, A.B.; Schwartz, D.E.; Sturrock, E.D.; Danilov, S.M. Fine epitope mapping of monoclonal antibodies 9B9 and 3G8, to the N domain of human angiotensin I-converting enzyme (ACE) defines a region involved in regulating ACE dimerization and shedding. *Tissue Antigens* **2010**, *75*, 136–150. [CrossRef]

39. Danilov, S.M.; Tikhomirova, V.E.; Metzger, R.; Naperova, I.A.; Bukina, T.M.; Goker-Alpan, O.; Tayebi, N.; Gayfullin, N.M.; Schwartz, D.E.; Samokhodskaya, L.M.; et al. ACE phenotyping in Gaucher disease. *Mol. Genet Metab.* **2018**, *123*, 501–510. [CrossRef]
40. Danilov, S.; Jaspard, E.; Churakova, T.; Towbin, H.; Savoie, F.; Wei, L.; Alhenc-Gelas, F. Structure-function analysis of angiotensin I-converting enzyme using monoclonal antibodies. *J. Biol. Chem.* **1994**, *269*, 26806–26814. [CrossRef]
41. Popova, I.A.; Lubbe, L.; Petukhov, P.A.; Kalantarov, G.F.; Trakht, I.N.; Chernykh, E.R.; Leplina, O.Y.; Lyubimov, A.V.; Garcia, J.G.; Dudek, S.M.; et al. Epitope mapping of novel monoclonal antibodies to human angiotensin I-converting enzyme. *Protein Sci.* **2021**, *30*, 1577–1593. [CrossRef]
42. Te Riet, L.; van Esch, J.H.; Roks, A.J.; van den Meiracker, A.H.; Danser, A.J. Hypertension: Renin-angiotensin-aldosteron system alterations. *Circ. Res.* **2015**, *116*, 960–975. [CrossRef]
43. Satou, R.; Penrose, H.; Navar, L. Inflammation as a regulator of the renin-angiotensin system and blood pressure. *Curr. Hypertens. Rep.* **2018**, *20*, 100. [CrossRef] [PubMed]
44. McDonagh, A.F. Ex uno plures; the concealed complexity of bilirubin species in the neonatal blood samples. *Pediatrics* **2006**, *118*, 1185–1187. [CrossRef] [PubMed]
45. Skirgello, O.E.; Balyasnikova, I.V.; Binevski, P.V.; Sun, Z.-L.; Baskin, I.I.; Palyulin, V.A.; Nesterovitch, A.B.; Albrecht, R.F.; Kost, O.A.; Danilov, S.M. Inhibitory antibodies to human angiotensin-converting enzyme: Fine epitope mapping and mechanism of action. *Biochemistry* **2006**, *45*, 4831–4847. [CrossRef] [PubMed]
46. Danilov, S.M.; Gordon, K.; Nesterovitch, A.B.; Lünsdorf, H.; Chen, Z.; Castellon, M.; Popova, I.A.; Kalinin, S.; Mendonca, E.; Petukhov, P.A.; et al. An angiotensin I-converting enzyme mutation (Y465D) causes a dramatic increase in blood ACE via accelerated ACE shedding. *PLoS ONE* **2011**, *6*, e25952. [CrossRef] [PubMed]
47. Li, L.-J.; Zhang, F.-B.; Liu, S.-Y.; Tian, Y.-H.; Le, F.; Wang, L.-Y.; Lou, H.-Y.; Xu, X.-R.; Huang, H.-F.; Jin, F. Human sperm devoid of germinal angiotensin-converting enzyme is responsible for total fertilization failure and lower fertilization rates by conventional in vitro fertilization. *Biol. Reprod.* **2014**, *90*, 125. [CrossRef]
48. Bosma, P.J. Inherited disorders of bilirubin metabolism. *J. Hepatol.* **2003**, *38*, 107–117. [CrossRef]
49. Sticova, E.; Jirsa, M. New insights in bilirubin metabolism and their clinical implications. *World J. Gastroenterol.* **2013**, *19*, 6398–6407. [CrossRef]
50. Memon, N.; Weinberger, B.J.; Hegy, T.; Aleksunes, L.M. Inherited disorders of bilirubin clearance. *Pediatric Res.* **2016**, *79*, 378–386. [CrossRef]
51. Minucci, G.C.A.; Zuppi, C.; Capoluongo, E. Gilbert and Crigler-Najjar syndromes: An update of the UDP-glucoronosyltransferase 1a1 (UGT1A1) gene mutation database. *Blood Cells Mol. Dis.* **2013**, *50*, 273–280.
52. Bosma, P.; Seppen, J.; Goldhoorn, B.; Bakker, C.; Elferink, R.O.; Chowdhury, J.; Jansen, P. Bilirubin UDP-glucuronosyltransferase 1 is the only relevant bilirubin glucuronidating isoform in man. *J. Biol. Chem.* **1994**, *269*, 17960–17964. [CrossRef]
53. Horsey, A.J.; Cox, M.H.; Sarwat, S.; Kerr, I.D. The multidrug transporter ABCG2: Still more questions than answers. *Biochem. Soc. Trans.* **2016**, *44*, 824–830. [CrossRef] [PubMed]
54. Vlaming, M.L.; Pala, Z.; van Esch, A.; Wagenaar, E.; de Waart, D.R.; van de Wetering, K.; van der Kruijssen, C.M.; Oude Elferink, R.P.; van Tellingen, O.; Schinkel, A.H. Functionally overlapping roles of Abcg2 (Bcrp1) and Abcc2 (Mrp2) in the elimination of methotrexate and its main toxic metabolite 7-hydroxymethotrexate in vivo. *Clin. Cancer Res.* **2009**, *15*, 3084–3093. [CrossRef] [PubMed]
55. Zámbó, B.; Bartos, Z.; Mózner, O.; Szabó, E.; Várady, G.; Poór, G.; Pálinkás, M.; Andrikovics, H.; Hegedűs, T.; Homolya, L.; et al. Clinically relevant mutations in the ABCG2 transporter uncovered by genetic analysis linked to erythrocyte membrane protein expression. *Sci. Rep.* **2018**, *8*, 7487. [CrossRef]
56. Nakagawa, H.; Wakabayashi-Nakao, K.; Tamura, A.; Toyoda, Y.; Koshiba, S.; Ishikawa, T. Disruption of N-linked glycosylation enhances ubiquitin-mediated proteasomal degradation of the human ATP-binding cassette transporter ABCG2. *FEBS J.* **2009**, *276*, 7237–7252. [CrossRef]
57. Stec, D.E.; John, K.; Trabbic, C.J.; Luniwal, A.; Hankins, M.W.; Baum, J.; Hinds, T.D., Jr. Bilirubin binding to PPARα inhibits lipid accumulation. *PLoS ONE* **2016**, *11*, e0153427. [CrossRef] [PubMed]
58. Zunszain, P.; Ghuman, J.; McDonagh, A.; Curry, S. Crystallographic analysis of human serum albumin complexed with 4Z,15E-bilirubin-IXalpha. *J. Mol. Biol.* **2008**, *381*, 394–406. [CrossRef]
59. Ahlfors, C.E.; Vreman, H.J.; Wong, R.J.; Bender, G.J.; Oh, W.; Morris, B.H.; Stevenson, D.K.; Subcommittee, T.P. Effects of sample dilution, peroxidase concentration, and chloride ion on the measurement of unbound bilirubin in premature newborns. *Clin. Biochem.* **2007**, *40*, 261–267. [CrossRef]
60. Wennberg, R.P.; Ahlfors, C.E.; Bhutani, V.K.; Johnson, L.H.; Shapiro, S.M. Toward understanding kernicterus: A challenge to improve the management of jaundiced newborns. *Pediatrics* **2006**, *117*, 474–485. [CrossRef]
61. McDonagh, A.F.; Vreman, H.J.; Wong, R.J.; Stevenson, D.K. Photoisomers: Obfuscating factors in clinical peroxidase measurements of unbound bilirubin? *Pediatrics* **2009**, *123*, 67–76. [CrossRef]
62. Kumagai, A.; Ando, R.; Miyatake, H.; Greimel, P.; Kobayashi, T.; Hirabayashi, Y.; Shimogori, T.; Miyawaki, A. A bilirubin-inducible fluorescent protein from eel muscle. *Cell* **2009**, *153*, 1602–1611. [CrossRef]
63. Huber, A.H.; Zhu, B.; Kwan, T.; Kampf, J.P.; Hegyi, T.; Kleinfeld, A.M. Fluorescence sensor for the quantification of unbound bilirubin concentration. *Clin. Chem.* **2012**, *58*, 869–876. [CrossRef] [PubMed]

64. Hegyi, T.; Kleinfeld, A.; Huber, A.; Weinberger, B.; Memon, N.; Shih, W.J.; Carayannopoulos, M.; Oh, W. Effects of Soybean Lipid Infusion on Unbound Free Fatty Acids and Unbound Bilirubin in Preterm Infants. *J. Pediatr.* **2017**, *184*, 45–50. [CrossRef] [PubMed]
65. Iwatani, S.; Yamana, K.; Nakamura, H.; Nishida, K.; Morisawa, T.; Mizobuchi, M. A Novel Method for Measuring Serum Unbound Bilirubin Levels Using Glucose Oxidase-Peroxidase and Bilirubin-Inducible Fluorescent Protein (UnaG): No Influence of Direct Bilirubin. *Int. J. Mol. Sci.* **2020**, *21*, 6778. [CrossRef]
66. Temme, E.H.; Zhang, J.; Schouten, E.; Kesteloot, H. Serum bilirubin and 10-year mortality risk in a Belgian population. *Cancer Causes Control.* **2001**, *12*, 887–894. [CrossRef] [PubMed]
67. Ip, S.; Chung, M.; Kulig, J.; O'Brien, R.; Sege, R.; Glicken, S.; Maisels, M.J.; Lau, J. An evidenced-based review of important issues concerning neonatal hyperbilirubinemia. *Pediatrics* **2004**, *114*, 130–153. [CrossRef] [PubMed]

Disclaimer/Publisher's Note: The statements, opinions and data contained in all publications are solely those of the individual author(s) and contributor(s) and not of MDPI and/or the editor(s). MDPI and/or the editor(s) disclaim responsibility for any injury to people or property resulting from any ideas, methods, instructions or products referred to in the content.

MDPI AG
Grosspeteranlage 5
4052 Basel
Switzerland
Tel.: +41 61 683 77 34

Biomedicines Editorial Office
E-mail: biomedicines@mdpi.com
www.mdpi.com/journal/biomedicines

Disclaimer/Publisher's Note: The title and front matter of this reprint are at the discretion of the Guest Editors. The publisher is not responsible for their content or any associated concerns. The statements, opinions and data contained in all individual articles are solely those of the individual Editors and contributors and not of MDPI. MDPI disclaims responsibility for any injury to people or property resulting from any ideas, methods, instructions or products referred to in the content.